Optical Spectroscopy of Low Dimensional Semiconductors

NATO ASI Series

Advanced Science Institutes Series

A Series presenting the results of activities sponsored by the NATO Science Committee, which aims at the dissemination of advanced scientific and technological knowledge, with a view to strengthening links between scientific communities.

The Series is published by an international board of publishers in conjunction with the NATO Scientific Affairs Division

A Life Sciences **B Physics**	Plenum Publishing Corporation London and New York
C Mathematical and Physical Sciences **D Behavioural and Social Sciences** **E Applied Sciences**	Kluwer Academic Publishers Dordrecht, Boston and London
F Computer and Systems Sciences **G Ecological Sciences** **H Cell Biology** **I Global Environmental Change**	Springer-Verlag Berlin, Heidelberg, New York, London, Paris and Tokyo

PARTNERSHIP SUB-SERIES

1. Disarmament Technologies	Kluwer Academic Publishers
2. Environment	Springer-Verlag / Kluwer Academic Publishers
3. High Technology	Kluwer Academic Publishers
4. Science and Technology Policy	Kluwer Academic Publishers
5. Computer Networking	Kluwer Academic Publishers

The Partnership Sub-Series incorporates activities undertaken in collaboration with NATO's Cooperation Partners, the countries of the CIS and Central and Eastern Europe, in Priority Areas of concern to those countries.

NATO-PCO-DATA BASE

The electronic index to the NATO ASI Series provides full bibliographical references (with keywords and/or abstracts) to more than 50000 contributions from international scientists published in all sections of the NATO ASI Series.
Access to the NATO-PCO-DATA BASE is possible in two ways:

– via online FILE 128 (NATO-PCO-DATA BASE) hosted by ESRIN, Via Galileo Galilei, I-00044 Frascati, Italy.

– via CD-ROM "NATO-PCO-DATA BASE" with user-friendly retrieval software in English, French and German (© WTV GmbH and DATAWARE Technologies Inc. 1989).

The CD-ROM can be ordered through any member of the Board of Publishers or through NATO-PCO, Overijse, Belgium.

Series E: Applied Sciences - Vol. 344

Optical Spectroscopy of Low Dimensional Semiconductors

edited by

Gerhard Abstreiter
Walter Schottky Institut,
Technische Universität München,
Germany

Atilla Aydinli
Department of Physics,
Bilkent University,
Ankara, Turkey

and

Jean-Pierre Leburton
University of Illinois,
Urbana, U.S.A.

Kluwer Academic Publishers

Dordrecht / Boston / London

Published in cooperation with NATO Scientific Affairs Division

Proceedings of the NATO Advanced Study Institute on
Optical Spectroscopy of Low Dimensional Semiconductors
Ankara and Antalya, Turkey
9–20 September 1996

A C.I.P. Catalogue record for this book is available from the Library of Congress.

ISBN 0-7923-4728-5

Published by Kluwer Academic Publishers,
P.O. Box 17, 3300 AA Dordrecht, The Netherlands.

Sold and distributed in the U.S.A. and Canada
by Kluwer Academic Publishers,
101 Philip Drive, Norwell, MA 02061, U.S.A.

In all other countries, sold and distributed
by Kluwer Academic Publishers,
P.O. Box 322, 3300 AH Dordrecht, The Netherlands.

Printed on acid-free paper

Printed in the Netherlands

TABLE OF CONTENTS

QUANTUM DOTS

PREFACE

Technological advances in semiconductor growth has opened a broad horizon for semiconductor physics and applications during the past 20 years. High quality two-dimensional systems are achieved with nearly atomic precision by direct epitaxial growth. Such structures led to novel applications like low noise high frequency modulation doped field effect transistors and quantum well lasers. Semiconductor heterostructures of lower dimensionality like quantum wires and quantum dots are not yet as mature, partly due to the lack in precision of lateral structuring technology. In recent years, however, there was an enormous progress in novel epitaxial growth methods. This opens a wide new area of basic and applied semiconductor physics with the hope of novel applications in near future making use of the advantageous properties of one- and zero-dimensional systems. Ideas for future device applications mainly stem from the altered density of states being discrete or atomic-like for quantum dots. Optical spectroscopy has played and is playing a crucial role in the advancement of this fascinating field of semiconductor physics.

The NATO school organized at Bilkent University in Ankara and in Antalya brought together experts in this field and newcomers, especially young Ph.D. students and postdocs, to learn about recent developments and to discuss open questions in the area of optical spectroscopy of low dimensional semiconductors. The school turned out to be extremely fruitful and there was a great enthusiasm among the lecturers and students during the whole two weeks.

The NATO Advanced Study Institute considered various aspects of optical spectroscopy of low dimensional semiconductor structures. This included basic physics aspects, novel technology and material fabrication tools, characterization methods and new devices with special emphasis on quantum wire and quantum dot lasers. The advance in special epitaxial growth techniques, especially on patterned substrates is remarkable and allows nowadays for fabrication of well defined 1- and 0-dimensional semiconductor systems. Based on the well controlled technology the realization of quantum wire and quantum dot lasers was demonstrated as one of the highlights of the school. Especially the self-ordering of semiconductor quantum dots in various materials was discussed during the whole school. Other major topics included so-called V-groove quantum wires, cleaved edge overgrowth, micro-cavities, electronic excitations in low dimensional systems, phonons and also dissipative transport in nanostructures. While most of the lectures on experimental results were devoted to III-V based semiconductor structures like GaAs, AlAs, InP and InAs, some talks also considered the improvement which was achieved in recent years with the group IV semiconductor microstructures based on Si, Ge and C and II-VI semiconductors based on ZnSe and CdSe. The theoretical lectures covered growth kinetics and electronic properties of low dimensional semiconductor structures including many-particle effects. Most of the aspects covered by the lectures are included in this book.

We are grateful to the NATO Scientific Affairs Division, to TÜBITAK, the Scientific and Technical Research Council of Turkey and to Bilkent University for making this meeting possible through their financial support and to B. Tanatar and A. Serpenguzel for their assistance in the preparation for the school. We are also grateful to the lecturers for their excellent presentations and to all participants for their contributions in oral talks, posters and discussions. They made the meeting to a lively and memorable event. We also thank the local organizers of the side programs which provided a lot of insight into the rich ancient and present culture, the beautiful scenery and the warm hospitality of the people in Turkey.

Gerhard Abstreiter
Atilla Aydinli
Jean-Pierre Leburton

MBE GROWTH AND OPTICAL PROPERTIES OF $Si_{1-y}C_y$ and $Si_{1-x-y}Ge_xC_y$ ALLOY LAYERS

K. Eberl and K. Brunner
Max-Planck-Institut für Festkörperforschung
Heisenbergstr. 1, D-70569 Stuttgart, Germany

1. Abstract

High quality pseudomorphic $Si_{1-y}C_y$ and $Si_{1-x-y}Ge_xC_y$ alloy layers with a carbon concentration up to 7% are prepared by solid source molecular beam epitaxy. Near band-edge photoluminescence (PL) is observed from $Si/Si_{1-y}C_y$ multiple quantum well (MQW) structures. The band gap in the pseudomorphic films is reduced by about 65 meV per percent C. The data for $Si/Si_{1-y}C_y$ MQW´s indicate a type I heterostructure with the band offset being mainly in the conduction band. In $Si_{1-x-y}Ge_xC_y$ MQW´s compressive strain caused by Ge is partially compensated by C alloying and the band gap increases with y. PL measurements from closely spaced $Si_{1-y}C_y/Si_{1-x}Ge_x$ layers show a lower transition energy than that of isolated $Si_{1-y}C_y$ and $Si_{1-x}Ge_x$ reference samples. This is attributed to spacially indirect PL transitions between the electrons confined in the $Si_{1-y}C_y$ layers and the heavy holes located in the $Si_{1-x}Ge_x$ layers. The PL is dominated by no-phonon recombination.

2. Introduction

Si/SiGe heterostructures are advantageous for device applications like heterobipolar transistors and photodetectors [1]. Additional flexibility in the design of device structures

G. Abstreiter et al. (eds.), Optical Spectroscopy of Low Dimensional Semiconductors, 1–20.

is achieved by using thick relaxed SiGe buffer layers as a virtual substrate, which allows one to adjust the strain. This concept opened the possibility to prepare high mobility electron and hole channels in tensily strained Si and compressively strained Ge-rich SiGe quantum wells, respectively [2]. The main problems of strain relaxed buffer layers, however, are the several μm total film thickness, the still very high defect density and the characteristic surface roughness.

A different concept towards strain adjustment was suggested by adding C into the Si/SiGe material system. Pseudomorphic $Si_{1-y}C_y$ layers with tensile strain and SiGeC films with reduced compressive strain directly on Si were reported in 1991 [3-5]. Before that, Posthill et al. published the first epitaxial growth of $Si_{1-y}C_y$ alloy layers on Si, but the films where not free of SiC precipitates since they used relatively high substrate temperatures during growth [6]. Since that time several groups started to prepare Si and SiGe films with substitutional C by MBE [7,11] and CVD [12-17] technique. A recent overview on structural and phonon properties was given by Jain et al. [18].

Pseudomorphic $Si_{1-y}C_y$ and $Si_{1-x-y}Ge_xC_y$ alloy layers on Si have a good crystal quality and are essentially free of extended lattice defects when prepared under appropriate conditions. Until very recently there was no experimental data on the fundamental band gaps and band alignments. The influence of local strain around the C atoms in the Si matrix is unclear. Demkov and Sankey discussed this effect on the basis of a first principles model calculation [19] and predicted that the fundamental band pap in unstrained $Si_{1-y}C_y$ is reduced for a small amount of C in the Si lattice. As a consequence of that the question arises what the effects on the electronic and optical properties are, and how the carrier mobilities are affected. This contribution provides the latest investigations of the optical properties of $Si_{1-y}C_y$ and $Si_{1-x-y}Ge_xC_y$ alloy layers.

3. Experimental conditions

The samples discussed in the following are prepared by solid source MBE on undoped Si (100) substrate with a resistance above 5000 Ωcm . A schematic drawing of the MBE system is shown in figure 1. A directly heated pyrolytic graphite filament is used for C sublimation with a typical growth rate of up to 0.002 nm/s. The hot filament is completely surrounded by graphite shieldings. In addition the shutter is lined with a graphite shielding to reduce problems with metal, CO and CO_2 contamination. Si is evaporated in an electron beam evaporator, which is feedback controlled by a mass-spectrometer. For Ge we use a special PBN crucible effusion cell with a Si orifice plate to reduce B background doping. Elemental B and a GaP decomposition source [20] are used for p- and n-type doping, respectively. Typically the active multilayer structures are grown on a 300 to 400 nm thick Si buffer layer. Optimized sample quality with respect to photoluminescence (PL) intensity for $Si/Si_{1-y}C_y$ QW structures is achieved at a substrate temperature of about 550 °C [21].

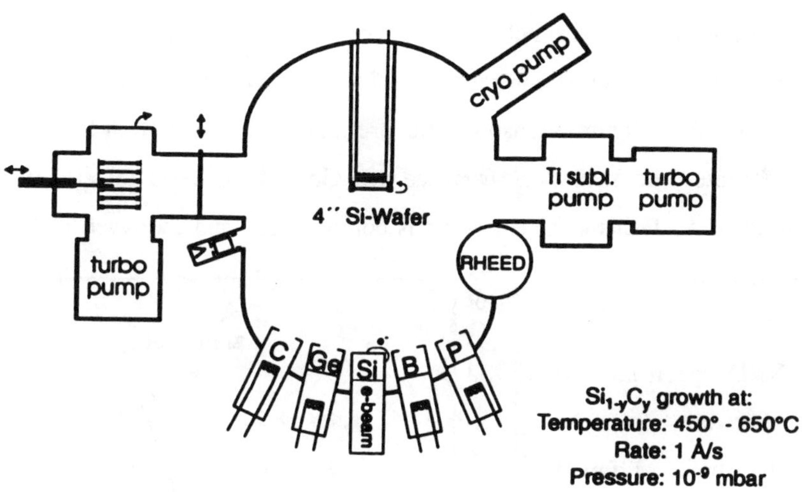

Figure 1: Schematic drawing of the SiGeC MBE system. The growth chamber has a base pressure below $3x10^{-11}$ mbar. During deposition the pressure is typically in the 10^{-9} mbar range. The load lock chamber is shown on the left side.

The structural properties of the samples were determined by double-crystal high resolution X-ray diffraction (XRD) and by comparing the data with dynamical simulation results, which are based on Vegard's law and linearly interpolated elastic constants. An example is shown in figure 2. The reproducibility for layer thickness, Ge and C concentration from sample to sample is about 5%. The C content measured by XRD agrees well with the C nominally deposited and calibrated by secondary-ion mass spectroscopy (SIMS) analysis. However, the absolute uncertainty in SIMS is about a factor of two due to the missing absolute reference samples for SIMS in the high C concentration regime. A pronounced dependence of the substitutional C incorporation on growth temperature is not observed within the temperature range from 400 to 600°C for C concentrations below 3%. Figure 2 shows an XRD spectrum from a sample with three different layers on Si (100). A 200 nm thick $Si_{0.992}C_{0.008}$ layer followed by a 200 nm $Si_{0.973}Ge_{0.063}$ and a 200 nm $Si_{0.93}Ge_{0.063}C_{0.008}$ layer. The layers are fully strained to match the Si substrate (*pseudomorphic growth*). The upper curve shows the measured spectrum and the lower curve shows the simulated data. The $Si_{0.992}C_{0.008}$ peak is shifted to larger angles indicating tensile strain, in contrast to the $Si_{0.973}Ge_{0.063}$ peak which indicates lateral compression in the film. Figure 3 illustrates the pseudomorphic growth of $Si_{1-y}C_y$ and $Si_{1-x}Ge_x$ on Si substrate. The XRD peak from the $Si_{0.93}Ge_{0.063}C_{0.008}$ layer falls together with the Si substrate peak. That means the strain is compensated and the layer has the same lattice constant as Si.

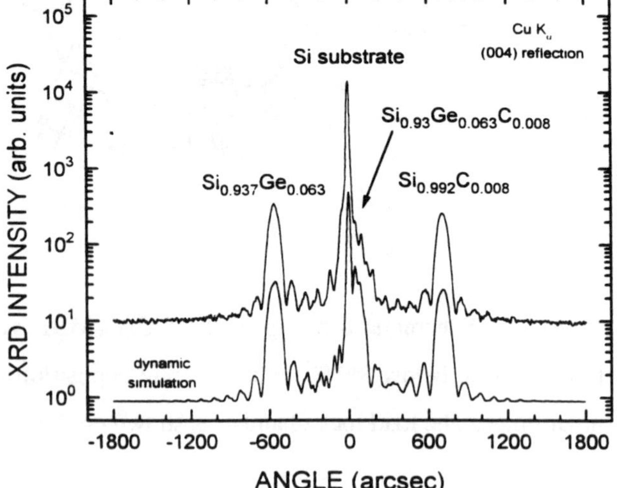

Figure 2: XRD spectrum and simulated spectrum from a sample with three different layers on Si (100). A 200 nm thick $Si_{0.992}C_{0.008}$ layer followed by a 200 nm $Si_{0.973}Ge_{0.063}$ and a 200 nm $Si_{0.93}Ge_{0.063}C_{0.008}$ layer is grown on a Si buffer layer.

For the PL studies the samples were excited by a 476 nm Kr^+ laser beam at a power density of about 0.2 W/cm^2. The temperature of the sample during measurement was about 8 K. PL is analyzed by a liquid-nitrogen cooled Ge detector using standard lock-in techniques.

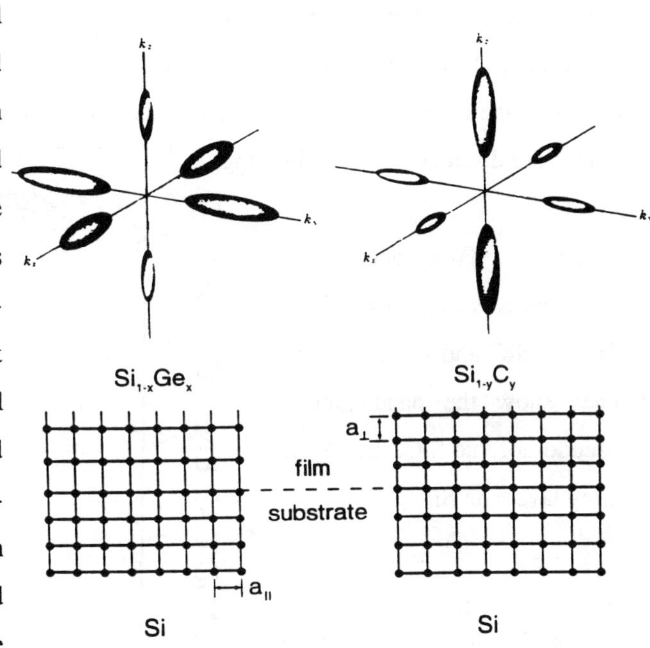

Figure 3: Two dimensional sketch of strained $Si_{1-x}Ge_x$ and $Si_{1-y}C_y$ on Si. The difference in lattice constant is accommodated by tetragonal deformation of the epilayer. The upper part shows surfaces of constant energy in k-space for the conduction band. It illustrates the influence of lateral strain on the 6-fold degenerated Δ-valleys in Si. The two Δ-minima in growth direction become the conduction band minima in $Si_{1-y}C_y$ due to the lateral tension.

4. $Si_{1-x}Ge_x$ and $Si_{1-y}C_y$ Alloy Layers

Figure 4 shows the band gap versus lattice constant for Si, Ge, cubic SiC and diamond. The solid curve between Si and Ge indicates the band gap for unstrained SiGe alloys with the characteristic kink at about 85% Ge, which is due to the crossover from the Si like 6-fold Δ minima to the Ge like 4-fold L minima in the conduction band (CB). The dashed line towards Ge shows the reduction of the band gap for pseudomorphic SiGe on Si, which is determined by the strain-degenerate 4-fold Δ minima in the CB and the heavy holes (hh) in the valance band (VB).

The lattice mismatch between Si and Ge is only 4% therefore the fundamental properties can be interpolated for SiGe alloys to a large degree. The band gap of $Si_{1-y}C_y$ alloys was expected to homogenously increase with the C content, because the band structures of Si, cubic SiC and diamond are very similar except the energy value of the band gap [23,24]. The PL data obtained recently indicate a strong band gap reduction for pseudomorphic $Si_{1-y}C_y$ alloy films as shown by the dashed line starting from Si towards SiC. Subtracting the effect of tensile strain reveals that unstrained $Si_{1-y}C_y$ alloys with a small C content have a smaller band gap than Si as indicated by the continuous curve. Further details are discussed in the following.

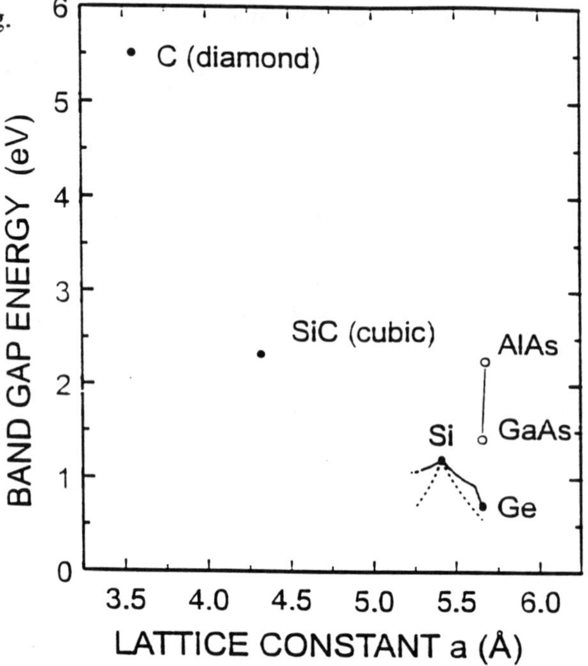

Figure 4: Fundamental band gap versus lattice constant for Si, Ge, SiC and C. The dashed lines show the band gap for pseudomorphic SiGe and $Si_{1-y}C_y$ alloy layers on Si.

Figure 5a shows a typical PL spectrum from a 30-period 5.2 nm $Si_{0.99}C_{0.01}$ / 15.6 nm Si multi-QW structure. The Si-TO phonon assisted PL at 1.1 eV originates from the Si substrate, buffer layer and 200 nm Si cap layer. The intense PL peaks from excitons confined in the QWs are the Si-Si TO phonon replica at 1.032 eV and the no-phonon (NP) transition at 1.089 eV. The linewidth of these peaks is about 12 meV. The weak PL shoulders labeled as QW-TA and QW-(TO+O^{Γ}) are probably TA phonon and TO plus Γ point optical phonon assisted recombinations of carriers confined in the $Si_{1-y}C_y$ quantum wells. All lines labeled with QW shift as a function of the C content y in the $Si_{1-y}C_y$ layers as shown for the no-phonon line in figure 5b. The PL energy of the NP line shifts linearly by $\Delta E^{NP} = -y\cdot(5.7\ eV)$. reflecting the decrease of the fundamental band gap in the MQW structure. For the layer thickness of 5.2 nm and a C content of about 2% or more we

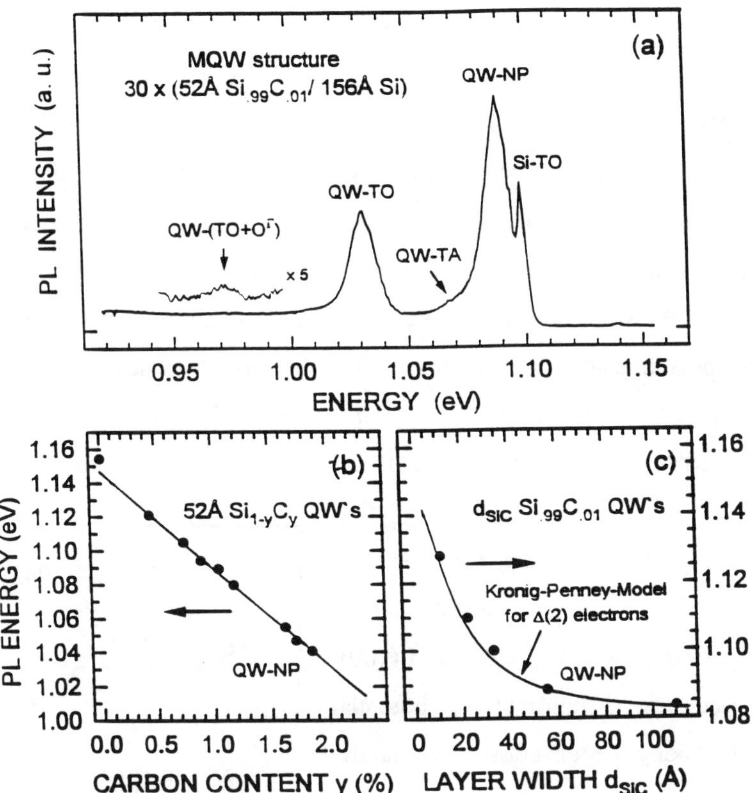

Figure 5: a) PL spectrum of a 30 period $Si_{0.99}C_{0.01}$/Si multi-QW well structure (8K). The Si and $Si_{1-y}C_y$ layers are 15.6 nm and 5.2 nm thick, respectively. b) Energy of the no-phonon PL peak as a function of the carbon content. c) Energy of the no-phonon PL peak as a function of the $Si_{0.99}C_{0.01}$ layers thickness. For smaller thickness the PL peak shifts to higher energies due to the quantum confinement in the $Si_{1-y}C_y$ quantum wells.

observe an increasing broad emission background at low energy, which is similar to the broad PL reported earlier [25]. In thinner $Si_{1-y}C_y$ multi-QWs we observe band-edge PL for a C content up to about y = 6.4% [21].

Figure 5c shows the energy dependence of the NP PL peak from 30-period $Si_{0.99}C_{0.01}$ / 15.6 nm Si multi-QWs as a function of the $Si_{0.99}C_{0.01}$ alloy layer thickness d_{SiC}. The peak is shifting to higher energy by reducing the QW width d_{SiC}. This proofs, that the PL lines originate from quantum confined levels and not from defect states related to C incorporation. In other words, it proofs, that $Si_{1-y}C_y$ alloy layers are a reasonable semiconductor material with a well-defined band gap. The measured data follow the curve, which represents a Kronig-Penney-Model calculation for the twofold degenerate conduction band-edge Δ(2) and thus using m = $0.92m_o$ as the effective mass for the electrons and neglecting any band offset in the valence band. This and more detailed PL studies on MQW structures with the $Si_{1-y}C_y$ layer thickness constant and varying the Si layer thickness indicate that the band offset in this heterostructure is mainly in the conduction band [22], which is complementary to the situation for pseudomorphic Si/SiGe heterostructures, where the band offset is mainly in the valence band.

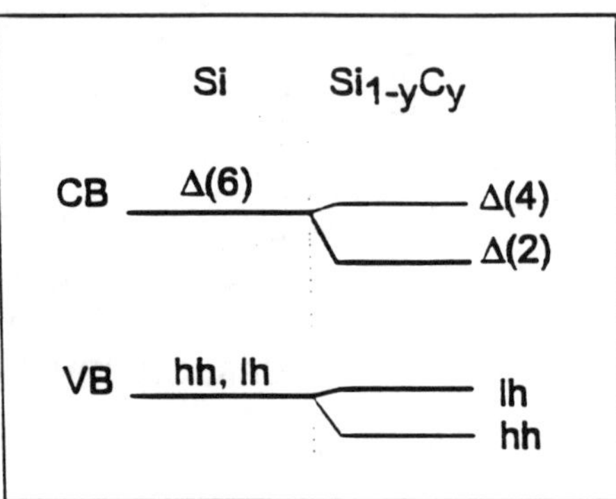

Figure 6: Schematic diagram of the conduction (CB) and valence-band-edge (VB) for pseudomorphic $Si_{1-y}C_y$ on Si (100). It is based on a deformation potential calculation taking the tensile strain within the $Si_{1-y}C_y$ layer into account and neglecting intrinsic effects of C on the band offset.

This model of the band alignment is also supported by just looking at the strain induced splitting of the electron and hole valleys. Figure 6 shows a schematic diagram of the conduction (CB) and valence band-edge (VB) for pseudomorphic $Si_{1-y}C_y$ layers on Si. It is based on a deformation potential calculation. The tensile biaxial strain introduced by substitutional C splits the electron valleys and thus, the twofold degenerate conduction band-edge Δ(2) is shifted down in energy by about $\Delta E^{\Delta(2)} = -y\cdot(4.6\ eV)$. In the valence band the light hole states are expected to increase only slightly in energy with increasing tensile strain $\varepsilon=0.35\cdot y$.

Summarizing the data for $Si_{1-y}C_y$ alloys we find the following: For small C contents of at least up to 2.5% the optical band gap is shrinking by about $\Delta E(y) = -y\cdot(6.5\ eV)$. This takes the electron confinement into account, which is about 8 meV for the 5.2 nm QWs discussed in figure 2. Since the tensile strain accounts for about $-y\cdot(4.6\ eV)$ we conclude that the intrinsic band gap of unstrained $Si_{1-y}C_y$ will decrease roughly by $\Delta E_g(y) = -y\cdot(1.9\ eV)$ for small C concentrations. Exciton binding of electrons and holes with a typical energy of 10 meV is neglected. The tentative change of the band gap for unstrained and strained $Si_{1-y}C_y$ alloys is schematically illustrated in figure 4 and more detailed in figure 9 by the continuous and the dashed curves starting from Si toward SiC, respectively.

5. $Si_{1-x-y}Ge_xC_y$ Alloy Layers

Figure 7 shows low temperature PL spectra from a $Si_{0.84}Ge_{0.16}$ and two $Si_{0.84-y}Ge_{0.16}C_y$ MQW structures with y=0.008 and y=0.01. The samples have 25 periods and the thickness of the $Si_{0.84}Ge_{0.16}$ and $Si_{0.84-y}Ge_{0.16}C_y$ layers are 4 nm. The QWs are separated by 17 nm thick Si layers. The main PL originating from the QWs are no-phonon and TO phonon lines of confined carriers. The intense peak at E=1.1 eV is the TO phonon replica from the thick Si layers. For compressively strained $Si_{1-x}Ge_x$ on Si the band offset is mainly in the valence band with the heavy hole (hh) shift being about $\Delta E_{hh}= x\cdot(0.72\ eV)$ [27]. In the conduction band there is a small energy lowering of the Δ(4) electron valleys [28]. The band alignment for Si/SiGe is schematically shown in the left part of figure 8. The PL from $Si_{0.842}Ge_{0.16}C_{0.008}$ multi-QWs is shifted to about 20 meV higher energy compared to the $Si_{0.84}Ge_{0.16}$ as shown in figure 7. The substitutional incorporation of 0.8% C into $Si_{0.84}Ge_{0.16}$ reduces the compressive strain from $\varepsilon= -0.64\%$ to -0.36%. The PL blueshift is attributed to the strain-induced increase and a minor intrinsic lowering of

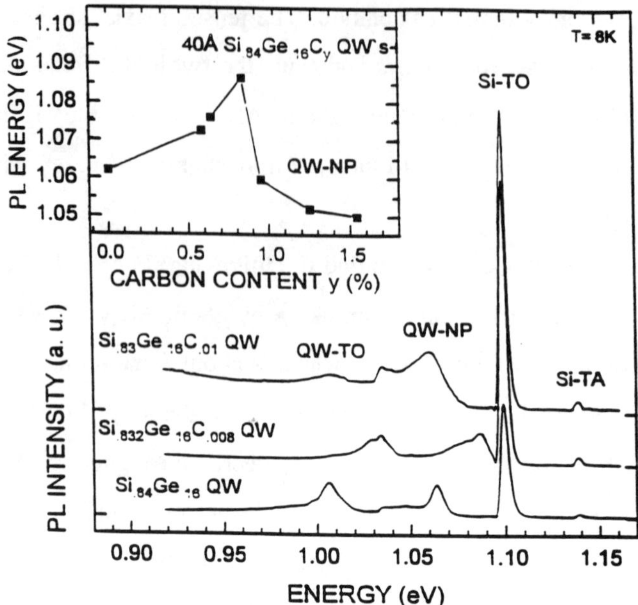

Figure 7: PL spectra from one 25 period $Si/Si_{0.84}Ge_{0.16}$ and two $Si/Si_{0.84-y}Ge_{0.16}C_y$ multi-QWs with y=0.008 and y=0.01. The Si and $Si_{1-x-y}Ge_xC_y$ layers are 20 nm and 4 nm thick, respectively. The TO phonon replica and the no-phonon (NP) PL peaks labeled with QW originate from the QWs. The low intensity shoulder close to E=1.03 eV is the TO+O(Γ) phonon assisted transition from the Si buffer layers. The insert shows the energy shift of the no-phonon PL energy as a function of the carbon concentration.

the fundamental band gap [29]. For C concentrations in $Si_{0.84-y}Ge_{0.16}C_y$ which are higher than 0.8% we find a downward shift of the PL energies as shown by the topmost spectrum in figure 7 with y=1% and more precisely in the insert.

The band gap increase in $Si_{1-x-y}Ge_xC_y$ is schematically illustrated in the right part of figure 8. Due to partial relief of compressive strain, the Δ(2) to Δ(4) and the hh to lh splittings are reduced and would be brought back to degeneracy in case of perfect strain compensation, which is achieved for about y=x/8.2. Additionally, there is an intrinsic effect of C on the band gap for relaxed $Si_{1-x-y}Ge_xC_y$ like for $Si_{1-y}C_y$ alloys which were discussed in the previous section.

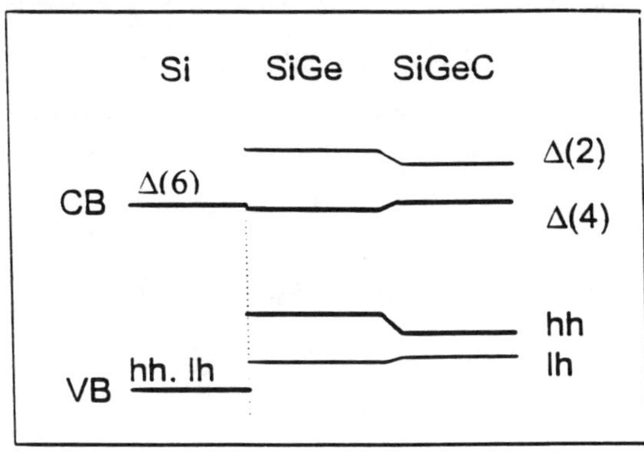

Figure 8: Schematic diagram of the conduction (CB) and valence-band edge (VB) for pseudomorphic SiGe on Si (100). The effects of substitutional C in SiGe are shown in the right side of the picture just by considering the effect of partial reduction of compressive strain. Intrinsic effects of C especially on the band offsets are not considered.

The band gap in pseudomorphic $Si_{1-x}Ge_x$ with x<25% is [30]: $E_g(x) = 1.171-1.01x+0.835x^2$ eV. Thus, the band gap of strained $Si_{0.84}Ge_{0.16}$ is 1.03 eV at 4.2K, which is 140 meV smaller than that of Si. We assume, that substitutional C increases the energy gap by about $\Delta E = y\cdot(0.24\ eV)$ based on the extrapolation of the increasing NP-PL energy shown in the insert of figure 4 for up to y=0.008. As a consequence, for strain compensated $Si_{0.82}Ge_{0.16}C_{0.02}$ on Si the band gap would still be about 100 meV smaller than in Si. For comparison, unstrained $Si_{0.84}Ge_{0.16}$ has a band gap of about 1.11 eV, whereas the relaxed $Si_{0.82}Ge_{0.16}C_{0.02}$ film would have a roughly 30 meV smaller band gap of about 1.07 eV. This is illustrated in figure 9. A similar result was reported for x=0.24 and 0.38 by Amour et al [29].

The drastic decrease of the NP-PL energy above y=0.008 (see inset of figure 7) has not been investigated in detail so far. However, part of it may be explained by a type I to type II band alignment transition. As more C is incorporated into SiGe the Δ(4) valleys are shifted upwards in energy and at some point the Δ(4) level probably crosses with the Δ(6) CB in the neighbouring Si layers. At this point the PL transition becomes spatially indirect from the Δ(6) valleys in Si to the hh band in SiGeC and one measures essentially the CB offset. Just considering the effects of strain reduction and neglecting intrinsic effects one would expect that the VB offset decreases with increasing C content (see figure 8). The continuous shift to lower energies for y > 0.01, however indicates an increasing VB offset in Si/SiGeC. This is also confirmed in measuements on coupled QWs discussed in figure 10 and 11.

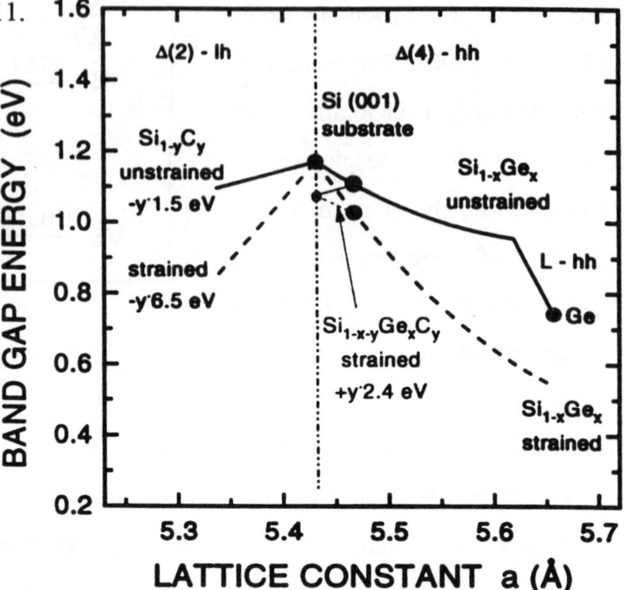

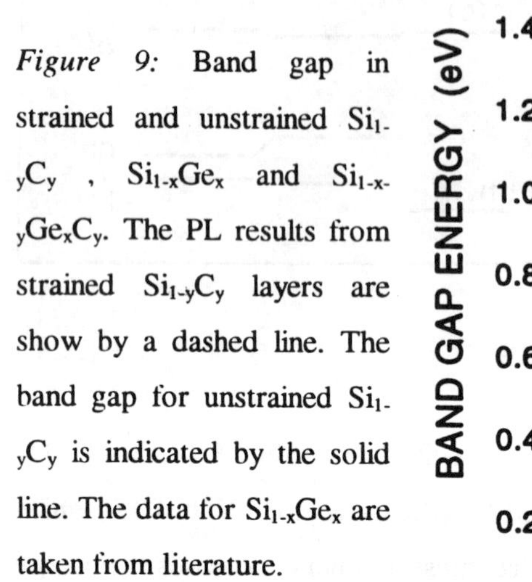

Figure 9: Band gap in strained and unstrained $Si_{1-y}C_y$, $Si_{1-x}Ge_x$ and $Si_{1-x-y}Ge_xC_y$. The PL results from strained $Si_{1-y}C_y$ layers are show by a dashed line. The band gap for unstrained $Si_{1-y}C_y$ is indicated by the solid line. The data for $Si_{1-x}Ge_x$ are taken from literature.

In summary, for SiGeC alloys we find a very similar downward bowing of the intrinsic energy gap like for $Si_{1-y}C_y$ discussed in figure 4 and 9. When starting from a certain unstrained SiGe alloy composition and adding substitutional C into the layer, the energy gap goes down by about 10 - 20 meV per percent C for unstrained alloys, at least for small C concentrations. For pseudomorphic layers, C reduces the strain in the films and increases the band gap by about $\Delta E = y \cdot (0.24\ eV)$, but this increase is only part of the one which would be obtained by reducing the Ge content to achieve similar strain reduction [29]. The experimental data suggest that SiGeC layers can be prepared on Si, which have small and adjustable lateral strain, a smaller band gap and a substantial band offset in the VB. The results discussed in section 4 and 5 are summarised in figure 9.

6. Coupled $Si_{1-x-y}Ge_xC_y$ / $Si_{1-z}C_z$ Layers

The starting idea of the experiments discussed in the following was to put a SiGe layer right next to a $Si_{1-z}C_z$ quantum well in order to achieve strong confinement for both holes

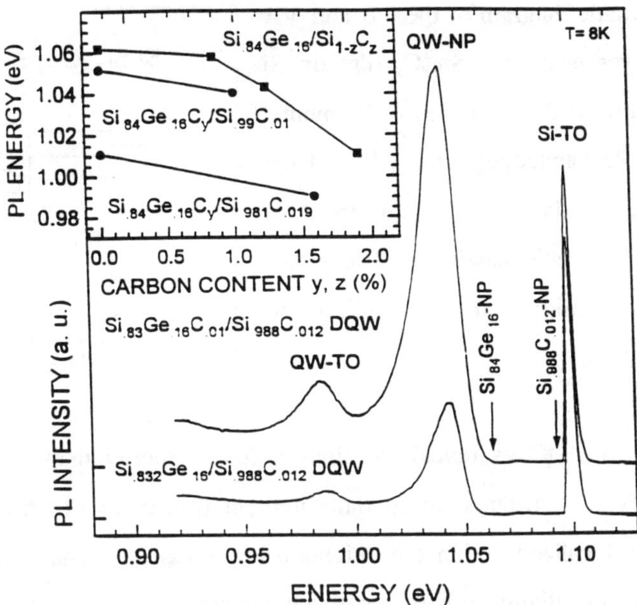

Figure 10: PL spectra of coupled $Si_{1-x-y}Ge_xC_y$ / $Si_{1-z}C_z$ quantum well structures with x=0.16, y=0 or 0.01, and z=0.012. The $Si_{1-x-y}Ge_xC_y$ and the $Si_{1-z}C_z$ layers are 4 nm and 3.3 nm thick, respectively. The coupled QWs are repeated 25 times with thick Si layers in between. The insert shows the energy of the no-phonon PL peak as a function of the carbon content z in $Si_{1-x}Ge_x$ / $Si_{1-z}C_z$ layers (upper curve).The lower two curves in the insert show the energy shift for $Si_{1-x-y}Ge_xC_y$ / $Si_{1-z}C_z$ double quantum wells for different C concentrations in the individual layers.

and electrons in the corresponding layers directly on Si. Similar concepts have been realized on SiGe buffer layers with a Si layer between compressively strained Ge quantum wells [31]. PL spectra of two coupled $Si_{1-x-y}Ge_xC_y$ / $Si_{1-z}C_z$ quantum well structures with x=0.16, y=0 or 0.01, and z=0.012 are shown in figure 10. The $Si_{1-x-y}Ge_xC_y$ and the $Si_{1-z}C_z$ layers are 4 nm and 3.3 nm thick, respectively. The double quantum wells (DQWs) are repeated 25 times with 17 nm thick Si layers in between. The arrows between the Si-TO line from the substrate and the QW-NP PL line from the DQWs mark the energetic position of the reference samples with only $Si_{0.84}Ge_{0.16}$ and

Figure 11: Schematic diagram of the CB and VB for a $Si_{1-x}Ge_x$ layer next to a $Si_{1-y}C_y$ film on Si (100). The continuous line indicates Δ(2) minima and the heavy hole bandedge in the CB and the VB, respectively. The dashed lines show the Δ(4) minima and the light hole bandedge. The arrow indicates the PL transition from Δ(2) in $Si_{1-y}C_y$ to the hh VB in $Si_{1-x}Ge_x$.

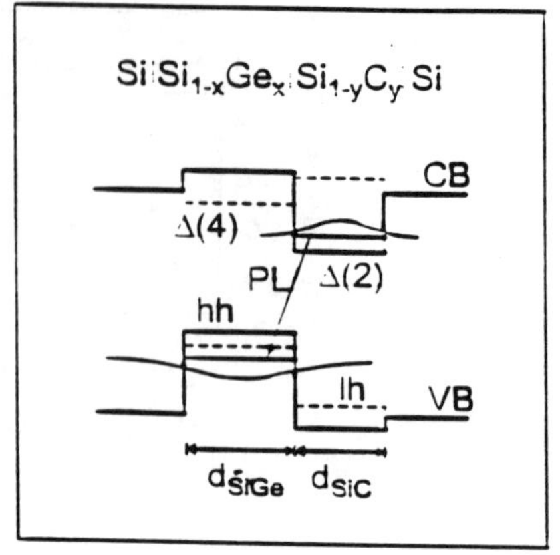

$Si_{0.988}C_{0.012}$ QWs. The DQWs reveal PL lines at lower energy than the corresponding single QWs. They are attributed to spatially indirect transitions of Δ(2) electron and heavy hole states localized within the neighbouring $Si_{0.988}C_{0.012}$ and $Si_{0.84}Ge_{0.16}$ layers, respectively. The band alignment is schematically illustrated in figure 11. The no-phonon line is strongly enhanced compared to the TO phonon line. The intensity gain observed for the DQW with respect to the corresponding single QWs is considerable. For optimized QW thicknesses with respect to maximum overlap of the wave functions for electrons in $Si_{0.988}C_{0.012}$ and holes in $Si_{0.84}Ge_{0.16}$ we achieve a three orders of magnitude intensity increase [26]. A strong dependence of the DQW PL on the spatial separation of carriers is demonstrated by a 40% intensity decrease observed for a DQW structure with a 0.4 nm thick Si layer in between the $Si_{0.84}Ge_{0.16}$ and $Si_{0.988}C_{0.012}$ layers. Adding C into the $Si_{0.84}Ge_{0.16}$ layers results in a further PL intensity enhancement and a slight downwards shift in energy.

The insert shows the energy of the no-phonon PL peak as a function of the carbon content in $Si_{0.84}Ge_{0.16}$ /$Si_{1-z}C_z$ layers with y varied from zero to 2% (upper curve). The lower two curves in the insert show the energy shift for $Si_{0.84-y}Ge_{0.16}C_y$ / $Si_{1-z}C_z$ double quantum wells as a function of y for different C concentrations in the $Si_{1-z}C_z$ layers. The behaviour of the upper curve is interpreted as follows: For y=0 we have a spatially direct, type I transition from the Δ(4) CB to the hh VB within the $Si_{0.84}Ge_{0.16}$ layer. Up to about z=0.8% there is hardly any shift of the NP-PL energy. As more and more C is incorporated into the adjacent $Si_{1-z}C_z$ layer, the Δ(2) CB shifts downwards and a crossing of the Δ(4) in SiGe and the Δ(2) level in $Si_{1-z}C_z$ takes place. For $z \geq 0.008$ optically excited electrons relax into the Δ(2) CB states in $Si_{1-z}C_z$ and thus the PL energy decreases as observed from $Si_{1-z}C_z$ single QW structures.

A small PL redshift of about -10 meV / %C is observed for the $Si_{0.84-y}Ge_{0.16}C_y$ / $Si_{1-z}C_z$ double quantum wells with respect to corresponding $Si_{0.84}Ge_{0.16}$/$Si_{1-z}C_z$ DQW structures. This can be assigned to an increase of the hh VB offset in a 4 nm $Si_{0.84-y}Ge_{0.16}C_y$ quantum well by C incorporation relative to Si. This VB shift is driven by intrinsic properties and not by the effect of strain reduction, which would rather decrease the VB offset as discussed above and shown in figure 8. However, more detailed studies are needed to get a reliable picture of the quantitative change of the VB offset, since other groups found a slight decrease of the VB offset in capacitance-voltage measurements from Si/SiGeC heterostructures [16]. One has to keep in mind that confinement effects, a modified excitonic binding energy in DQWs, inhomogenities within the layers or impurities may affect the quantitative PL shifts with C content given in figure 10.

In conclusion, we can describe the energy shifts of the PL lines for $Si_{0.84}Ge_{0.16}$/$Si_{1-z}C_z$ and $Si_{0.84-y}Ge_{0.16}C_y$ / $Si_{1-z}C_z$ DQWs considering strain and intrinsic effects of C on the alloy layers within the picture of the CB and VB alignments shown in figure 11. An improved PL efficiency of the DQWs embedded in Si is observed compared to single QWs. Further enhanced no-phonon transitions and even more efficient capture of excited carriers may be expected for larger Ge and C contents. It is also interesting that the DQW structures presented here are nearly symmetric in strain, which means that there are no strong overall thickness limitations for the superlattice. This makes this concept interesting for future applications in waveguides, infrared detectors, modulators and electroluminescence devices grown directly on Si.

7. Summary

In pseudomorphic $Si_{1-y}C_y$ layers on Si the band gap is reduced by about 65 meV/%C. Considering deformation potentials we find the band gap of unstrained $Si_{1-y}C_y$ alloys with small carbon concentration to be samller than in Si by about 15 meV/%C. The band gap in pseudomorphic SiGeC is increased by substitutional C incorporation by about 24 meV/%C. The band alignment in $Si/Si_{1-y}C_y$ is most probably type I with the offset mainly in the conduction band. A strong electron and hole confinement is achieved directly on Si in coupled $Si_{1-x}Ge_x/Si_{1-y}C_y$ quantum wells, which results in a strong no-phonon PL intensity.

Hall measurements on homogenously and modulation doped $Si_{1-y}C_y$ and $Si_{1-x-y}Ge_xC_y$ alloys have been published in Ref. [32,33]. The results are summarised as follows: Enhanced electron mobilities are observed in thick pseudomorphic n-type doped $Si_{1-y}C_y$ layers at low temperatures, which has been theoretically predicted [34] and is attributed to the splitting of the Δ valleys with the $\Delta(2)$ valleys being the CB minima. In a modulation doped p-type $Si_{0.49}Ge_{0.49}C_{0.02}$ QW we observe an improved hole mobility at room temperature and 77 K compared to a corresponding sample without C, which is a consequence of the reduced strain in the layer due to substitutional C.

The PL investigations and the Hall measurements presented provide a picture of the influence of substitutional C on the band gap and the band alignment. The results demonstrate that $Si_{1-y}C_y$ and $Si_{1-x-y}Ge_xC_y$ alloys are useful semiconductor materials which offer more flexibility in the design of device structures directly on Si avoiding the complicated and thick relaxed SiGe buffer layers.

8. Acknowledgement

This work has been supported financially by the Bundesministerium für Bildung und Forschung within the „Si Nanoelektronik" project. The authors gratefully acknowledge the excellent technical work of W. Winter and the continuous support of K. von Klitzing.

References:

1. E. Kasper, Properties of Strained and Relaxed Silicon Germanium, EMIS Datareviews Series No 12, (1995).

2. K. Ismail, F.K. LeGoues, K.L. Saenger, M. Arafa, J.O. Chu, P.M. Mooney and B. S. Meyerson, Identification of a mobility-limiting scattering mechanism in modulation-doped Si/SiGe heterostructures, Phys. Rev. Lett., **73**, 3447 (1994).
3. S.S. Iyer, K. Eberl, M. Goorsky, F.K. LeGoues, J.C. Tsang, and F. Cardone, Synthesis of $Si_{1-y}C_y$ alloys by molecular beam epitaxy, Appl. Phys. Lett. **60**, 356 (1992).
4. K. Eberl, S.S. Iyer, J.C. Tsang, M.S. Goorsky and F.K. LeGoues, The growth and characterization of $Si_{1-y}C_y$ alloys on Si(001), J. Vac. Sci. Technol. B. **10** 934 (1992).
5. K. Eberl , S.S. Iyer, S. Zollner, J.C. Tsang, and F.K. LeGoues, Growth and strain compensation effects of the ternary SiGeC alloy system, Appl. Phys. Lett. 60, 3033 (1992).
6. J.B. Posthill, R.A. Rudder, S.V. Hattanggady, C.G. Fountain and R.J. Markunas, On the fesibility of growing dilute $C_{1-x}Si_x$ epitaxial alloys, Appl. Phys. Lett. **56**, 734 (1990).
7. H.J. Osten, E. Bugiel and P. Zaumseil, Growth of an inverse tetragonal distorted SiGe layer on Si(001) by adding small amounts of carbon, Appl. Phys. Lett. **64**, 3440 (1994); and H. Rücker, M. Methfessel, E. Bugiel and H.J. Osten, Strain-stabilized highly concentrated pseudomphic $Si_{1-y}C_y$ layers on silicon, Phys. Rev. Lett. **72**, 3578 (1994).
8. J. Kolodzey, P.R. Berger, B.A. Orner, D. Hits, F. Chen, A. Khan, S. Shao, M.M. Waite, S. Ismat Shah, C.P. Swann and K.M. Unruh, Optical and electronic properties of SiGeC alloys grown on Si substrate, J. Crystal Growth **157**, 386 (1995).
9. W. Faschinger, S. Zerlauth, G. Bauer and L. Palmetshofer, Electrical properties of $Si_{1-x}C_x$ alloys and modulation doped Si / $Si_{1-x}C_x$ /Si structures, Appl. Phys. Lett. **67**, 3933 (1995).
10. P.O. Pettersson, C.C. Ahn, T.C. McGill, E.T. Croke and A. T. Hunter, Sb-surfactant-mediated growth of Si/$Si_{1-x}C_x$ superlattices by molecular beam epitaxy, Appl. Phys. Lett. **67**, 2530 (1995).

11.G. He, M. D. Savellano and H. A. Atwater, Synthesis of dislocation free $Si_y(Sb_xC_{1-x})_{1-y}$ alloys by molecular beam epitaxy and solid phase epitaxy, Appl. Phys. Lett. **65**, 1159 (1995).

12.J.W. Strane, H.J. Stein, L.R. Lee, B.L. Doyle, S.T. Picraux and J.W. Mayer, Metastable SiGeC formation by solid phase epitaxy, Appl. Phys. Lett. **63**, 2786 (1993).

13.J.L. Regolini, F. Gisbert, G. Dolino and P. Boucaud, Growth and characterization of strain compensated SiGeC epitaxial layers, Mater. Lett. **18**, 57 (1993).

14.Z. Atzomon, A.E. Bair, E.J. Jaquez, J.W. Mayer, D. Chandrasekhar, D.J. Smith, R.L. Hervig and McD. Robinson, Chemical vapor deposition of heteroepitaxial SiGeC films on (100) Si substrates, Appl. Phys. Lett. **65**, 2559 (1994).

15.L.D. Lanzerotti, A. S. Amour, C.W. Liu and J. C. Sturm, Tech. Dig. IEDM, 930, (1994)

16.K. Rim, S. Takagi, J.J. Welser, J.L. Hoyt and J.F. Gibbons, Capacitance-Voltage characterization of p-Si/SiGeC MOS capacitors, Mat. Res. Soc. Symp. Proc. Vol. **397**, 327 (1995).

17.J. Mi, P. Warren, P. Letourneau, M. Judelewicx, M. Gailhanou, M. Dutoit, C. Dubois and J.C. Dupuy, High quality SiGeC epitaxial layers grown on (100) Si by rapid thermal chemical vapor depositon using methylsilane, Appl. Phys. Lett, **67**, 259 (1995).

18.S.C. Jain, H.J. Osten, B. Dietrich and H. Rücker, Growth and properties of strained SiGeC layers, Semicond. Sci. Technol. 10, 1289 (1995).

19.A.A. Demkov and O.F. Sankey, Theoretical investigation of random Si-C alloys, Phys. Rev. B **48**, 2207 (1993).

20.G. Lippert, H.J. Osten, D. Krüger, P. Gaworzewski and K. Eberl, Heavy phosphorus doping in molecular beam epitaxy grown silicon with a GaP decomposition source, Appl. Phys. Lett. **66**, 3197 (1995).

21.K. Brunner, K. Eberl, W. Winter and N.Y. Jin-Phillipp, Photoluminescence study of $Si_{1-y}C_y$ /Si quantum well structures grown by molecular beam epitaxy, Appl. Phys. Lett. **69** ,91 (1996).

22.K. Brunner, K. Eberl and W. Winter, Near Band-edge Photoluminescence from pseudomorpnic $Si_{1-y}C_y$ /Si quantum well structures, Phys. Rev. Lett. **76**, 303 (1996).

23.R.A. Soref, Optical bandgap of the ternary semiconductor SiGeC, J. Appl. Phys. **70**, 2470 (1991).

24.A.R. Powell, K. Eberl, B.A. Ek and S.S. Iyer, SiGeC growth and properties of the ternary system, J. Cryst. Growth 127, 425 (1993).

25.P. Boucaud, C. Francis, A. Larré, F.H. Julien, J.M. Lourtioz, D. Bouchier, S. Bonar and J.T. Regolini, Photoluminescence of strained $Si_{1-y}C_y$ alloys grown at low temperature, Appl. Phys. Lett. **66**, 70 (1995).

26.K. Brunner, W. Winter and K. Eberl, Spatially indirect radiative recombination of carriers localized in SiGeC/$Si_{1-y}C_y$ double quantum well structure on Si substrate, Appl. Phys. Lett. 69, 1279 (1996).

27.J.C. Sturm, H. Manoharan, L.C. Lenchyshyn, M.L.W. Thewalt, N.L. Rowell, J.-P. Noel and D.C.Houghton, Well-resolved band-edge photoluminescence of excitons confined in strained SiGe quantum wells, Phys. Rev. Lett. **66**, 1362 (1991).

28.D.C. Houghton, G.C. Aers, S.-R. E. Yang, E. Wang, and N.L. Rowell, Type I band alignment in SiGe/Si(001) quantum wells: Photoluminescence under applied [110] and [100] uniaxial strain, Phys. Rev. Lett. **75**, 866 (1995).

29.A. S. Amour, C.W. Liu, J. C. Sturm, Y. Lacroix and M. L. W. Thewalt, Defect-free band-edge photoluminescence and band gap measurement of pseudomorphic SiGeC alloy layers on Si, Appl. Phys. Lett. **67**, 3915 (1995).

30.D. Dutartre, G. Bremond, A. Souifi, T. Benyattou, Excitonic photoluminescence from Si-capped strained SiGe layers, Phys. Rev. B **44**, 11525, (1991).

31.M. Gail, G. Abstreiter, J. Olajos, J. Engvall, H. Grimmeis, H. Kibbel and H. Presting, Room-temperature photoluminescence of $Ge_mSi_nGe_m$ structures, Appl. Phys. Lett. **66**, 2978 (1995).

32.K. Eberl, K. Brunner and W. Winter, Pseudomorphic $Si_{1-y}C_y$ and SiGeC alloy layers on Si, Thin Solid Films (1996) in press.

33.K. Brunner, K. Eberl, W. Winter, N.Y. Jin-Phillipp and F. Phillipp, Fabrication and band alignment of pseudomorphic $Si_{1-y}C_y$ and SiGeC and coupled $Si_{1-y}C_y$ / SiGeC quantum well structures on Si substrate, J. Crystal Growth (1996) in press.

34.M. Ershov and V. Ryzhii, Monte Carlo study of electron transport in strained silicon-carbon alloy, J. Appl. Phys. 76, 1924 (1994).

OPTICAL PROCESSES IN TWO-DIMENSIONAL II-VI SYSTEMS

R.CINGOLANI
*Istituto Nazionale di Fisica della Materia- Unita' di Lecce,
Dipartimento di Scienza dei Materiali, Universita' degli Studi di Lecce,
via Arnesano , 73100 Lecce (Italy)*

1. Introduction and Background

Since the first observation of excitonic gain in ZnSe-based multiple quantum wells (MQWs) [1, 2, 3], much attention has been devoted to the radiative recombination processes underlying stimulated emission in blue-green quantum well lasers. Unlike the case of III-V heterostructures, excitons are very stable in II-VI quantum wells and their binding energy can easily exceed the energy of the longitudinal optical (LO) phonon. It is therefore expected that excitons remain stable even in the presence of high carrier densities and at high temperatures, thus playing an important role in the operation of blue-green lasers and modulators. The understanding of laser action and of the optical non- linearities in these materials is thus complicated by interplay of the exciton and free-carrier populations, so that lasing can be generated either by free-carrier or excitons. The former mechanism involves free carriers distributed on a three level system comprised of the electron and hole quasi-Fermi levels, and the renormalized band gap from which stimulated emission originates. The latter mechanism involves the stimulated recombination of free excitons, and may occur provided that an intermediate level - a localized exciton state or the final state of a scattering process exists to form the three level system. Since the demand for efficient blue-green light generators is increasing, due to the appealing application to high density optical data storage

G. Abstreiter et al. (eds.), Optical Spectroscopy of Low Dimensional Semiconductors, 21–40.

exploiting the diffraction-limited focussing of short wavelengths , it is now important to understand deeply the physics of these processes. Similar considerations hold for wide gap III-V materials (Nitrides) which have had an impressive progress in the last two years (thus becoming natural competitors of II-VI materials [4]). To date the situation in the literature reflects the difficult assessment of the different recombination mechanisms in wide gap lasers. Different authors have associated lasing of II-VI quantum wells to excitonic or free-carrier recombination, so that the physics of gain in blue-gree lasers remains somewhat controversial [1, 5, 6]. In Nitrides the situation is similar due to the large polarity of the structures, though the physical understanding is even more difficult due to the lack of information on the excitonic properties of GaN (dependence on politypes, binding energy etc.). In this contribution we will review recent works on the physics of II-VI quantum wells, and discuss the relevant concepts of lasing in wide band gap heterostructures. The main conclusion is that ZnCdSe/ZnSe quantum wells represent a prototype system where excitonic recombination and free-carrier recombination compete in determining lasing. In particular, depending on the degree of confinement of the exciton wavefunction, identical injection conditions can create either an exciton-rich or a free-carrier-rich phase. This in turn contributes to determining the main stimulated emission mechanism in quantum wells of different width and composition. Throughout the manuscript we will discuss several experiments performed on $Zn_{1-x}Cd_xSe/ZnSe$ quantum wells grown by MBE, similar to those investigated in the pioneering works of the Brown-Purdue group [7]. These structures usually consist of ten-periods of Cd content ranging between 0.1 and 0.26, and well width $< L_z$ in the 3 to 20 nm range. This choice of compositional and structural parameters of the wells allowed us to tune the exciton binding energy (E_b) from about 20 meV to about 38 meV [8]. This, in turn, permits the investigation of samples in which the exciton binding energy can be either larger or smaller than the LO-phonon energy, with a correspondingly large variation in the thermal and electrostatic stability of the excitonic particle. For brevity, in what follows we will define quantum wells shallow (or Cd-poor) if the binding energy of the exciton is substantially lower than the LO-phonon energy, and deep (or Cd-rich), if the binding energy is higher [8].

2.Linear Optical Properties

In Fig.1 we show representative absorption (solid line) and PL spectra (dashed line) of selected ZnCdSe/ZnSe MQW samples. Distinct excitonic resonances are observed in the low energy range of each absorption spectrum. The MQW-related absorption edge is found to blue-shift in narrow wells, due to confinement, and to red-shift with increasing Cd-content, due to the decrease of the energy gap of the strained ternary alloys. Higher-energy features at a constant position reflect the excitonic and interband absorption edge of the ZnSe barrier and buffer layers. The PL spectra exhibit negligible Stokes-shift in shallow quantum wells, the emission being of excitonic origin. Conversely, an increasing Stokes shift is observed in deep (Cd-rich) MQWs. This normally reflects a dominant exciton localization at sample inhomogeneities, either interface-related, or due to compositional fluctuations within the quantum well [9]. The temperature dependence of the absorption spectra clearly indicates the dramatic difference in the stability of the excitonic resonances of deep (Fig.2a) and shallow quantum wells (Fig.2b). The deep quantum wells exhibit well resolved excitonic bands up to room temperature, consistent with the large binding energy. Conversely, thermal ionization of the weakly bound exciton occurs above 200 K in the shallow wells. In order to compare these excitonic properties with the well known case of III-V materials, we plot in Fig.3a the well-width dependence of the exciton binding energy of $Zn_{0.77}Cd_{0.23}Se/ZnSe$, $GaAs/Al_{0.3}Ga_{0.7}As$ and $In_{0.11}Ga_{0.89}As/GaAs$ MQWs. In Fig.3b we also plot the compositional dependence of the exciton binding energy of

ZnCdSe/ZnSe MQWs for a fixed well width ($L_z = 3nm$). The results of Figs.3a and 3b clearly indicate that the excitonic properties of II-VI ternary QWs can be taylored by a proper combination of well width and composition, giving an unprecedented

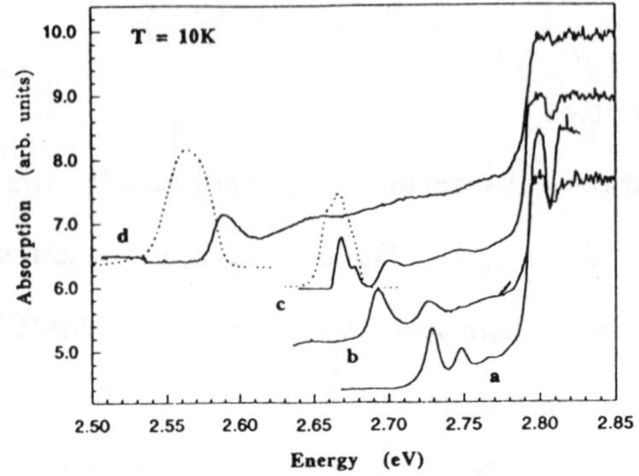

Fig.1 - Absorption (continuous lines) and photoluminescence (dashed lines) of $Zn_{(1-x)}Cd_xSe/ZnSe$ multiple quantum well structures of different composition (x) and thickness (L_w). The sample parameters are: (a) x=0.11 and L_w=3 nm, (b) x=0.16 and L_w=3nm, (c) x=0.11 and L_w=7 nm, (d) x=0.23, L_w=3 nm. The temperature is 10 K.

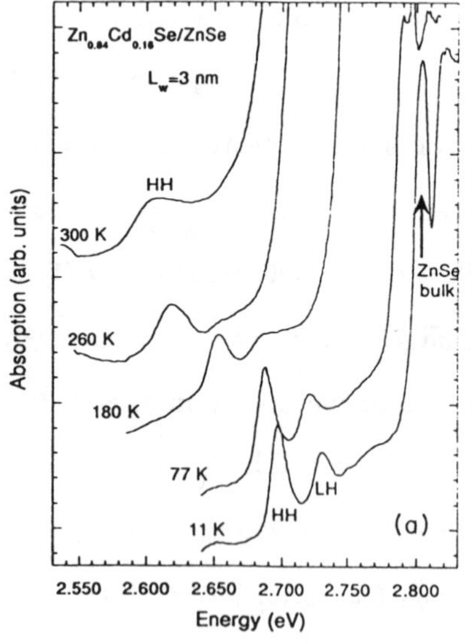

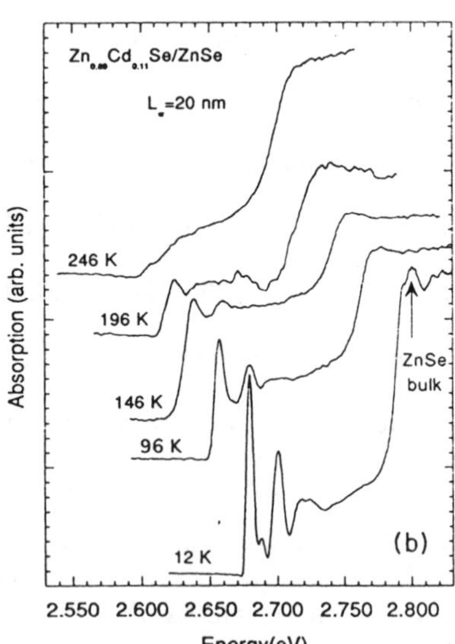

Fig.2 - Temperature dependence of the absorption of a deep (a) and a shallow (b) quantum well sample.

flexibility to. this material system (also, the energy gap of the structure can be tuned from the green to the blue in the range of parameters selected in Figs.3).

3. Non-linear Optical Properties

Such a strong dependence of the excitonic properties on the quantum well parameters raises the important question about the excitonic stability in real devices. This can be studied by pump and probe transmission spectra, eventually collecting simultaneously the stimulated emission excited by the pump and emitted from the sample edge, in order to correlate the occurrence of lasing with the actual strength of the excitonic resonance in the absorption spectra [10]. This experiment has been performed in many samples with binding energies varying between 20 meV and 40meV. In Fig.4 we show the pump and probe spectra of a deep quantum well (E_b=37 meV). Different symbols correspond to different pump intensities. A clear saturation of the excitonic absorption with increasing pump power can be observed. The systematic analysis of these experiments demonstrates that in deep and narrow quantum wells (large binding energy) excitons are very stable, and can survive up to comparatively high photogeneration rates (up to about $I_{sat} = 50kWcm^{-2}$). Conversely, shallow quantum wells exhibit bulk- like excitonic stability, and the excitonic resonance disappears at relatively low power densities ($I_{sat} = 2kWcm^{-2}$). The stimulated emission threshold of these samples is of the order of $5kWcm^{-2}$, regardless of sample composition and well width. Threfore lasing in deep quantum wells occurs when the excitonic resonance is still present in the absorption spectrum, whereas in shallow wells it occurs when the exciton absorption is already bleached. This suggests that the lasing mechanism is different in MQWs with substantially different exciton binding energy.

This situation can be modeled assuming the coexistence of the exciton and electron-hole gas within the quantum well. The key ideas of the model are the following:

(i) Due to the strong exciton stability in II-VI quantum wells, excitons and free carriers coexist even at very high density, under steady-state excitation conditions. The hot-carriers generated by the photoexcitation can form excitons during the electron-hole pair lifetime leading to a mixed gas of bound and unbound electron-hole pairs.
(ii) The relative density of the exciton and free-carrier gas is determined by a phase diagram, modeled by a mass-action law. At a given temperature and total density, the equilibrium between the two phases is reached when the exciton formation rate balances the ionization rate.
(iii) The phase diagram primarily reflects the renormalization of the exciton binding energy that is determined by many-body effects and by the temperature of the hot-carriers photogenerated by the pump. At steady-state we assume the same temperature for excitons, electrons and holes. In optical experiments, the total density and the gas temperature are determined by the photogeneration rate (i.e. the pump power) and by the excess energy of the electron-hole pair (i.e. the excitation wavelength), respectively.
(iv) Due to the large density of excitons coexisting with the carrier plasma, both the self-energy correction induced by free-carriers and that induced by excitons have to be included in the renormalization of the exciton binding energy. This is unlike the case of III-V materials, in which excitons are readily ionized at relatively low injection rates.
(v) The many-body energy renormalization is used to quantitatively analyse the non-linear pump and probe spectra. From the analysis, we obtain in turn the total density of the photogenerated elementary excitations at different pump intensities and therefore the exciton/free-carrier phase diagram.
The main drawback of this treatment is the neglect of exciton localization (particularly important in Cd-rich quantum wells as exemplified by the increased Stokes shift in Fig. 1) which might raise considerably the screening threshold of the exciton leading to an even stabler exciton resonance.

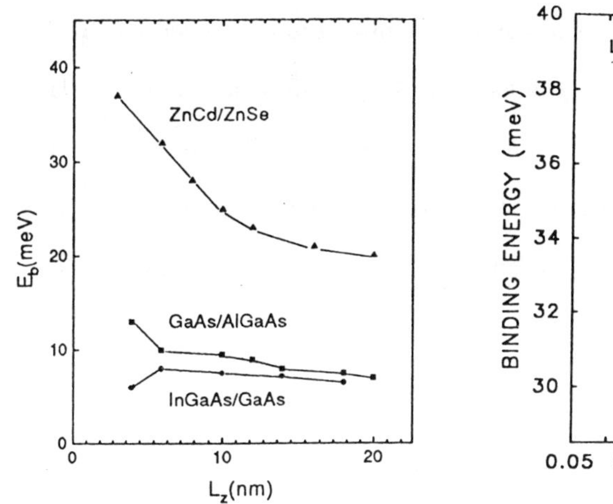

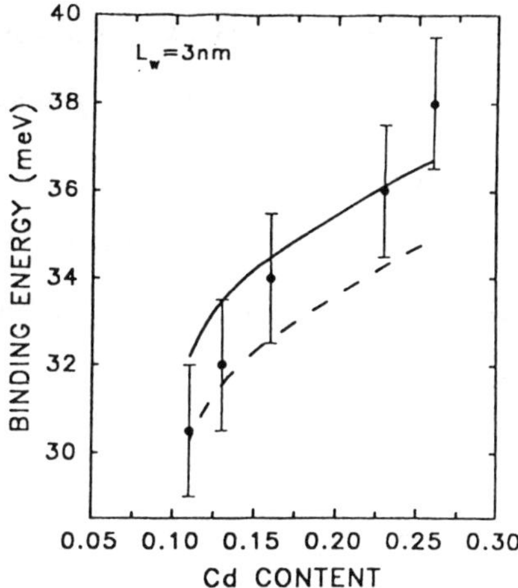

Fig.3a - Well width dependence of the exciton binding energy of different MQW systems. The compositions are x=0.23 for the ZnCdSe, x=0.3 for the AlGaAs,and x=0.1 for the InGaAs.
3b - Composition dependence of the exciton binding energy of a ZnCdSe/ZnSe quantum well of width L_z= 3nm. The curves are the results of variational calculations performed with slightly different values of the dielctric constant.

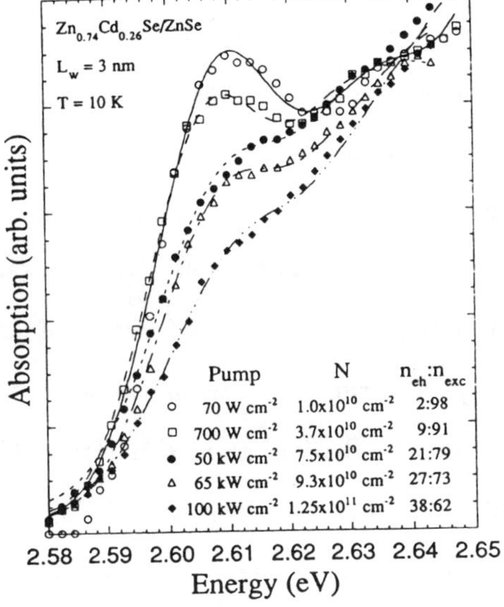

Fig.4 - Experimental (symbols) and calculated (curves) pump and probe transmission spectra of 3nm wide $Zn_{0.77}Cd_{0.23}Se/ZnSe$ quantum well samples measured at 10 K and under different pump power densities. The table indicates the pump power density (pump), the corresponding total density of photogenerated elementary excitations (N) obtained from the self-consistent line shape calculation, and the ratio of the free-carrier to exciton population ($n_{eh} : n_{exc}$).

The total density of elementary excitations N in our model is expressed as the sum of the free-carrier (n_{eh}) and exciton (n_{exc}) densities. The equilibrium between the two phases established by the mass-action law depends on the equilibrium temperature of the whole system and on the density-dependent exciton binding energy $E_b'(n_{eh}, n_{exc})$. This is expressed by a set of coupled equations for the total density of photogenerated elementary excitations

$$N = n_{eh}(T) + n_{exc}(T) \tag{1}$$

and for the relative balance of free-carriers and excitons:

$$\frac{n_{eh}^2}{n_{exc}} = \frac{m_e m_h k_b T}{2\pi\hbar^2(m_e + m_h)} \cdot exp(\frac{-E_b'[n_{exc}, n_{eh}]}{k_b T}) \tag{2}$$

The renormalized exciton binding energy E_b' can be written as:

$$E_b'(N) = E_b - |\Delta E_g(n_{eh})| - |\Delta E_g(n_{exc})| - |\Delta E_{exc}(n_{eh})| - |\Delta E_{exc}(n_{exc})| \tag{3}$$

where E_b is the unperturbed quasi-two-dimensional binding energy of the exciton [11], ΔE_g is the band gap renormalization, and ΔE_{exc} is the correction to the exciton energy including both phase space filling and exchange. Due to the coexistence of a dense exciton and free-carrier gas, we have to consider that both the exciton population and the electron-hole populations induce many-body corrections. The band gap renormalization as well as the exciton energy corrections are therefore a function of both the free-carrier and exciton populations. The explicit forms of the renormalisation terms in Eq.(3) are calculated in ref.[10] as first-order (perturbative) approximations which are reasonably valid for carrier densities smaller than the saturation density of the exciton. By this method it is possible to have a very physical description of the excitonic and free-carrier many-body interaction which is extremely useful for the understanding of the blue-green device operation.

A more precise description has to resort to a computationally intensive calculation in the frame of the random phase approximation or by the Green function formalism . To this aim the reader is demainded to ref. [12] (and references therein).

The above model is used to calculate the non-linear pump and probe spectra of Fig.4. The non-linear absorption profile is given by:

$$\alpha(\hbar\omega, N) = C \cdot (\frac{1}{1+\frac{N}{N_{sat}}}) \cdot G\{\hbar\omega - [E_g - E_b'(n_{exc}, n_{eh})], \Gamma(N), \gamma\} + C' \cdot D[\hbar\omega - E_g'(n_{eh}, n_{exc}, \Gamma(N), \gamma)] \cdot S(\hbar\omega - E_b) \cdot [f_e(n_{eh}, T) - f_h(n_{eh}, T)] \tag{4}$$

The first factor in Eq.(4) accounts for the saturation of the oscillator strength of the excitonic transition through a critical saturation density N_{sat} [13]. The exciton density of states is modeled by a Gaussian function G whereas the Θ function in Eq.(4) models the step-like density of states. Both these quantities contain a density-dependent lifetime broadening $\Gamma(N)$ [12] and an inhomogeneous broadening γ to account for inhomogenieties in the ternary well (typically $\pm$ 1% in the Cd content). C and C' are amplitude constants containing the matrix element, evaluated in ref.[12]. Unlike the case of III-V quantum wells [12], we find that the Lorentzian description of the exciton fails in these II-VI quantum wells even at relatively low temperature [9]. In Eq. (4) E_g' is the renormalized band gap energy given by $E_g'(N) = E_g - \Delta E_g(n_{eh}, n_{exc})$. Coulomb coupling to the continuum, though very weak in these structures, was included using the Sommerfeld factor $S(\hbar\omega)$, according to Ref.[14].

The calculation of the non-linear absorption spectra has to be performed self-consistently, for a given total density, until the relative exciton-plasma population converged to a value left unvaried from one iteration to the next one.

The calculated PP spectra are shown (solid line) superimposed to the data (symbols) in Fig.4, and show excellent agreement with the experimental spectra. The hot-carrier temperature was directly obtained from the slope of the high-energy tail of the spontaneous luminescence spectra recorded under the same experimental conditions of the pump and probe measurements. Typical values ranged between 2 meV and 5 meV in the investigated excitation intensity range, and have been found to vary roughly sub-linear with the total density N. The phase diagrams obtained from the analysis of the pump and probe spectra are displayed in Figs.5 for a shallow and for a deep MQW sample. A number of important results are obtained from this analysis.

(i) The phase transition from exciton gas to free-carrier gas (cross-over of the curves in Figs.5) is found to occur at $7 \cdot 10^{10} cm^{-2}$ and $1.3 \cdot 10^{11} cm^{-2}$ in the shallow and in the deep sample, respectively, consistent with the different exciton binding energies.

(ii) The exciton oscillator strength vanishes at N-values close to the phase transition in all samples, indicated by the white arrows in Figs.5. Such condition corresponds to an exciton density to free-carrier density ratio approximately equal to one, thus demonstrating the impact of the exciton/plasma coexistence on the non-linear optical properties of ZnSe-based quantum wells

(iii) The boundary between the exciton phase and the plasma phase changes dramatically with the hot-carrier temperature. The smaller the temperature, the higher the critical density for the transition.

(iv) The phase transition is smooth and continuous, reflecting the progressive transformation of the Coulomb-correlated electron-hole pair into an unbound couple of carriers. In shallow quantum wells, the free-carrier phase prevails at injected densities of the order of $8 \cdot 10^{10} cm^{-2}$, consistent with the observation of exciton bleaching in the PP spectra. In depeer wells the exciton phase dominates up to higher N- values.

The results of the theory-experiment comparison also provide the density dependence of the exciton oscillator strength in II-VI quantum wells.

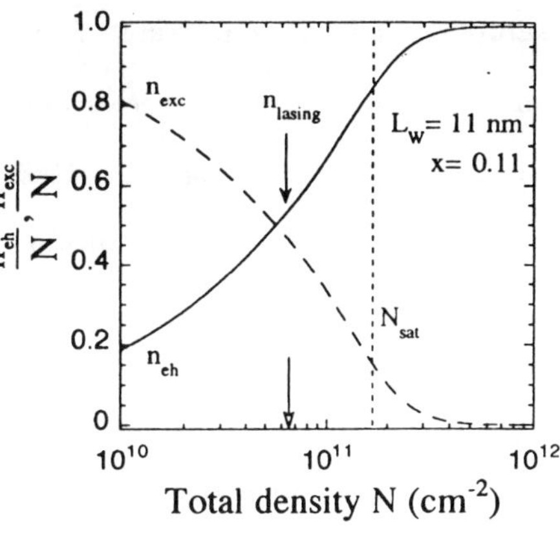

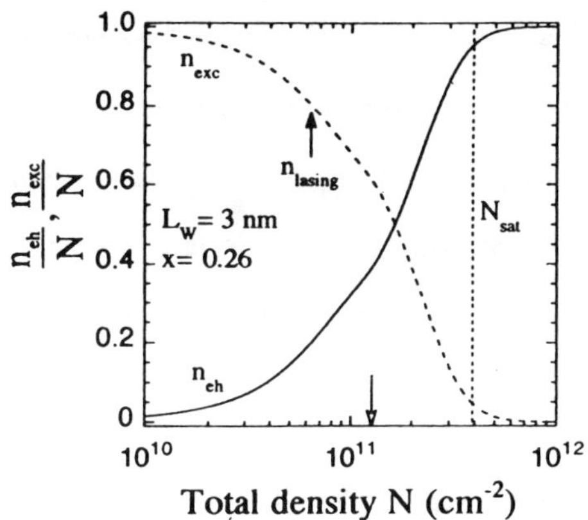

Fig.5 - Phase-diagrams of the exciton-free carrier gas for a shallow (a) and a deep (b) MQW sample. The dashed vertical line indicates the theoretical value of N_{sat} calculated including only free-carrier effects. . n_{exc} and n_{eh} are the exciton and electron-hole pair densities (long-dashed and continuous lines, respectively). n_{lasing} indicates the carrier density corresponding to the threshold intensity for stimulated emission. The white arrow indicates the total density at which the exciton oscillator strength vanishes, as obtained from the line shape analysis of the pump and probe spectra. Note that in shallow quantum wells lasing occurs in the free-carrier-rich side of the phase diagram (a), when the exciton resonance is bleached, whereas in deep quantum wells lasing occurs in the exciton-rich-side of the phase diagram, well below the total bleaching of the exciton (b).

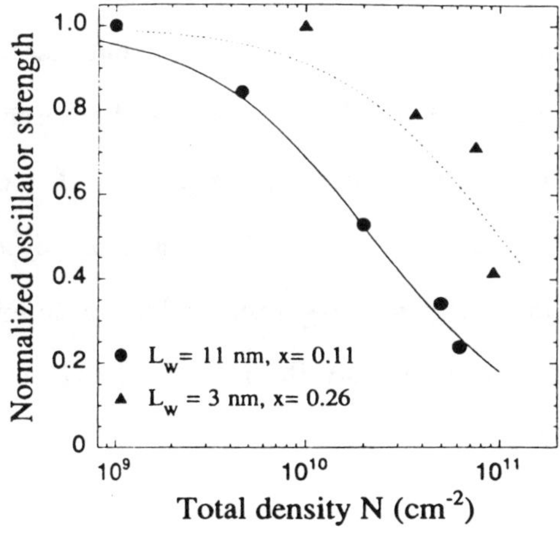

Fig.6 - Density dependence of oscillator strength of the heavy-hole exciton for the sample investigated in Figs.2a (dots) and 2c (triangles). The lines are calculated by means of eq.(14). N'_{sat} corresponds to the N-value for which the normalized oscillator strength is 0.5. The oscillator strength vanishes almost completely at N- values corresponding to the phase transition.

This is shown in Fig.6 for two representative samples. The data obtained from the measured integrated intensity of the excitonic absorption resonance (circles and triangle) in the PP spectra are shown superimposed to the corresponding calculated oscillator strength (lines), from the area of the calculated exciton peak. The oscillator strength follows the expected saturation behavior

$$f(N) = \frac{f(0)}{1 + \frac{N}{N'_{sat}}} \tag{5}$$

where the N'_{sat} obtained from the experiment is considerably smaller than the N_{sat} obtained by neglecting excitonic interactions [13]. This is a clear consequence of the complex interplay between exciton and free-carrier many- body processes and of the assumptions made about the hot-carrier distribution in the well under stationary conditions.

A direct test of these results comes from the analysis of the absorption spectra of Modulation Doping Quantum Wells (n-type MDQWs), in which a controlled density of electrons is inserted in the welll by doping the barriers [15]. In this case the barrier doping forms a one-component electron-plasma which screens the exciton approximately with one-half the strength of the photogenerated two-component electron-hole plasma. The absorption spectra of a series of ZnCdSe/ZnSe MDQWs of different carrier density is displayed in Fig.7a, showing a clear exciton bleaching. In this case the density dependence of the oscillator strength obtained for Cd-rich and Cd-poor structures (see Fig.7b) is determined experimentally, directly from the known doping densities, and is found to be in excellent agreement with that obtained by the theoretical analysis of the non-linear absorption spectra.

4. Lasing

The quantitative description of the exciton-free carrier phase diagram can be correlated to the stimulated emission threshold I_{lasing}. In comparing the stimulated emission threshold with the pump and probe experiments, it is clear that I_{lasing} can be either larger or smaller than the pump intensity corresponding to the saturation density, depending on the exciton stability in the investigated sample (i.e. on the size and composition of the quantum well). The deep and narrow quantum wells ($E_b \geq 32$ meV) exhibit stimulated emission at power densities such that distinguishable excitonic features can still be observed in the pump and probe spectra, i.e. $I_{lasing} < I_{sat}$, suggesting that excitons can play a role in the lasing process. On the contrary, in shallow and wide quantum wells ($E_b \leq$25 meV), lasing occurs for $I_{lasing} > I_{sat}$, when the exciton resonance is partially or even totally saturated. In this case, a dominant free-carrier recombination is expected. In the wide range of intermediate size and composition, I_{sat} and I_{lasing} are comparable, so that the exciton or free-carrier character of the emission should depend on wether the photogenerated N-value falls below or above the phase transition density.

The carrier density (n_{lasing}) corresponding to the stimulated emission threshold, obtained from the line shape analysis of the PP spectra, are plotted on the phase diagram of Figs.5. The values fall into the exciton-rich phase for the deep quantum wells and in the free-carrier-rich phase for the shallow samples. For intermediate well sizes and compositions, the actual lasing mechanism can be tuned by varying the initial excess energy of the photoinjected carriers (hot carrier temperature) or the excitation intensity, leading to a subtle competition between the two recombination mechanisms. In this case, the phase-transition between the exciton gas and the free-carrier gas determines the dominant recombination process occurring in II-VI quantum well lasers. The intriguing suggestion about lasing at the exciton to free-carrier phase transition, can be independently tested by measuring the diamagnetic or the Landau-like shift of the stimulated emission in high magnetic field [5]. A weak diamagnetic (parabolic) shift is expected for excitonic recombination (of the order of 2 $\mu eV \cdot T^{-2}$), whereas a

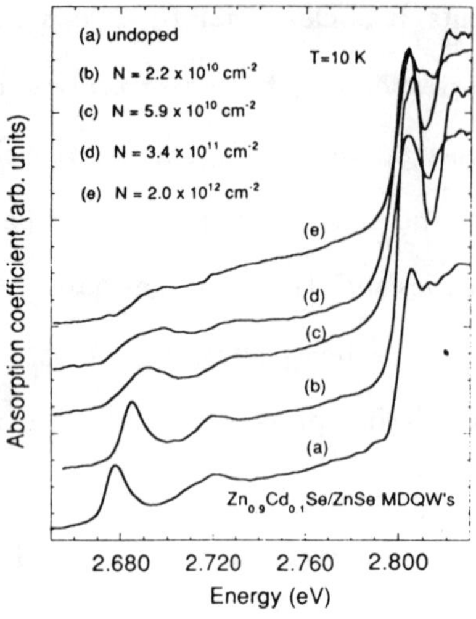

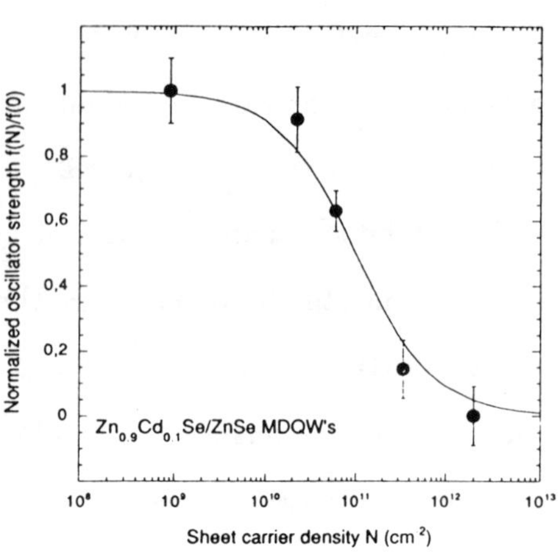

Fig.7a - Absorption spectra of n-type $Zn_{0.9}Cd_{0.1}Se/ZnSe$ modulation doping quantum wells of different doping density at 10 K. The well width is L_z=3nm. The doping densities are: a) $2.2 \cdot 10^{10}cm^{-2}$, b)$5.9 \cdot 10^{10}cm^{-2}$, c) $3.4 \cdot 11^{10}cm^{-2}$ and d)$2 \cdot 10^{12}cm^{-2}$.

7b - density dependence of the oscillator strength of the heavy-hole exciton exciton. The line is calculated by means of eq.(5).

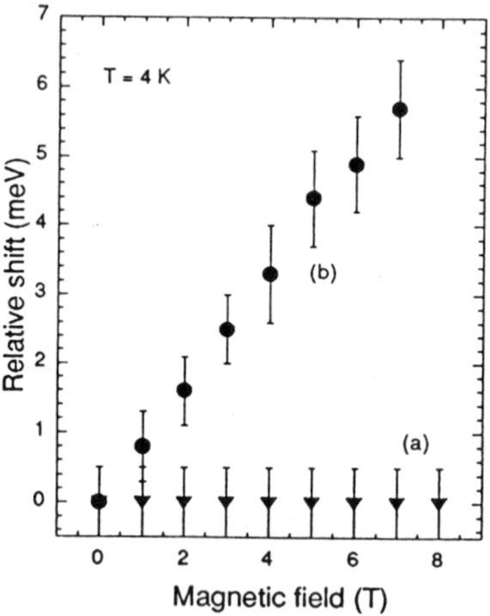

Fig.8 - Shift of the stimulated emission in magnetic field for two $Zn_{1-x}Cd_xSe/ZnSe$ multiple quantum well samples of well width L_w= 7nm and x =0.23 (a) and x=0.11 (b).

linear Landau shift (of the order of 0.5 meV/T) should be observed when free-carrier recombination dominates. In Fig.8, we display representative data for two MQWs of well width of 7 nm and Cd content 11 % and 23 %. Accordingly to the respective phase diagrams, the two saples around lasing threshold exhibit a completely different behavior. The deep MQW structure is indeed characterized by a diamagnetic exciton shift, whereas the shallow sample clearly shows a free- carrier like Landau shift.

Though this is a convincing proof of the phase diagram model, one should be aware that exciton localisation at compositional fluctuations or disorder causes the enhancement of the exciton stability due to the further shrinkage of the excitonic wavefunction. This might cause the phase transition (from excitons to free-carriers) to occur at densities larger than expected. An indirect confirmation of important localization comes from the observation of increasing Stokes shift between absorption and luminescence in Cd-rich samples (see Fig.1). More unambiguous evidence for exciton localization is obtained by means of time-resolved PL experiments. In these experiments, photoluminescence is excited in backscattering configuration under injection rates ranging from well below ($I \simeq 0.3 I_{lasing}$) to well above ($I \simeq 3 I_{lasing}$) the stimulated emission threshold. The excess energy of the off-resonant pumping (λ=390 nm, $\hbar\omega$=3.18 eV) is expected to be dissipated within a few hundred fs through exciton-LO phonon interaction, resulting in a quasi-equilibrium distribution of hot-excitons in the well. In Figs.9a-9b, we display the spectrally resolved decay time and rise time of a shallow and a deep quantum well sample, measured below and above I_{lasing}. As a general trend, we see long rise and decay times of the photoluminescence in the low energy tail of the exciton resonance, due to trapping of excitons at local potential fluctuations [16]. This temporal evolution changes with excitation intensity and quantum well parameters. In shallow quantum wells both the rise time and the decay time get shorter with increasing pump power. This change reveals the saturation of the available density of localization states and

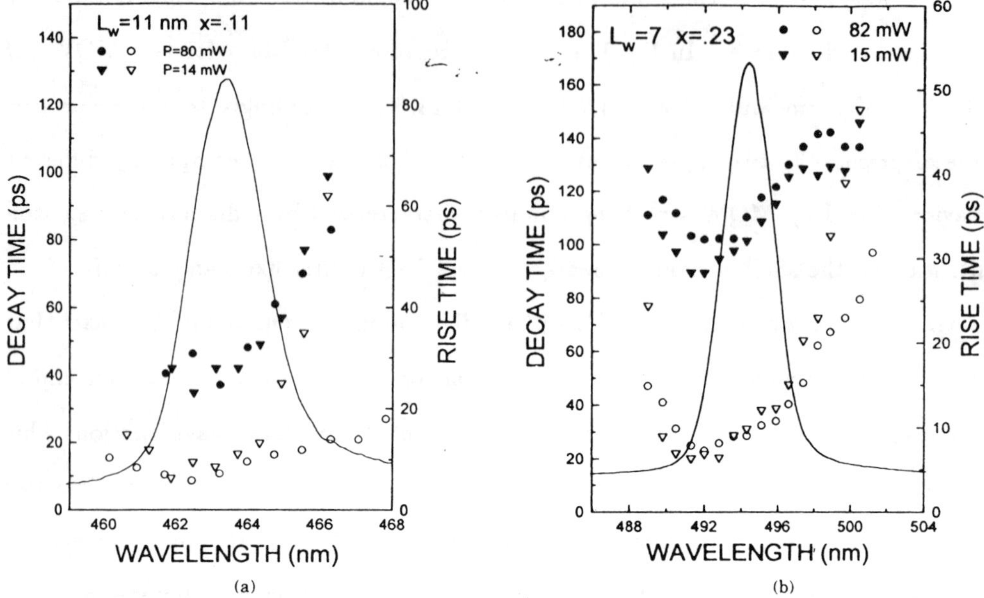

Fig.9 - (a) Spectrally resolved decay times (full symbols and left-hand scale) and rise times (empty symbols and right-hand scale) measured in a shallow and wide quantum well sample. The triangles are obtained at a power density of the order of 1.5 I_{lasing}, whereas the dots were obtained below threshold (around 0.3 I_{lasing}). P labels the integrated power of the picosecond laser.
(b) The same as in Fig.9(a), but for a deep and narrow quantum well.Note that the rise time remains long even above I_{lasing}. The solid lines represent the time integrated photoluminescence spectra.

the recovery of the intrinsic free-exciton recombination. A further increase of the excitation level generates a rather high carrier density resulting in the almost constant luminescence intensity during the first 150 ps, followed by a longer decay time. Under this condition, the excitonic population is fed by the carrier plasma as long as excess carriers are available to form new excitons, resulting in the "pleateau" region on the time-resolved curve. Conversely, no recovery of the intrinsic free-exciton recombination is observed in Cd-rich MQWs (Fig.9b), even at pump intensities comparable with I_{lasing}, indicating a dominant recombination of localized excitons even under stimulated emission conditions.

The above result is very important to understand the lasing mechanism. The stimulated emission intensity in deep wells is usually much smaller than that required to saturate localization. This supports the model of optical-gain from localized exciton states proposed by Ding et al.[1]. Conversely, the lasing threshold in shallow quantum wells (Fig. 9a) corresponds to excitation intensity levels such that both localized excitons and free-excitons are saturated, leaving free-carrier recombination as the main lasing mechanism [5].

5. Conclusions

In conclusion, we have shown that a smooth transition from excitonic lasing to free-carrier lasing occurs in II-VI quantum wells. This phenomenon is determined by the phase diagram of the exciton/plasma gas, which is described by a self-consistent mass-action law including hot-carrier effects and many-body renormalization of the exciton binding energy. The quantitative analysis of the non-linear pump and probe spectra allowed us to derive a free-exciton/plasma phase diagram and to interprete the basic physical mechanism underlying laser action in these materials. While free-carrier lasing is predominantly observed in quantum wells with bulk-like exciton binding energy, a strong excitonic character of the laser action is observed in strongly confined quantum wells. This is a consequence of the modified stability of the exciton particle further enhanced by exciton localization at local inhomogeneities within the wells.

Acknowledgments

The results presented in this work are the outcome of a collaboration between the Optoelectronic group at the University of Lecce (Material Science Department) and the TASC-INFM laboratory of Trieste (Italy). Special thanks are due to Prof. A.Franciosi for the fruitful collaboration.

References

[1] J.Ding, H.Jeon, T.Ishiara, M.Hagerott, A.V.Nurmikko, H.Luo, N .Samarth, and J.Furdyna, Excitonic gain and laser emission in ZnSe based quantum wells, *Phys. Rev. Lett.* **69**, 1707 (1992)

[2] J.Ding, M.Hagerott, T.Ishihara, H.Jeon, A.V.Nurmikko, ZnCdSe/ZnSe quantum well lasers: Excitonic gain in an inhomogeneously broadened quasi-two-dimensional system, *Phys. Rev.* **B47**, 10528 (1993)

[3] Y.Kuroda, I.Suemune, Y.Fujii, and M.Fujimoto, Ultraviolet stimulated emission and optical gain in CdZnS/ZnS strained-layer superlattices, *Appl. Phys. Lett.* **61**, 1182 (1992)

[4] I.Akasaki and H.Amano, Current status of III-V Nitrides research, *Proc. Int. Symposium on Blue Lasers and Light Emitting Devices, Chiba, Japan*, page 11 (1996)

[5] R. Cingolani, R. Rinaldi, L. Calcagnile, P. Prete, P. Sciacovelli, L. Tapfer, L. Vanzetti, F. Bassani, L. Sorba, and A. Franciosi, Recombination mechanism and lasing in shallow ZnCdSe/ZnSe quantum well structures, *Phys. Rev.* **B49**, 16769 (1994).

[6] Y.Kawakami, I.Hauksonn, H.Stewart, J.Simpson, I. Galbraith, K.A.Prior, and B.C.Cavenett, Exciton-related lasing mechanism in ZnSe-ZnCdSe multiple quantum wells, *Phys. Rev.* **B48**, 11994 (1993)

[7] A.Nurmikko and R.Gunshor, II-VI lasers - Opportunities and new directions, *Proc. Int. Symposium on Blue Lasers and Light Emitting Devices, Chiba, Japan*, page 3 (1996)

[8] R.Cingolani, P.Prete, D.Greco, P.V.Giugno, M.Lomascolo, R. Rinaldi, L.Calcagnile, L.Vanzetti, L.Sorba and A.Franciosi, Exciton spectroscopy of ZnCdSe/ZnSe multiple quantum wells, *Phys. Rev.* **B51**, 5176 (1995) and references therein

[9] R.Cingolani and K.Ploog, Frequency and density dependent radiative recombination processes in III-V semiconductor quantum wells and superlattices, *Adv. Phys.* **40**, 535 (1991)

[10] R.Cingolani, L.Calcagnile, G.Coli', R.Rinaldi, M.Lomascolo, M.DiDio, A. Franciosi, L.Vanzetti, G.C.LaRocca, and D.Campi, Radiative recombination processes in II-VI wide-gap quantum wells: the interplay between excitons and free-carriers *J.Opt. Soc. Am.* **B13**, 1268 (1996)

[11] We point out that E_b represents the unperturbed binding energy of the free exciton, and neglects any contribution of localization at sample inhomogeneities. We emphasize that although localization is not accounted for neither in the mass-action law nor in the resulting distribution function, we will account for localization effects in the calculation of the nonlinear absorption profiles by broadening the Gaussian exciton density of states and the quantum well (step function) density of states using the usual inhomogenous line-broadening parameter gamma.

[12] D.Campi and C.Coriasso, Optical non-linearities in multiple quantum wells: generalized Elliott formula, *Phys. Rev.* **B51**, 10719 (1995)

[13] S.Schmitt-Rink, D.S.Chemla, and D.A.B.Miller, Theory of transient exciton non-linearities in semiconductor quantum well structures *Phys. Rev.* **32**, 6601 (1985)

[14] D.S.Chemla, D.A.B.Miller, P.W.Smith, A.C.Gossard, and W.Wiegmann, Room temperature excitonic non-linearity and refraction in GaAs/AlGaAs quantum well structures, *IEEE J.Quantum El.* **QE-20**, 265 (1984)

[15] L.Calcagnile, R.Rinaldi, P.Prete, C.J.Stevens, R.Cingolani, L.Vanzetti, L. Sorba, and A.Franciosi, Free carrier effects on the excitonic absorption of n-type ZnCdSe/ZnSe modulation doping quantum wells, *Phys. Rev.* **B52**, 17248 (1995)

[16] M.Lomascolo, M.DiDio, R.Cingolani, L.Calcagnile, L.Sorba and A.Franciosi, Free versus localized exciton recombination in ZnCdSe/ZnSe multiple quantum wells, *Appl. Phys. Lett.* **69**, 1145 (1996)

OPTICAL INTERSUBBAND ABSORPTION AND EMISSION IN QUANTUM STRUCTURES

F. H. JULIEN and P. BOUCAUD
Institut d'Electronique Fondamentale, URA 22 CNRS
Bât. 220, Université Paris XI
91405 Orsay, France.

ABSTRACT

Specific infrared properties of intersubband transitions in quantum confined semiconductor structures are reviewed. A simple model is used to derive the basic optical properties of these transitions. We illustrate how optical intersubband spectroscopy can be a powerful tool for probing the confinement energies as well as the subband dispersions using various examples either in the conduction or the valence band of quantum well structures. Recent achievements on intersubband emission are then discussed with special emphasis on Quantum Cascade lasers and optically pumped systems.

1. Introduction

Intersubband transitions which occur between confined quantum states or bound-to-continuum states either in the valence band or in the conduction band are specific of quantum semiconductor structures. The quantum confinement of the carriers enhances the interaction between subbands leading to intersubband transitions with a narrow bandwidth (typically 4-10 meV) along with a large oscillator strength. The resonance wavelength which depends on both physical (effective masses) and structural (width, composition) parameters of the quantum structure, exhibits a large range of tunability from near-infrared to far-infrared. Intersubband optical transitions were first observed in the 2D electron gas formed in Si inversion layers [1]. Further evidences of intersubband transitions were obtained in GaAs/AlGaAs heterostructures [2] and quantum wells [3] using resonant Raman scattering techniques. The first direct observation of IR absorption between conduction subbands of n-doped GaAs/AlGaAs quantum wells (QW) was reported in 1985 [4]. It was confirmed that intersubband transitions between electronic states of QWs are strongly polarized along the confinement potential direction. This polarization selection rule, i.e. the light must have a polarization component perpendicular to the QW layers, has strong consequences for practical applications since normal incidence absorption is usually forbidden. However, as shown by Nee *et al.* [5] in Si(110) MOS inversion layers, intersubband absorption at normal incidence is possible in the presence of some anisotropy of the effective mass tensor. In-plane polarized

G. Abstreiter et al. (eds.), Optical Spectroscopy of Low Dimensional Semiconductors, 41–61.

absorption is also allowed for hole intersubband transitions due to strong band-mixing effects in the valence band [6, 7].

Intersubband transitions have now been observed in many other material systems such as InGaAs/InAlAs [8], InGaAsP/InP [9], GaAs/InGaP [10], Si/SiGe [11], SiGeC/Si [12], HgCdTe [13]. Various 2D confinement potentials have been investigated including step quantum wells [14] and coupled quantum wells [15]. Interminiband absorption in the conduction band of GaAs/AlGaAs superlattices has also been reported [16, 17]. It was shown that intersubband absorption spectroscopy is a valuable tool for probing the dispersion of superlattice minibands. In the quest for normal-incidence absorbing systems, n-doped vertical AlGaAs QWs fabricated by self-organized growth over a grooved substrate have been investigated [18]. Huge intersubband absorption was effectively measured at normal incidence but the broadening of the intersubband resonance was found to be somewhat larger than for regular planar QWs because of width inhomogeneities during the vertical growth. Recently, intersubband spectroscopy of GaAs/AlGaAs quantum wires grown on V-groove GaAs substrates has been reported [19]. Among many favorable properties related to the reduced dimensionality, one expects 1D structures to exhibit normal-incidence intersubband absorption for light polarized normal to the wire axis.

Intersubband transitions in quantum wells have attracted a lot of interest in recent years for their potential applications to long-wavelength infrared photodetection [20] as a promising alternative to II-VI bulk alloys. The performances of single-element quantum-well infrared photoconductors (QWIP) in terms of specific peak detectivity and temperature for background-limited infrared photodetection (BLIP) are remarkable but still somewhat smaller than those of HgCdTe photodetectors [21]. However, large focal plane arrays of GaAs/AlGaAs QWIPs have been fabricated benefiting from the maturity of the GaAs technology. Noise equivalent temperature differences (NEDT) as low as 10 mK have been demonstrated reflecting the high spatial uniformity of the pixels [22]. Because of their large dipole moment, intersubband transitions are also good candidates for non-linear optical devices [23-25]. For example, it was demonstrated that the second-harmonic generation susceptibility in QWs could be increased by three orders of magnitude over that of bulk GaAs due to resonant enhancement by intersubband transitions [26]. Many other applications of intersubband transitions have been investigated such as infrared light modulators [27, 28] or up-conversion devices [29]. However, one of the most spectacular achievements in recent years was the first demonstration of unipolar lasers based on intersubband emission, the so-called Quantum Cascade lasers [30].

This paper is organized as follows. In section 2, we first derive the basic optical properties of intersubband transitions such as the transition selection rules, the dipole moment or the oscillator strength. Optical spectroscopy of intersubband absorptions in various material systems and confinement potentials is illustrated in section 3 with examples of intersubband transitions in the conduction band or in the valence band using different spectroscopic techniques. We then discuss in section 4 the basic features associated with the difficulty to achieve population inversion between subbands as well as the recent achievements on intersubband emission with special emphasis on unipolar injection lasers and optically pumped systems.

2. Optical properties of intersubband transitions

In this section, we show that a simple model can be used to derive the basic optical properties of intersubband transitions by considering the case of a single quantum well sketched in Figure 1.

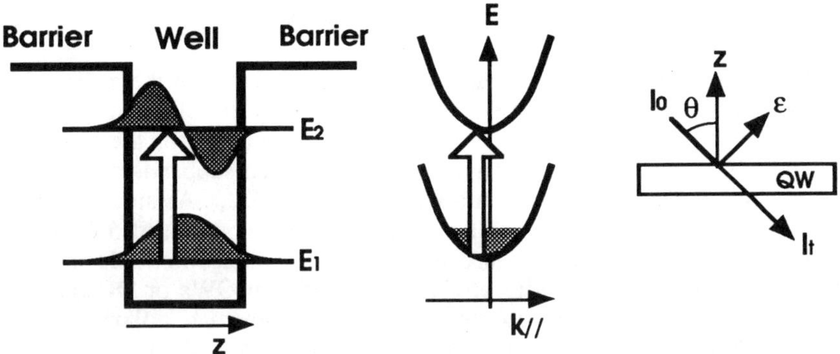

Figure 1. Confinement potential and in-plane dispersion diagram for a quantum well.

2.1. SELECTION RULES

The optical matrix element for a transition between subbands i and j is given by:

$$p_{ij} = < \psi_j \mid \varepsilon.\mathbf{p} \mid \psi_i > \tag{1}$$

where ε is the polarization vector of the light electric field, **p** is the momentum operator and ψ_i , ψ_j are the wave functions of the initial and final states which can be expressed within the envelope wave function approximation as [31]:

$$\psi_i(r) = u_i(r)\, F_i(r) = u_i(r) \frac{1}{\sqrt{S}} \exp(i\, \mathbf{k}_{//}.\mathbf{r}_{//})\, \phi_i(z) \tag{2}$$

where $u_i(r)$ is the periodic part of the Bloch functions at the band extremum, $\mathbf{k}_{//}$ and $\mathbf{r}_{//}$ are the wave and position vectors in the QW layer plane, $\phi_i(z)$ is the envelope function for subband i in the z confinement direction and S is the area. Accounting for the rapid variations of the Bloch functions over $1/k_{//}$ and over the spatial extent of the envelope wave functions

$$p_{ij} \approx \varepsilon. < u_j \mid \mathbf{p} \mid u_i > < F_i(r) \mid F_j(r) > + < u_j \mid u_i > < F_j(r) \mid \varepsilon.\mathbf{p} \mid F_i(r) > \tag{3}$$

The first term on the right-hand side is the optical matrix element for interband transitions which gives rise to the band-to-band selection rules. The second term is the intersubband contribution since by definition the Bloch functions are identical for both subbands ($< u_j \mid u_i > = \delta_{ij}$ and $< u_i \mid \mathbf{p} \mid u_i > = 0$). The intersubband optical matrix element is equal to:

$$<F_j(\mathbf{r}) \mid \varepsilon.\mathbf{p} \mid F_i(\mathbf{r}) > = \frac{i\,(E_j - E_i)\,m_0}{\hbar\, e}\,\mu_{ij}$$

$$\mu_{ij} = e < f_j(z) \mid z \mid f_i(z) > \varepsilon.\mathbf{z} \tag{4}$$

μ_{ij} is the intersubband dipole moment, **z** is the unit vector in the z direction, e is the electron charge, E_i is the confinement energy and m_0 the free electron mass. As seen, the dipole moment only involves the envelope wave functions for the two subbands. In symmetric QWs, since z is odd, only transitions between subbands with opposite parity of the envelope wave functions are allowed: $i\text{-}j = \pm 1, \pm 3, \ldots$. As an example, transitions from the ground state to the first excited state are allowed but transitions from the ground state to the second excited state are forbidden. This selection rule is of course not relevant for asymmetric potential profiles, such as step QWs or DC-biased QWs, for which all transitions become allowed. The dipole moment is polarized normal to the layers, i.e. along the z confinement direction. Using the so-called infinite quantum well approximation, i.e. assuming an infinite barrier potential on both sides of the well, one can easily derive the following expression for the dipole:

$$\mu_{ij} = \frac{8}{\pi^2}\,\frac{i\,j}{(j^2 - i^2)^2}\; e\, t_W \sin(\theta) \tag{5}$$

where t_w is the well width and θ is the angle between the direction of incident light and the direction perpendicular to the layers. Therefore, for the transition between the ground state and the first excited state, the dipole for light polarized perpendicular to the layers ($\theta=\pi/2$) is:

$$\mu_{12} = \frac{16}{9\,\pi^2}\, e\, t_W \approx 0.18\, e\, t_W \tag{6}$$

The equivalent dipole length of intersubband transitions is giant, of the order of 18% of the well width. The simple model presented above holds very well for bound-to-bound intersubband transitions in the conduction band. It is clearly a gross approximation for intersubband transitions in the valence band since the hole wave function must account for the strong coupling at $k\neq 0$ between the heavy hole, light hole and spin-orbit subbands. Normal incidence excitation of hole intersubband transitions is allowed because of this coupling [32].

The treatment can be extended to lower dimensional systems. In the case of a 1D confinement, one can show that for electronic transitions between 1D subbands there is only one forbidden direction for the polarization vector which is parallel to the wire axis. For 0D systems, there is *a priori* no forbidden direction but the actual polarization of intraband transitions between confined levels will depend on the spatial wave function symmetry of the involved states.

2.2. OSCILLATOR STRENGTH

The oscillator strength of the intersubband transition between the ground state and the first excited state writes:

$$f = \frac{2\, m_0\, E_{21}}{e^2\, \hbar^2}\, \mu_{12}^2 \approx 0.96\, \frac{m_0}{m^*} \tag{7}$$

where m* is the effective mass and the intersubband energy $E_{21} \approx 3\,\hbar^2\,\pi^2/(2\,m^*\,t_W^2)$. As seen, the oscillator strength of intersubband transitions does not depend on the energy of the transition, i.e. on the width of the well, but only depends on the carrier effective mass which is material dependent. Lower effective masses give larger oscillator strengths. As an example, in the Γ conduction band, f≈14 in GaAs QWs ($m_c^*/m_0 \approx 0.067$) and f≈42 in InAs QWs ($m_c^*/m_0 \approx 0.023$). It can be shown that this giant magnitude of the oscillator strength of intersubband transitions is in fact comparable to that of interband transitions [33].

2.3. ABSORBANCE

The intersubband absorbance per QW for the 1→2 transition can be expressed as:

$$a(\hbar\omega) = \mathrm{Ln}(I_0/I_t) = \sigma(\hbar\omega)\,(n_1 - n_2)$$

$$\sigma = \frac{h\, e^2}{4\, m_0\, \tilde{n}\, c\, \varepsilon_0}\, \frac{\sin^2(\theta)}{\cos(\theta)}\, f_{12}\, \frac{\hbar\Gamma / \pi}{(E_{21} - \hbar\omega)^2 + (\hbar\Gamma)^2} \tag{8}$$

where I_0, I_t are the incident and transmitted light intensities, n_i is the bidimensional density of carriers in subband i, σ is the absorption cross-section, c is the speed of light, ε_0 is the permittivity constant, ñ is the index of refraction and a Lorentzian lineshape has been assumed with a full width at half maximum (FWHM) equal to $2\hbar\Gamma$.

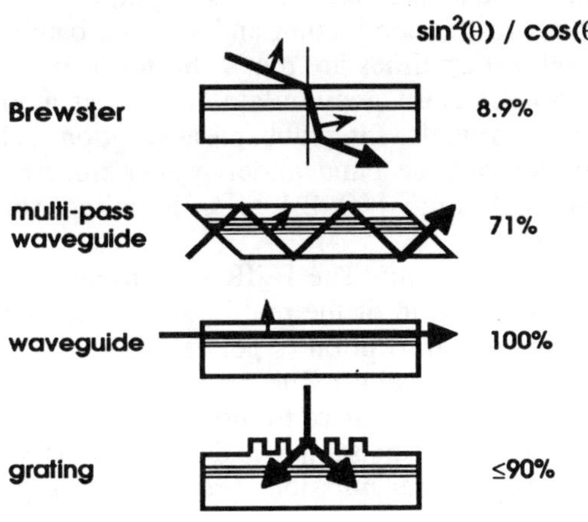

Figure 2. Experimental configurations for observing intersubband transitions. The coupling efficiency is indicated.

As revealed by the geometrical factor $\sin^2(\theta)/\cos(\theta)$ in Equ. (8), care must be taken when choosing the experimental configuration in order to ensure a good coupling efficiency of the light with the intersubband transition. Some of the usual configurations are sketched in Figure 2. When the light is incident at Brewster's angle ($\tan(i)=\tilde{n}$), the coupling efficiency is rather small and a large number of QWs are required to get significant absorption [4]. The multi-pass waveguide configuration [20] where the light enters the sample at normal incidence on a facet polished at 45° angle, is widely used for spectroscopic measurements since it allows large absorptions due to the excellent coupling of p-polarized light and to multiple passes within the active layer. In slab waveguides the absorption of light propagating with a polarization perpendicular to the layer plane (TM mode) can be made arbitrarily large by adjusting the waveguide length [34]. The grating configuration which consists of a linear or crossed grating fabricated on top of the sample, is mainly used for QWIP detectors since it allows normal incidence irradiation [35].

3. Absorption spectroscopy

Observation of intersubband absorption requires the ground subband to be populated. Epitaxial doping of the layers is widely used since large carrier densities can be achieved in most semiconductors. In addition, the choice of the doping impurity, either n-type or p-type, allows intersubband transitions to be studied in the conduction or valence band [4, 6]. However, band-bending effects due to the charge distribution may induce large shifts of the intersubband energy. Additional shifts may occur at large carrier densities due to depolarization shifts and to many-body effects such as exchange-correlation [3, 36, 37]. Modulation doping of the barrier material is usually preferred to doping of the well material since the latter results in extra broadening of the intersubband resonance due to scattering by the impurities [38]. Spectroscopic measurements on doped samples have been carried out using a variety of techniques such as inelastic light scattering [2, 3], infrared transmission spectroscopy [34], Fourier Transform Infrared (FTIR) spectroscopy [4] or photocurrent spectroscopy [39].

An alternate technique for populating the subbands relies on optical pumping of interband transitions to generate photo-carriers in the conduction and valence bands. Because the intersubband and intrasubband relaxation times are much shorter than the band-to-band recombination times, the photocarriers accumulate in the ground conduction and valence subbands of the quantum wells. Intersubband absorption can then take place in the conduction band or the valence band under proper infrared excitation. This photo-induced absorption technique [40-42] is well suited for intersubband spectroscopy of undoped structures.

Figure 3 gives an example of the experimental set-up. The FTIR spectrometer is operated in the step-scan mode. A polarizer placed ahead of the multi-pass waveguide sample allows to probe either the in-plane polarized absorption (s-polarization) or the absorption polarized perpendicular to the layers (p-polarization). The Ar^{++} or Ti:Saphirre pump laser is focused onto the sample surface. A chopper is used to modulate the pump intensity. The resulting changes in the IR transmission of the sample is detected using an IR detector and a lock-in amplifier. The photoinduced transmission spectrum ($\Delta T/T$) is obtained after Fourier transform. Intersubband absorptions as low as 10^{-5} can be detected using set-up.

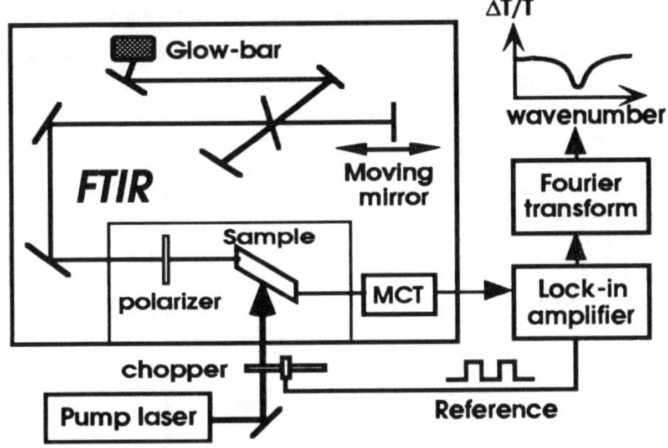

Figure 3. Experimental set-up for photo-induced intersubband absorption spectroscopy.

3.1. CONDUCTION-BAND INTERSUBBAND TRANSITIONS

Figure 4 gives an example of photo-induced absorption spectroscopy of a non-intentionally-doped multi-quantum well containing 100 periods of 7.8 nm thick GaAs wells and 10 nm thick $Al_{0.3}Ga_{0.7}As$ barriers. Measurements were performed at room-temperature.

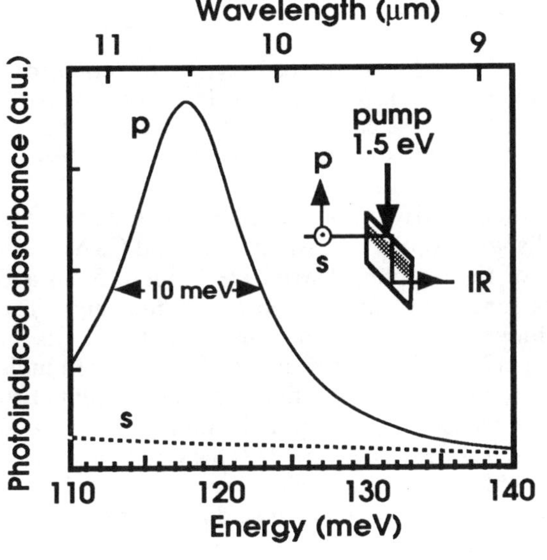

Figure 4. Photo-induced absorption spectroscopy at room-temperature of undoped $GaAs/Al_{0.3}Ga_{0.7}As$ multi-quantum wells. Interband excitation is set at 1.5 eV (4 W/cm^2).

The resonance peaked at 118 meV in the p-polarized spectrum is the $e1 \rightarrow e2$ intersubband absorption in the conduction band. The resonance vanishes in the s-polarized spectrum as expected from the polarization selection rule for conduction intersubband transitions (see 1.1). The residual s-polarized absorption arises from free-carriers photo created in the GaAs substrate. The FWHM of the intersubband resonance, of the order of 10 meV at 300 K, is typical of a sample containing a large number of QWs, part of the broadening being attributed to interface roughness as well as monolayer fluctuations of the different well thicknesses during the MBE growth.

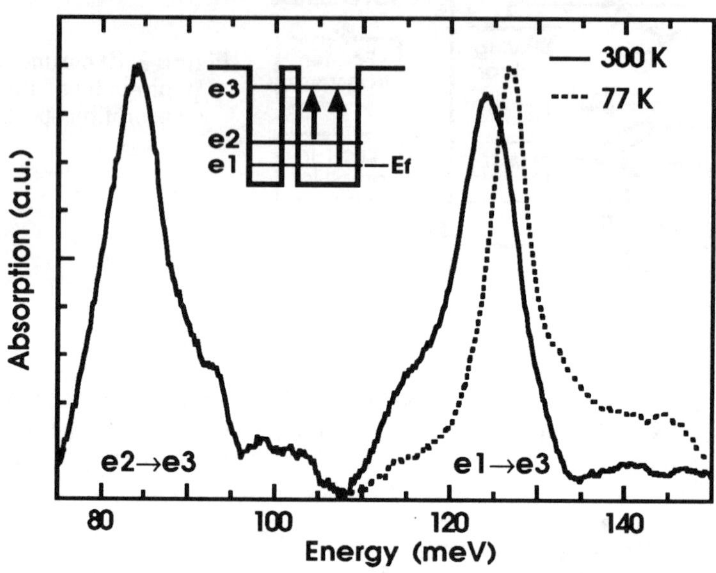

Figure 5. FTIR absorption spectroscopy of a modulation-doped asymmetric coupled quantum well structure at 300 K (full line) and at 77 K (dotted line). The energy spacing between the e1 and e2 subbands deduced from the energy difference between the resonances at 300 K is 39.8 meV which is close to the LO-phonon energy in GaAs.

Figure 5 shows FTIR absorption spectra of a MBE-grown sample containing 100 n-doped asymmetric coupled quantum wells with $Al_{0.22}Ga_{0.78}As$ barriers and GaAs wells [43]. The thicknesses of the wells and of the coupling barrier are 7.5 nm, 5 nm and 1.7 nm, respectively, while the barriers separating each period are 20 nm thick. The barriers are modulation doped to achieve a 2D electron density in the wells of $1x10^{11}$ cm^{-2}. FTIR measurements were performed at 300 K and 77 K using the multi-pass waveguide configuration. The room-temperature absorption spectrum exhibits two peaks under p-polarized irradiation, which vanish when the polarization of the infrared beam is rotated parallel to the layer planes. The resonance at 124.0 meV with a FWHM of 8.5 meV is related to the e1→e3 intersubband transition while the peak at 84.2 meV with a FWHM of 9 meV can be assigned to the e2→e3 intersubband transition which is allowed because of the asymmetric potential (see 1.1). By lowering the temperature down to 77 K, the e1→e3 intersubband resonance occurs at 126.6 meV with a reduced FWHM of 6.2 meV. Thus, although the band gap of GaAs has increased by 95 meV between 300K and 77K, the intersubband energy is only blue-shifted by 2.6 meV. This weak sensitivity of the energy to temperature is characteristic of intersubband transitions, in sharp contrast with interband transitions [44]. Another important feature at low temperatures is that the e2→e3 absorption is no more observable. This behavior is clearly related to the thermal population of the first excited subband. Indeed, the occupation factor of the e2 subband ≈ $\exp(-E_{21}/kT)$, where kT is the thermal energy, is expected to drop from ≈ 22% at room temperature down to ≈ 0.2% at 77K. It is interesting to note in Fig. 5 that despite the lower e2 population, the e1→e3 and e2→e3 integrated absorbances at room temperature are of similar amplitudes, which is illustrative of a larger oscillator strength of the e2→e3 transition.

3.2. VALENCE-BAND INTERSUBBAND TRANSITIONS

Intersubband transitions in the valence band have been studied in various material systems such as p-doped GaAs/AlGaAs [22], InGaAsP/InP [9] or InGaAs/AlGaAs [45] as well as indirect-gap compounds like Si/SiGe [46-50] or SiGeC/Si [12]. For SiGe(C) quantum wells grown on Si(100), the band discontinuity mainly occurs in the valence band. Although most studies were performed using FTIR absorption spectroscopy, it was shown that the photo-induced IR absorption spectroscopy is well suited to study indirect-gap materials since large photocarrier densities can be achieved in the ground hole subband due to the rather long band-to-band recombination times [50].

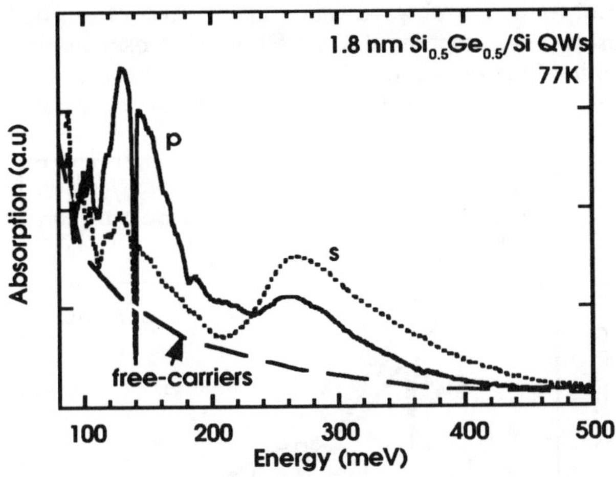

Figure 6. Photo-induced intersubband absorption spectroscopy at 77 K of undoped 1.8 nm thick $Si_{0.5}Ge_{0.5}$ wells with Si barriers for p-polarization (full line) and s-polarization (dotted line). The free-carrier contribution is indicated by the dashed line.

Figure 6 shows the photo-induced absorption spectroscopy of a multi-pass waveguide sample containing 20 undoped periods of 1.8 nm thick $Si_{0.5}Ge_{0.5}$ wells and 32 nm thick Si barriers. The pump excitation is provided by an Ar^{++} laser. The spectra for both p- and s-polarizations show resonances characteristic of hole intersubband absorptions in addition to significant free-carrier absorption. Energy calculations show that the only bound levels in the structure are the heavy-hole, light hole and spin-orbit states, hh1, lh1 and so1, respectively. The resonance at 115 meV is attributed to the hh1→lh1 transition while the s-polarized peak at 265 meV corresponds to the hh1→so1 transition. Both transitions are allowed for carriers away from Brillouin zone center. As seen significant s-polarized intersubband absorption is observed. However, one can notice that the p-polarized intersubband absorption is larger at low energies. The intersubband resonances in the valence band are much broader than those observed in the conduction band of GaAs QWs because the mixing between heavy-hole and light-hole bands away from Brillouin zone center leads to significant non-parabolicity of the hole subbands [51].

In p-doped SiGe/Si quantum wells, the photo-induced intersubband absorption (PIA) at low temperatures differs from the results obtained using FTIR absorption spectroscopy. This is illustrated in Figure 7 which shows the p-polarized absorption using both techniques for a multi-pass waveguide sample containing 50 periods of 3 nm thick $Si_{0.8}Ge_{0.2}$ wells and 20 nm thick Si barriers. The wells are uniformly doped at $p=4.2x10^{18}$ cm^{-3}. The resonance in both spectra is attributed to the hh1→hh2 transition which is found to be mainly p-polarized. Bound-to-continuum transitions from the hh1 subband contribute to the broadening at large energies. As seen in Fig. 7, the hh1→hh2 resonance is significantly shifted to lower energies in the photo-induced spectrum as compared to the FTIR measurements. Indeed, one expects at very low temperatures the photo-created holes to be thermalized in the hh1 ground subband above the Fermi energy. The photo-induced technique will then probe absorption by holes with in-plane wave vector superior to the Fermi wave vector k_F, i.e. away from Brillouin zone center. In contrast, the absorption as measured using FTIR spectroscopy involves carriers close to zone center with in-plane momentum $k \leq k_F$. The shift observed in the photo-induced spectrum is therefore a direct manifestation of the subband non-parabolocities.

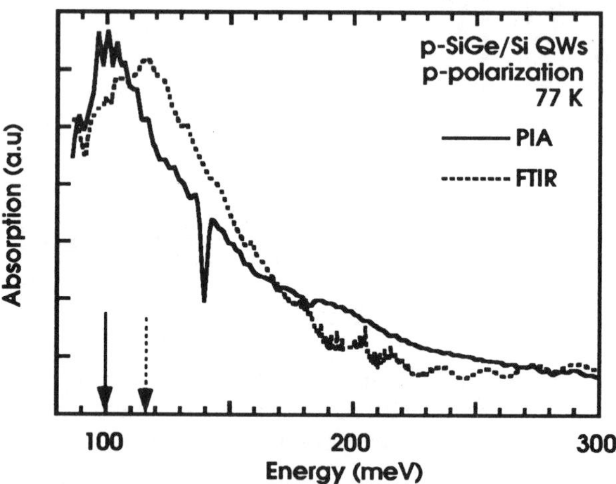

Figure 7. Photo-induced absorption spectroscopy (PIA, full line) and direct absorption (FTIR, dotted line) spectroscopy of p-doped 3 nm thick $Si_{0.8}Ge_{0.2}$/Si quantum wells in p-polarization at 77 K.

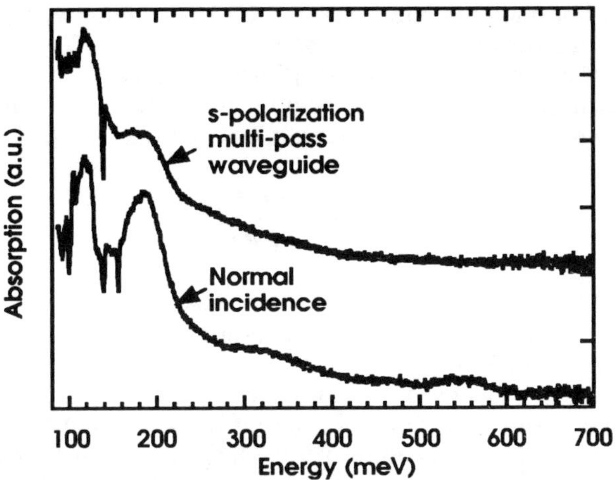

Figure 8. In-plane polarized absorption spectrum of p-doped 3 nm thick $Si_{0.8}Ge_{0.2}/Si$ quantum wells in a multi-pass waveguide and at normal incidence.

Figure 8 shows the in-plane polarized photo-induced absorption of the same sample measured using a multi-pass waveguide (s-polarization) and at normal incidence. The peaks at 120 meV and 190 meV correspond respectively to the hh1→so1 transition and to the transition from hh1 to the continuum. As seen, although the resonances are located at the same energies in both spectra, the absorption lineshape can be rather different at normal incidence. This is due to Fabry-Perot interferences in the multi-quantum well layer, which are clearly evidenced by the oscillations at large energies, and to the large variations of the index of refraction resulting from the anomalous dispersion associated with the intersubband absorptions [52].

4. Intersubband emission

The possibility of intersubband emission has been theoretically investigated as early as 1971 by Karazinov and Suris [53]. The original design was based on a photon-assisted tunneling process in a biased superlattice. However, it took 17 years before the first experimental evidence by M. Helm and co-workers that a spontaneous intersubband emission could occur in quantum wells under application of an electric field in the plane the layers [54] or perpendicular to the layers [55]. In the latter case, it was found that the far-infrared emissions in the biased superlattice do not result of a photon-assisted tunneling process but of resonant tunneling of electrons from well to well. Resonant tunneling structures have been predicted to exhibit population inversion as well as large optical gains at far-infrared wavelengths for most studies [56-64].

Indeed, for intersubband energies well below the optical phonon energy (λ>35 μm), the non-radiative relaxation between subbands is dominated by the relatively slow emission of acoustic phonons, while at energies larger than the optical phonon energy the emission of LO-phonons leads to very short lifetimes of electrons in the excited subbands. Typical non-radiative times (τ_{nr}) in GaAs quantum wells ranges from ≈ 0.3-2 ps for mid-infrared transitions (LO-phonon emissions) to ≈ 100-400 ps for far-infrared transitions (acoustic-phonon emissions) [65-68]. However, because intersubband transitions lie in the infrared spectral region, the radiative lifetime (τ_r), although inversely proportional to the oscillator strength, can be rather long:

$$\tau_r = \frac{3\, m_o\, \varepsilon_o\, c}{2\, \pi\, \tilde{n}\, e^2} \frac{\lambda^2}{f} \approx \frac{3\, m^*\, \varepsilon_o\, c}{2\, \pi\, \tilde{n}\, e^2} \lambda^2 \tag{9}$$

since it exhibits a λ^2 evolution. Typical values in GaAs are of the order of 15 ns at $\lambda = 4$ μm and 9 μs at $\lambda = 100$ μm. Materials with smaller effective mass give shorter radiative lifetimes which is beneficial for emission.

As seen, the balance between the radiative and non-radiative processes is not favorable for spontaneous emission. Depending on wavelength, the efficiency of an intersubband spontaneous emission, $\eta = \tau_{nr}/(\tau_r+\tau_{nr})$, can be by 3 to 6 orders of magnitude smaller than the corresponding value for band-to-band luminescence in direct gap semiconductors. Therefore, light emitting devices based on intersubband luminescence should exhibit very low efficiencies. However, if population inversion can be achieved between subbands, large stimulated gains are expected because of the large oscillator strength of intersubband transitions. The condition for population inversion only requires the non-radiative lifetime in the upper state to be longer than in the lower state. In tunneling structures, this condition imposes the necessity to both provide a high injection current density and quickly remove electrons from the ground subband within a time shorter than the non-radiative lifetime of the excited subband. A proper design of the tunneling structure should also be aimed at reducing the leakage tunneling current from the excited state since it may dramatically lower the non-radiative lifetime of the upper state. It was shown that biased coupled quantum wells placed between two n+ layers which act as injector/collector regions could be good candidates for intersubband emission even at short infrared wavelengths [64].

4.1. QUANTUM CASCADE LASERS

Intersubband electroluminescence has effectively been observed in coupled quantum wells at mid-infrared wavelengths ≈ 4-5 μm in the AlInAs/GaInAs material system [69-71]. The radiative efficiency was found to be somewhat larger at shorter infrared wavelengths since it benefits from a large decrease of the radiative lifetime along with an enhanced LO-phonon scattering time. Indeed, the LO-phonon scattering time scales like the square of the momentum required for relaxation and is significantly increased for large subband separations. Careful design of the structure involved the use of a three coupled well active region which acts as an interference filter for preventing electrons to escape from the excited state, and of graded injector regions for efficient population of the excited state and quick removing of carriers from the lower states. The injector region consists of an n-doped AlInAs/GaInAs superlattice with constant period and varying duty cycle to obtain a graded gap pseudoquaternary alloy. By embedding the whole structure repeated many times within an infrared waveguide, the first intersubband Quantum Cascade laser (QC) emitting at 4.2 μm has been demonstrated [30, 72]. The active coupled-well region is essentially a three-level system where population inversion is achieved between the two excited states ($t_{32}^{nr} \approx 4.3$ ps) and strong inelastic relaxation with nearly zero momentum transfer ($t_{21}^{nr} \approx 0.6$ ps) occurs between the first excited and the ground state which are separated by an energy close to the LO-phonon energy. The emission is diagonal in real space since it occurs between states belonging to different wells. Pulsed operation was achieved up to 125 K [73] and the threshold current density was found to vary exponentially with temperature ($\exp(T/T_0)$) from 6.0 kA/cm^2 at 50 K to 9.3 kA/cm^2 at 125 K with a $T_0 \approx 112$ K. Peak powers as large as 32 mW were obtained.

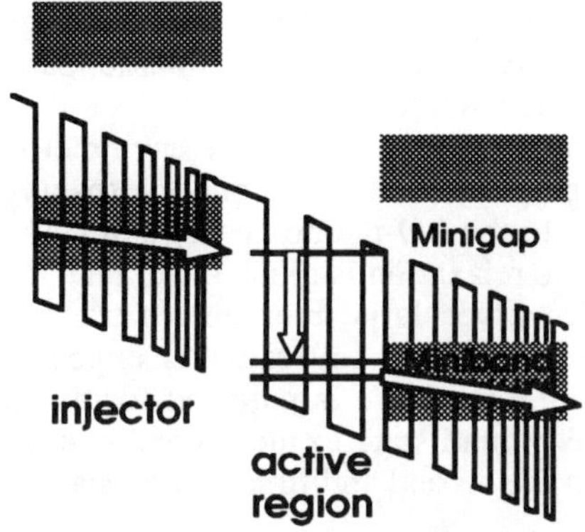

Figure 9. Conduction band diagram of a Quantum Cascade laser with vertical emission.

It was soon made clear that, instead of diagonal emissions, vertical transitions in two coupled wells could bring a major improvement of the threshold current due to narrower gain spectrum, provided that electrons in the upper state could be prevented from escaping into the continuum [74]. Following earlier work on strong localization of electrons in the presence of Bragg reflectors [75], the superlattice injector was designed to fulfill the Bragg reflection conditions for the upper state while allowing fast emptying of the two lower states through miniband transport (see Figure 9). Continuous wave operation at $\lambda = 4.6$ µm could be achieved up to 85 K [76]. Pulsed operation up to room-temperature has been recently demonstrated at $\lambda \approx 5$ µm [77]. By adjusting the layer thicknesses and using an optimized waveguiding structure, electroluminescence at 8-13 µm wavelengths has also been reported [78], shortly followed by the demonstration of QC lasing operation at 8.4 µm [79]. Continuous wave operation at $\lambda = 7.4$-8.6 µm at 110 K has already been achieved [80].

QC lasers are unipolar devices whose performances in terms of emitting powers, operating temperatures and T_0 already compare favorably with competing interband lasers made of small-gap III-V or II-VI materials [77]. Unlike those interband laser diodes, QC lasers are not limited by the Auger recombination processes and the emission wavelength exhibits only weak temperature dependence.

4.2. OPTICALLY PUMPED SYSTEMS

Optical pumping of electrons has also been proposed to achieve population inversion between excited subbands at far-infrared wavelengths [81-83]. Experimentally, Bales *et al.* have shown that optical excitation at 10 μm of a coupled quantum well structure resulted in a spontaneous far-infrared emission at low temperatures [84]. Recently, an alternate design have been proposed for achieving intersubband emission at mid-infrared wavelengths based on optical pumping of a three-bound-state coupled quantum well structure [85]. By carefully engineering the LO-phonon scattering rates between subbands through a proper choice of the quantum well composition and layer thicknesses, it was shown that the conditions for population inversion could be fulfilled. Since no electron transport is required for emission, the optically pumped structures relie on a much simpler design as compared to QC emitters. Moreover, because metallic contacts or heavily doped layers are not necessary, free-carrier and plasmon losses can be minimized which is beneficial for long-wavelength operation. However, although large optical gains have been predicted [85, 86], only observations of intersubband luminescences at $\lambda \approx$ 14-15 μm have been reported so far [87, 88, 43].

The principle of operation is illustrated in the inset of Figure 10. The asymmetric coupled quantum well structure (ACQW) is designed to present an energy spacing between the first excited and ground subbands close to the LO-phonon energy in GaAs (36 meV). In such a situation, the e2→e1 non-radiative relaxation is much faster than the e3→e2 non-radiative relaxation, since the latter involves large momentum transfer between subbands. As a result, almost complete population inversion is expected between the two excited states. The asymmetry of the structure is required for direct optical pumping at λ = 10 μm of electrons from the ground state to the second excited state. The intersubband emission takes place between the second and first excited states.

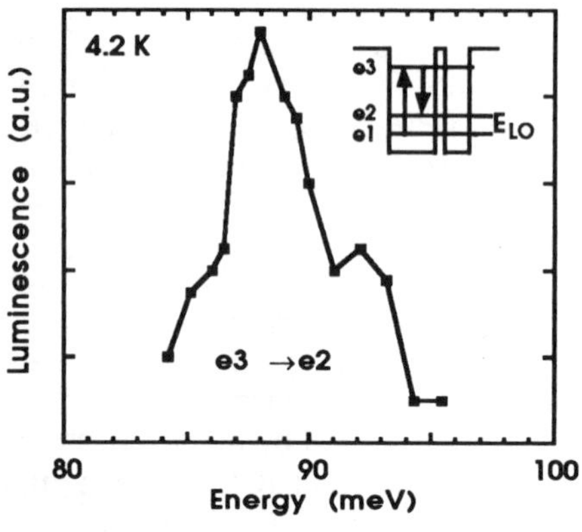

Figure 10. Intersubband emission spectrum at 4.2 K of GaAs/$Al_{0.22}Ga_{0.78}As$ asymmetric coupled quantum wells under optical pumping at 9.8 μm.

Figure 10 shows the intersubband emission spectrum at 4.2 K for the GaAs/AlGaAs ACQW structure described in section 2.1. As seen, the emitted photon energy ≈ 88 meV is down-converted from the pump photon energy by an energy close to the LO-phonon energy. The broadening of the emission resonance is rather small (FWHM ≈ 4 meV), as expected from a vertical emitting transition. The efficiency at 4.2 K is of the order of 60 nW per Watt of pump power. The emission is found to persist up to room-temperature with a maximum efficiency around 190 K [88]. Very recently, intersubband luminescence under band-to-band optical pumping has also been observed [89] and large optical gains have been predicted [85]. Interband pumping could be of major practical interest for intersubband lasers since compact-sized highly-efficient diode lasers can be used to optically pump the quantum wells.

5. Conclusion

We have attempted to give a review on some specific infrared properties of intersubband transitions. These transitions which are direct manifestation of the quantum confinement of carriers, are very sensitive to the nature and the shape of the confinement potential. Optical intersubband absorption spectroscopy has been shown to be a valuable tool for probing the confinement energies as well as the subband dispersions either in the conduction or the valence band. Examples have been given for different materials either doped or undoped under various confinement situations. Although most reported work has been devoted to 2D confined systems, infrared absorption spectroscopy could be of great help for disentangling the energy structure and wave function symmetry of lower dimensional systems such as quantum wires and quantum dots. The recent demonstration of intersubband stimulated emission in Quantum Cascade structures has triggered new interest for intersubband emitting devices. It was shown that engineering of the phonon scattering rate can lead to intersubband population inversion in optically pumped structures. Inversion-less intersubband lasers based on the non-parabolicity of conduction subbands have also been reported [90]. The recent demonstration of Fano-like interferences in the intersubband absorption spectrum of quantum wells strongly coupled to a continuum could lead to new inversion-less emitting situations [91]. Finally, 0D confined structures may be good candidates for infrared emission between discrete levels because of the quenching of the inter-level optical phonon scattering expected in such structures.

Acknowledgements

We acknowledge V. Berger, J.-P. Leburton, J.-M. Lourtioz, Z. Moussa, J. Nagle, A. Sa'ar, R. Planel, P. Vagos and L. Wu for their contributions to the results presented here.

References

1. A. Kamgar, P. Kneschaurek, G. Dorda and J.F. Koch (1974) Resonant spectroscopy of electronic levels in a surface accumulation layer, Phys. Rev. Lett. **32**, 1251.
2. G. Abstreiter and K. Ploog (1979) Inelastic light scattering from a quasi-two-dimensional electron system in GaAs-AlGaAs heterojunctions, Phys. Rev. Lett. **42**, 1308.
3. A. Pinczuk, H.L. Störmer, R. Dingle, J.M. Worlock, W. Wiegmann and A.C. Gossard (1979) Observation of intersubband excitations in a multilayer two dimensional electron gas, Solid State Commun. **32**, 1001.
4. L. C. West and S. J. Eglash (1985) First observation of an extremely large dipole infrared transition within the conduction band of a GaAs quantum well, Appl. Phys. Lett. **46**, 1156.
5. S.M. Nee, U. Claessen and F. Koch (1984) Subband resonance of electrons in Si(110), Phys. Rev. **B 29**, 3449.

6. B. F. Levine, S. D. Gunapala, J. M. Kuo, S. S. Pei, and S. Hui (1991) Normal incidence hole intersubband long wavelength GaAs/AlGaAs quantum well infrared photodetectors, Appl. Phys. Lett. **59**, 1864.
7. J.S. Park, R.P.G. Karunasiri and K.L. Wang (1991) Si1-xGex/Si multiple quantum well infrared detector, Appl. Phys. Lett. **59**, 2588.
8. H. Asai and Y. Kawamura (1990).Well-width dependence of intersubband absorption in InGaAs/InAlAs multiquantum wells, Appl. Phys. Lett. **56**, 1149.
9. S. D. Gunapala, B. F. Levine, D. Ritter, R. Hamm and M.B. Panish (1991) InGaAs/InP long wavelength quantum well infrared photodetectors, Appl. Phys. Lett. **58**, 2024.
10. C. Francis, M. Bradley, P. Boucaud, F.H. Julien, and M. Razeghi (1993) Intermixing of GaInP/GaAs multiple quantum wells, Appl. Phys. Lett. **62**, 178.
11. H. Hertle, G. Schuberth, E. Gornik, G. Abstreiter, F. Schäffler (1991) Intersubband absorption in the conduction band of Si/Si1-xGex multiple quantum wells, Appl. Phys. Lett. **59**, 2977.
12. P. Boucaud, J.M. Lourtioz, F.H. Julien, P. Warren and M. Dutoit (1996) Intersubband absorption in Si/Si1-x-yGexCy quantum wells, Appl. Phys. Lett. **69**, 1734.
13. C.R.M. de Oliveira, A.M. de Paula, C.L. Cesar, L.C. West, C. Roberts, R.D. Feldman, R.F. Austin, M.N. Islam and G.E. Marques (1995) Photoinduced intersubband transition in undoped HgCdTe multiple quantum wells, Appl. Phys. Lett. **66**, 2998.
14. Y.J. Mii, K.L. Wang, R.P.G. Karunasiri and P.F. Yuh (1990) Observation of large oscillator strngths for both 1-2 and 1-3 intersubband transitions of step quantum wells, Appl. Phys. Lett. **56**, 1046.
15. C. Sirtori, F. Capasso, D.L. Sivco, S.N.G. Chu and A.Y. Cho (1991) Observation of large second-order susceptibility via intersubband transitions at $\lambda \approx 10\mu m$ in asymmetric coupled AlInAs/GaInAs quantum wells, Appl. Phys. Lett. **59**, 2302.
16. A. Katalsky, T. Duffield, S.J. Allen and J.P. Harbison (1988) Photovoltaic detection of infrared light in a GaAs/AlGaAs superlattice, Appl. Phys. Lett. **52**, 1320.
17. M. Helm, W. Hilber, T. Fromherz, F.M. Peeters, K. Alavi, and R.N. Pathak (1993) Infrared absorption in superlattices: a probe of the miniband dispersion and the structure of the impurity band, Phys Rev. **B 48**, 1601.
18. V. Berger, G. Veirmeire, P. Demeester and C. Weisbuch (1995) Normal incidence intersubband absorption in vertical quantum wells, Appl. Phys. Lett. **66**, 218.
19. A. Sa'ar, A. Givant, S. Calderon, O. Ben-Shalom, E. Kapon, A. Gustafsson, D. Oberli, C. Caneau (1996) Intersubband optical transitions in one dimensional quantum wire structures, Superlattices and Microstruct. **19**, 217.
20. B. F. Levine, K. K. Choi, C. G. Bethea, J. Walker, and R. J. Malik (1987) New 10 μm infrared detector using intersubband absorption in resonant tunneling GaAlAs superlattices, Appl. Phys. Lett. **50**, 1092.
21. I. Gravé and A. Yariv (1992) Fundamental limits in quantum well intersubband detection, in E.Rosencher, B. Vinter and B.Levine (eds.), *Intersubband Transitions in Quantum Wells*, Plenum Press, NewYork, 15.

22. B.F. Levine (1992) Recent progress in quantum well infrared photodetectors, in E.Rosencher, B. Vinter and B.Levine (eds.), *Intersubband Transitions in Quantum Wells*, Plenum Press, NewYork, 43.
23. M. M. Fejer, S. J. B. Yoo, R. L. Byer, A. Harwitt, J. S. Harris (1989) Observation of extremely large quadratic susceptibility at 9.6-10.8 μm in electric-field biased AlGaAs quantum wells, Phys. Rev. Lett. **62**, 1041.
24. C. Sirtori, F. Capasso, D. Sivco and A.Y. Cho (1992) Giant, triply resonant, third-order nonlinear susceptibility $\chi^3(3\omega)$ in coupled quantum wells, Phys. Rev. Lett. **68**, 1010.
25. A. Sa'ar, N. Kuze, J. Feng, I. Gravé and A. Yariv (1992) Observation of third-order intersubband dc Kerr effect at midinfrared wavelengths in GaAs quantum wells, Appl. Phys. Lett. **61**, 1263.
26. P. Boucaud, F. H. Julien, D. D. Yang, J.-M. Lourtioz, E. Rosencher, P. Bois, and J. Nagle (1990) Detailed analysis of second-harmonic generation near 10.6 μm in GaAs/AlGaAs asymmetric quantum wells, Appl. Phys. Lett. **57**, 215.
27. A. Harwitt and J. S. Harris (1987) Observation of Stark shifts in quantum well intersubband transitions, Appl. Phys. Lett. **50**, 685.
28. F.H. Julien, P. Vagos, J-M. Lourtioz, D.D. Yang and R. Planel (1991) Novel all-optical 10 μm waveguide modulator based on intersubband absorption in GaAs/AlGaAs quantum wells, Appl. Phys. Lett. **59**, 2645.
29. P. Vagos, P. Boucaud, F. H. Julien, J.-M. Lourtioz and R. Planel, Photoluminescence up-conversion induced by intersubband absorption in asymmetric coupled quantum wells (1993) Phys. Rev. Lett. **70**, 1018.
30. J. Faist, F. Capasso, D. Sivco, C. Sirtori, A.L. Hutchinson, S.N. G. Chu and A.Y. Cho (1994) Quantum cascade laser, Science **264**, 553.
31. G. Bastard (1988) *Wave Mechanics Applied to Semiconductor Heterostructures,* Les Editions de Physique, Les Ulis,.
32. Y.C. Chang and R.B. James (1989) Saturation of intersubband transition in p-type semiconductor quantum wells, Phys Rev. **B 39**, 12672.
33. J. Khurgin (1992) Comparative analysis of the intersubband versus band-to-band transitions in quantum wells, Appl. Phys. Lett. **62**, 1390.
34. D. D. Yang, F. H. Julien, P. Boucaud, J.-M. Lourtioz, and R. Planel (1990) Intersubband absorption of GaAs/AlGaAs quantum wells in MBE grown mid-infrared slab waveguides, IEEE Photon. Technol. Lett. **2**, 181.
35. J.Y. Andersson and L. Lundqvist (1992) Coupling of radiation into quantum well infrared detectors by the use of reflection grating and waveguide structures, in E.Rosencher, B. Vinter and B.Levine (eds.), *Intersubband Transitions in Quantum Wells*, Plenum Press, NewYork, 1.
36. M. Ramsteiner, J.D. Ralston, P. Koidl, B. Dischler, H. Biebl, J. Wagner and H. Ennen (1990) Doping density dependence of intersubband transitions in GaAs/AlGaAs quantum-well structures, J. Appl. Phys. **67**, 3900.
37. T. Ando, A.B. Fowler and F. Stern (1982) Electronic properties of 2D systems, Rev. Mod. Phys. **54**, 437.
38. E.B. Dupont, D. Delacourt, D. Papillon, J.P. Schnell and M. Papuchon (1992) Influence of ionized impurities on the linewidth of intersubband transitions in GaAs/GaAlAs quantum wells, Appl. Phys. Lett. **60**, 2121.

39. E. Martinet, F. Luc, E. Rosencher, P. Bois, E. Costard, S. Delaître and E. Böckenhoff (1992) Electric field effects on bound to quasibound intersubband absorption and photocurrent in GaAsAlGaAs quantum wells, in E.Rosencher, B. Vinter and B.Levine (eds.), *Intersubband Transitions in Quantum Wells*, Plenum Press, NewYork, 299.
40. M. Olszakier, E. Ehrenfreund, E. Cohen, J. Bajaj, and J. Sullivan (1989) Photoinduced intersubband absorption in undoped multi-quantum-well structures, Phys. Rev. Lett. **62**, 2997.
41. D. D. Yang, F. H. Julien, J.-M. Lourtioz, P. Boucaud, and R. Planel (1990) First demonstration of room-temperature intersubband-interband double-resonance spectroscopy of GaAs/AlGaAs quantum wells, IEEE Photon. Technol. Lett. **2**, 398.
42. F.H. Julien (1992) Room-temperature photo-induced intersubband absorption in GaAs/AlGaAs quantum wells, in E.Rosencher, B. Vinter and B.Levine (eds.), *Intersubband Transitions in Quantum Wells*, Plenum Press, NewYork, 163.
43. F.H. Julien, Z. Moussa, P. Boucaud, Y. Lavon, A. Sa'ar, J. Wang, J.-P. Leburton, V. Berger, J. Nagle and R. Planel (1996) Intersubband mid-infrared emission in optically pumped quantum wells, Superlattices and Microstruct. **19**, 69.
44. P. Von Allmen, M. Berz, G. Petrocelli, F.-K. Reinhart and G. Harbeke (1988) Intersubband absorption in GaAs/AlAs quantum wells between 4.2 K and room-temperature, Semicond. Sci. Technol. **3**, 1211.
45. E.L. Martinet, B.J. Vartanian, G.L. Woods, H.C. Chui, J.S. Harris, M.M. Fejer, B.A. Richman and C.A. Rella (1994) Applications of high indium content InGaAs/AlGaAs quantum wells in the 2-7 µm regime, in H. C. Liu, B. F. Levine, J. Y. Andersson (eds.), *Quantum Well Intersubband Transitions: Physics and Device*, Kluwer Academic Publishers, Dordrecht, 261.
46. J. S. Park, R. P. G. Karunasiri, K. L. Wang (1992) Normal incidence infrared detector using p-type SiGe/Si multiple quantum wells, Appl. Phys. Lett. **60**, 103.
47. R. People, J. C. Bean, S. K. Sputz, C. G. Bethea, L. J. Peticolas (1992) Normal incidence hole intersubband quantum well infrared photodetector in pseudomorphic GeSi/Si, Thin Solid Films **222**, 120.
48. M. Seto, M. Helm, Z. Moussa, P. Boucaud, F. Julien, J.M. Lourtioz, J.F. Nützel and G. Abstreiter (1994) Second-harmonic generation in asymmetric Si/SiGe quantum wells, Appl. Phys. Lett. **65**, 2969.
49. S. Zanier, J.M. Berroir, Y. Guldner, J.P. Vieren, I. Sagnes, F. Glowacki, Y. Campidelli and P.A. Badoz (1995) Free-carrier and intersubband infrared absorption in p-type SiGe/Si multiple quantum wells, Phys. Rev. **B 51**, 14311.
50. P. Boucaud, L. Gao, Z. Moussa, F. Visocekas, F. H. Julien, J.-M. Lourtioz, I. Sagnes, Y. Campidelli, P.-A. Badoz (1995) Photo-induced intersubband absorption in Si/SiGe quantum wells, Appl. Phys. Lett. **67**, 2948.
51. E. Corbin, K.B. Wong and M. Jaros (1994) Absorption in p-type Si-SiGe quantum well structures, Phys. Rev. **B 50**, 2339.
52. L. Wu, P. Boucaud, J.-M. Lourtioz, F. H. Julien, I. Sagnes, Y. Campidelli, P.-A. Badoz (1995) Absorption and resonant dispersion associated with normal incidence intersubband transitions in Si/SiGe quantum wells, Appl. Phys. Lett. **67**, 3462.

53. R.F. Kazarinov and R.A. Suris (1971) Possibility of amplification of electromagnetic waves in a semiconductor with a superlattice, Sov. Phys. Semicond. **5**, 707.
54. M. Helm, E. Colas, P. England, F. DeRosa and S.J. Allen (1988) Observation of grating-induced intersubband emission from GaAs/AlGaAs superlattices, Appl. Phys. Lett. **53**, 1714.
55. M. Helm, P. England, E. Colas, F. DeRosa, and S. J. Allen, Jr. (1989) Intersubband emission from semiconductor superlattices excited by sequential resonant tunneling, Phys. Rev. Lett. **63**, 74.
56. P. Yuh and K. L. Wang (1987) Novel infrared band-aligned superlattice laser, Appl. Phys. Lett. **51**, 1404.
57. S. Borenstain and J. Katz (1988) Intersubband Auger recombination and population inversion in quantum-well subbands, Phys. Rev. B **39**, 10852.
58. H. C. Liu (1988) A novel infrared source, J. Appl. Phys. **63**, 2856.
59. S. I. Borenstain and J. Katz (1989) Evaluation of the feasability of a far-infrared laser based on intersubband transitions in GaAs quantum wells, Appl. Phys. Lett. **55**, 654.
60. M. Helm and S. J. Allen (1990) Can barriers with inverted tunneling rates lead to subband population inversion ?, Appl. Phys. Lett. **56**, 1368.
61. J.-W. Choe, A. G. U. Perera, M. H. Francombe and D. D. Coon (1991) Estimates of infrared intersubband emission and its angular dependence in GaAs/AlGaAs multiquantum well structures, Appl. Phys. Lett. **59**, 54.
62. J. P. Lohr, J. Singh, R. K. Mains and G. I. Haddad (1991) Theoretical studies of the applications of resonant tunneling diodes as intersubband laser and interband excitonic modulators, Appl. Phys. Lett. **59**, 2070.
63. Q. Hu and S. Feng (1991) Feasability of far-infrared lasers using multiple semiconductor quantum wells, Appl. Phys. Lett. **59**, 2923.
64. A. Katalsky, V. J. Goldman and J. H. Abeles (1991) Possibility of infrared laser in a resonant tunneling structure, Appl. Phys. Lett. **59**, 2636.
65. M. C. Tatham, J. F. Ryan and C. T. Foxon (1989) Time-resolved Raman measurements of intersubband relaxation in GaAs quantum wells, Phys. Rev. Lett. **63**, 1637.
66. D. Y. Oberli, D. R. Wake, M. V. Klein, J. Klem, T. Henderson and H. Morkoc (1987) Time-resolved Raman scattering in GaAs quantum wells, Phys. Rev. Lett. **59**, 696.
67 F. Julien, J.-M. Lourtioz, N. Herschkorn, D. Delacourt, J.P. Pocholle, M. Papuchon, R. Planel and G. Le Roux (1993) Optical saturation of intersubband absorption in GaAs/AlGaAs quantum wells, Appl. Phys. Lett. **62**, 2289.
68. J. Faist, C. Sirtori, F. Capasso, L. Pfeiffer and K. West (1994) Phonon-limited intersubband lifetimes and linewidths in a two-dimensional electron gas, Appl. Phys. Lett. **64**, 872.
69. J. Faist, F. Capasso, C. Sirtori, D. L. Sivco, A. L. Hutchinson, S.-N. G. Chu and A. Y. Cho (1993) Quantum-well intersub-band electroluminescent diode at $\lambda = 5$ μm, Electron. Lett. **29**, 2230.
70. J. Faist, F. Capasso, C. Sirtori, D. L. Sivco, A. L. Hutchinson, S.-N. G. Chu and A. Y. Cho (1994) Mid-infrared field-tunable intersubband electroluminescence

at room-temperature by photon-assisted tunneling in coupled-quantum wells, Appl. Phys. Lett. **64**, 1144.
71. J. Faist, F. Capasso, C. Sirtori, D. L. Sivco, A. L. Hutchinson, S.-N. G. Chu and A. Y. Cho (1994) Narrowing of the intersubband electroluminescence spectrum in coupled quantum well heterostructures, Appl. Phys. Lett. **65**, 94.
72. J. Faist, F. Capasso, D. L. Sivco, C. Sirtori, A. L. Hutchinson and A. Y. Cho (1994) Quantum cascade laser: an intersubband semiconductor laser operating above liquid nitrogen temperature, Electron. Lett. **30**, 865.
73. J. Faist, F. Capasso, D. L. Sivco, A. L. Hutchinson, C. Sirtori, S. N. G. Chu and A. Y. Cho (1994) Quantum cascade laser: temperature dependence of the performance characteristics and high To operation, Appl. Phys. Lett. **65**, 2901.
74. J. Faist, F. Capasso, C. Sirtori, D. L. Sivco, A. L. Hutchinson and A. Y. Cho (1995) Vertical transition quantum cascade laser with Bragg confined excited state, Appl. Phys. Lett. **66**, 538.
75. C. Sirtori, F. Capasso, J. Faist, D.L. Sivco, S.N.G. Chu and A. Y. Cho (1992) Quantum wells with localized states at energies above the barrier height: a Fabry-Perot electron filter, Appl. Phys. Lett. **61**, 898.
76. J. Faist, F. Capasso, C. Sirtori, D.L. Sivco, A.L. Hutchinson and A.Y. Cho (1995) Continuous wave operation of a vertical transition cascade laser above 80 K, Appl. Phys. Lett. **67**, 3057.
77. J. Faist, F. Capasso, C. Sirtori, D.L. Sivco, A.L. Hutchinson and A.Y. Cho (1996) Room-temperature mid-infrared quantum cascade lasers, Electron. Lett. **32**, 560.
78. C. Sirtori, F. Capasso, J. Faist, D. L. Sivco, A. L. Hutchinson and A. Y. Cho (1995) Quantum cascade unipolar intersubband light emitting diodes in the 8-13 μm wavelength region, Appl. Phys. Lett. **66**, 4.
79. C. Sirtori, J. Faist, F. Capasso, D. L. Sivco, A. L. Hutchinson and A. Y. Cho (1995) Quantum cascade laser with plasmon-enhanced waveguide operating at 8.4 μm, Appl. Phys. Lett. **66**, 3242.
80. C. Sirtori, J. Faist, F. Capasso, D. L. Sivco, A. L. Hutchinson, S. N. G. Chu and A. Y. Cho (1995) Quantum cascade laser with plasmon-enhanced waveguide operating at 8.4 μm, Appl. Phys. Lett. **66**, 3242.
81. G. Sun and J. B. Khurgin (1993) Optically pumped four-level infrared laser based on intersubband transition in multiple quantum wells: feasibility study, IEEE J. Quantum Electron. **29**, 1104.
82. V. Berger (1994) Three-level laser based on intersubband transitions in asymmetric quantum wells: a theoretical study, Semicond. Sci. Technol. **9**, 1493.
83. A. Afzali-Kushaa, G. I. Haddad and T. B. Norris (1995) Optically pumped intersubband lasers based on quantum wells, IEEE J. Quantum Electron. **31**, 135.
84. J. W. Bales, K. A. McIntosh, T. C. L. G. Sollner, W. D. Goodhue and E. R. Brown (1990) Observation of optically pumped intersubband emission from quantum wells, SPIE Vol. 1283, 74.
85. F.H. Julien, A. Sa'ar, J. Wang and J.-P. Leburton (1995) Optically pumped intersubband emission in quantum wells, Electron. Lett. **31**, 838.
86. J. Wang, J.-P. Leburton, F.H. Julien and A. Sa'ar (1996) Design and performance optimization of optically-pumped mid-infrared intersubband semiconductor lasers, IEEE Photonics Technol. Lett. **8**, 1001.

87. Z. Moussa, P. Boucaud, F.H. Julien, Y. Lavon, A. Sa'ar, V. Berger, J. Nagle, N. Coron (1995) Observation of infrared intersubband emission in optically pumped quantum wells, Electron. Lett. 31, 912.
88. Y. Lavon, A. Sa'ar, Z. Moussa, F. H. Julien, R. Planel (1995) Observation of optically pumped midinfrared intersubband luminescence in a coupled quantum well structure, Appl. Phys. Lett. **67**, 1984.
89. P. Boucaud, F.H. Julien, V. Berger and J. Nagle (1996) unpublished.
90. J. Faist, F. Capasso, C. Sirtori, D.L. Sivco, A.L. Hutchinson, M.S. Hybertsen and A.Y. Cho (1996) Quantum cascade lasers without intersubband population inversion, Phys. Rev. Lett. **76**, 411.
91. J. Faist, F. Capasso, C. Sirtori, D.L. Sivco, A.L. Hutchinson and A.Y. Cho (1996) Tunable Fano interference in intersubband absorption, International Conf. on Intersubband Transitions in Quantum Wells: Physics and Applications, Ginosar 1995, Proc. in Superlattices and Microstruct. **19**.

INELASTIC LIGHT SCATTERING BY ELECTRONS IN LOW-DIMENSIONAL SEMICONDUCTORS

A. PINCZUK, B.S. DENNIS, L.N. PFEIFFER, and K.W. WEST
Bell Laboratories, Lucent Technologies
Murray Hill, New Jersey 07974-0636, USA

VITTORIO PELLEGRINI
Scuola Normale Superiore
Piazza dei Cavalieri 7, I-56126, Pisa, Italy

A.S. PLAUT
Department of Physics, University of Exeter
Exeter, EX4 4QL, UK

1. Introduction

Free carriers in semiconductor heterojunctions, quantum wells and superlattices have behaviors in common with those of an ideal two-dimensional electron gas. The greatly enhanced carrier mobilities in modulation doped heterostructures have stimulated research on the electron gas that is at the frontiers of condensed matter physics. Fundamental electron-electron interactions in the 2D electron gas manifest in remarkable phenomena like the fractional quantum Hall effect and novel electron solid phases[1-3]. The fractional quantum Hall effect(FQHE) may be considered as an archetype of the class of quantum liquids that emerge in electron systems of reduced dimensions. In such electron liquid states experimental research in conjunction with theoretical studies of electron correlation uncover the intriguing physics of the 2D systems in semiconductor quantum structures.

Resonant inelastic light scattering is a powerful method to study the elementary excitations of low-dimensional electron systems in artificial semiconductor quantum structures. Interest in such optical studies was stimulated by a proposal of Burstein et al.[4], presented at the 14th International Conference on the Physics of Semiconductors. The intervening 18 years have seen extensive light scattering research on the electron systems that reside in semiconductor quantum structures. Light scattering studies of

G. Abstreiter et al. (eds.), Optical Spectroscopy of Low Dimensional Semiconductors, 63–82.

modulation doped semiconductors have been considered in several reviews[5-7].

The resonant inelastic light scattering method is currently being used in numerous studies of novel behavior of electrons in quantum wells, quantum wires and quantum dots. Some of the most important applications of the light scattering method follow from capabilities to measure collective spin-density and charge-density excitations, as well as excitations that display single-particle behavior[8]. These capabilities offer superb opportunities to carry out unique spectroscopic determinations of the structure of energy levels and of fundamental interactions in the electron systems of semiconductor quantum structures.

Resonant inelastic light scattering has been employed in studies of the 2D electron gas embedded in a large magnetic field. This research enables new studies of the intriguing physics of electrons in systems of reduced dimensions. Of great current interest are studies in the quantum Hall regimes, where light scattering offers insights into electron behaviors that are not accessible by the conventional magneto-transport experiments. The impact of light scattering experiments is most significant in the regime of the FQHE, where current theories predict that the incompressible states should exhibit collective gap excitations with dispersions that are characteristic of the electron quantum fluid[1-3]. In fact, by revealing the existence of such dispersive excitations light scattering experiments at filling factor $\nu=1/3$ have offered the first direct evidence that this FQHE state supports the predicted collective gap modes[9-12].

This paper presents a short review of recent light scattering research in the quantum Hall regimes. These low temperature experiments demonstrate the fundamental physics that may be uncovered in light scattering studies of low-dimensional electron systems in semiconductor quantum structures. In the next Section we consider several aspects of the applications of the light scattering method in studies of the excitations of electrons in quantum Hall states. In Section 3 we consider studies of gap excitations of the fractional quantum Hall state with Landau level filling factor $\nu=1/3$[9-12]. In Section 4 we briefly consider the recent observations of *unstable* spin-density excitations of coupled electron double-layers at even integer values of ν[13].

2. Resonant Inelastic Light Scattering in the Quantum Hall Regimes

The transport anomalies seen at the "magic" Landau level filling factors of the integer and fractional quantum Hall effects can be regarded as a sequence of metal-insulator transitions. The observed insulating behaviors indicate that quantum Hall states are incompressible and that an energy gap separates the ground state from its excited states[1-3]. At integer values of ν, energy gaps arise from quantization of kinetic energy into Landau levels and the Zeeman splitting of the spin states[1,2,14]. Given the absence of single-particle gaps at fractional values of ν, the FQHE reveals the remarkable behaviors that emerge from electron interactions in two dimensions. The gaps of the FQHE are novel phenomena due to condensation of the electron system into incompressible quantum fluids[1-3,15,16].

Energy gaps of quantum Hall states are represented by neutral density excitations of the 2D electron gas in the perpendicular magnetic field. These dispersive collective modes are constructed with neutral pair-states in which a particle(or quasiparticle) is promoted to an excited state leaving behind a hole(or quasihole). The Landau gauge is used when one of the quantum numbers is the excitation wavevector. For wavevector $\vec{q}$ there is a displacement between the centers of the cyclotron orbits of the two single-particle states in the neutral pairs. The displacement is given by

$$x_o = q\ell_o^2 \tag{1}$$

Here $\ell_o = (\hbar c/eB)^{1/2}$ is the magnetic length and the direction of the displacement is orthogonal to that of $\vec{q}$. The collective excitations are constructed with these neutral pair states as a basis. The excitations can be regarded as magnetic excitons with average dipole length x_o[17-19]. The magnetic exciton concept is consistent with the classical description in which the Coulomb attraction is balanced by Lorentz forces, and there is linear motion of the center of mass of the pair with a constant velocity inversely proportional to the separation between the two particles[19].

Measurements of thermal activation of magnetotransport yield gap energies in the quantum Hall regimes[20]. These determinations are interpreted as energies required to create widely separated quasiparticle-quasihole pairs. It follows from Eq. (1) that such neutral pairs can be regarded as gap excitations in the limit $q \to \infty$. Measurements of the magnetic field dependence of acceptor luminescence have also been used to extract energies of widely separated neutral pairs[21]. The study of long wavelength($q \to 0$) collective gap excitations and their wavevector dispersions is the major goal of spectroscopy in the quantum Hall regimes. This quest is particularly important in the regime of the FQHE, where optical experiments could offer significant tests of the theories of electron quantum fluids in 2D systems.

The collective gap excitations of the FQHE states are intra-Landau-level modes (i.e. without changes in Landau quantum number). The neutral pairs are associated here with fractionally charged quasiparticles that obey fractional statistics[16,22,23]. The dispersive modes were calculated by means of numerical evaluations of small finite systems and within a single-mode-approximation(SMA)[24,25]. The calculated wavevector dispersions exhibit distinctive *magnetoroton* minima at wavevectors $q \gtrsim 1/\ell_o$. The roton, caused by the dependences of quasiparticle-quasihole interactions on q(and on x_o according to Eq. (1)), is a characteristic signature of electron interactions in large perpendicular magnetic fields[16,18,19].

Figure 1 displays the first light scattering observations of gap excitations of the FQHE state with ν=1/3. The mode is the sharpest peak in the spectra(full width at half maximum FWHM=0.04meV). The broader, but quite sharp, bands labeled L_o and L_o' are due to luminescence from recombination of the electron gas with weakly photoexcited holes in the highest Landau level of the GaAs quantum well. Strong luminescence occurs in these spectra because in such resonant inelastic light scattering

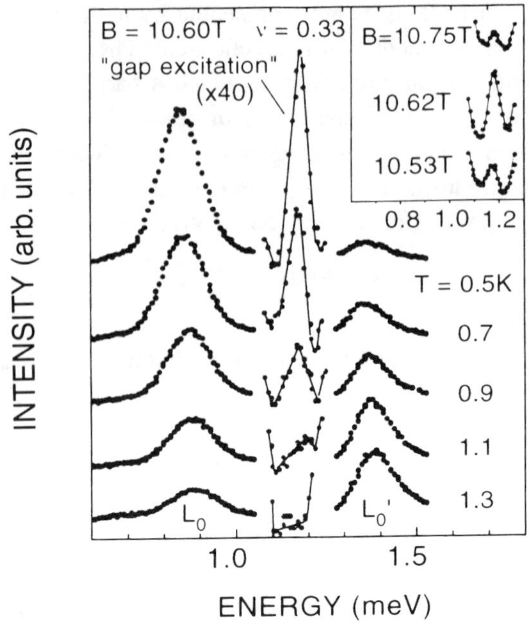

Figure 1. Temperature dependence of resonant inelastic light scattering spectra at the magnetic field of the FQHE state with ν=1/3. The sharp peak at 1.18meV is interpreted as the q=0 collective gap excitation of the electron quantum liquid state. The sample is a 250Å $GaAs/Al_{0.1}Ga_{0.9}As$ single quantum well with electron density $n=8.5x10^{10}cm^{-2}$. The low temperature electron mobility is $\mu=3x10^{6}cm^{2}/Vs$. The inset shows the magnetic field dependence of the T=0.5K spectra. The temperature dependence of the L_o and L_o' intensities is an optical anomaly which is characteristic of the ν=1/3 state[27].(After Ref. [9]).

experiments the photon energies overlap with the fundamental optical transitions of the quantum structure that hosts the 2D electron system. Given that the photon wavevectors are much smaller than $1/\ell_o$, the sharp light scattering peak in Fig. 1 is interpreted as a $q\rightarrow 0$ collective gap excitation of the state at ν=1/3.

The light scattering measurements of gap excitations of the FQHE state are surprising. At long wavelengths the dynamical structure factors, S(q,ω), which enter into conventional expressions of inelastic scattering cross-sections, vanish as q^4 for the intra-Landau-level transitions associated with gap excitations of the FQHE[16,26]. The light scattering experiments are also intriguing because the character of q=0 gap excitations is not clearly specified in current theories. In fact, Girvin et al.[26] speculated that two gap excitations, each near the magnetoroton minimum at

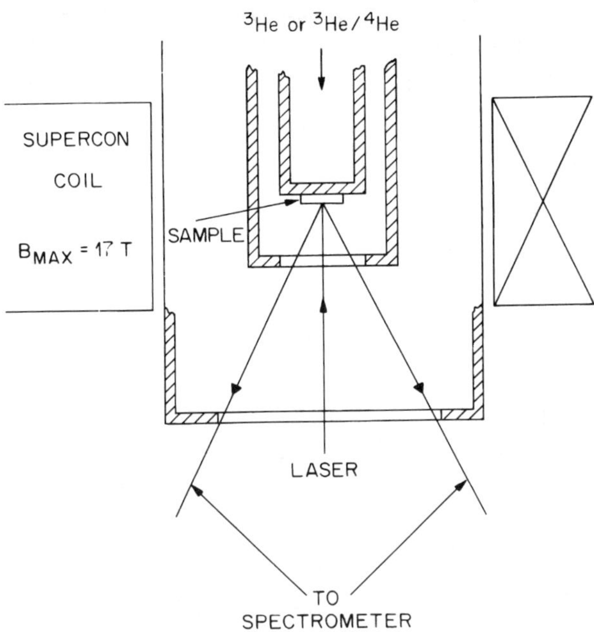

Figure 2. Schematic representation of inelastic light scattering experiments in the quantum Hall regimes. The magnetic field is here perpendicular to the quantum structure sample as well as to the 2D electron layer.(After Ref. [11]).

wavevectors close to $1/\ell_0$ could pair to produce a two-roton state with q=0. It was also proposed that the q=0 gap excitation could consist of two dipoles in a configuration that has a quadrupole moment but no net dipole moment[28]. Within the conceptual frameworks defined by such hypotheses, light scattering by the q=0 gap excitation at ν=1/3 might not be related to S(0,ω). Instead, the resonant light scattering cross-sections by two-roton states may be similar to second-order Raman scattering by optical phonons. These considerations stimulated the resonant light scattering experiments that culminated in the observations of q=0 gap excitations.

The experiments considered here employ the backscattering geometry shown in Fig. 2. In this setup ^{3}He or dilution ^{3}He/^{4}He cryostats are inserted in the cold bore of a superconductive magnet. The assembly has silica windows for optical access. To be compatible with the low temperatures required for measurements of low-energy collective gap excitations, the power densities of incident tunable dye laser light are kept low. Such power levels are in the range of $2\text{x}10^{-5}$ to 10^{-3}Wcm^{-2}. This experimental configuration enables low temperature measurements of inelastic light

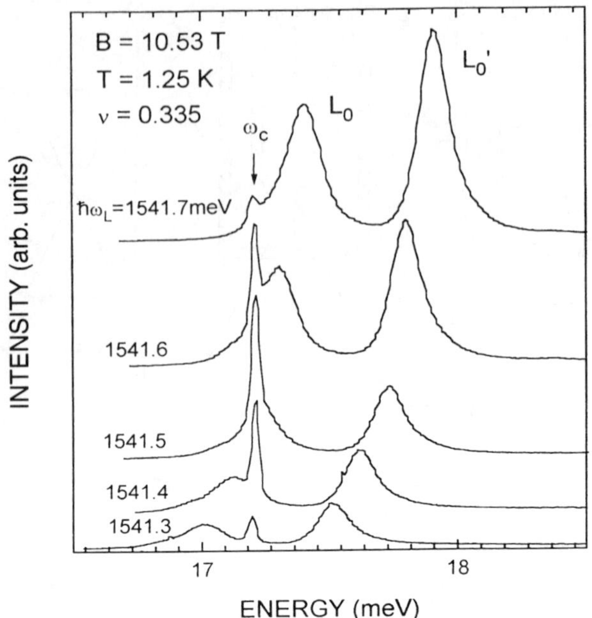

Figure 3. Spectra of resonant inelastic light scattering by the long wavelength charge-density inter-Landau-level excitation at $\nu=1/3$. The spectra show the very sharp light scattering peak by the q=0 magnetoplasmon at ω_c as well as the L_o and L_o' luminescence bands. The sample is the 250Å GaAs single quantum well of Fig. 1. $\hbar\omega_L$ is the incident photon energy. (After Ref. [10]).

scattering spectra at energy shifts that are smaller than 0.1meV[13].

For the backscattering geometry of Fig. 2 the scattering wavevector is $\vec{\kappa}=\vec{k}_L-\vec{k}_S$, where $\vec{k}_L$ and $\vec{k}_S$ are the wavevectors of the incident and scattered light. In this scattering geometry $\vec{\kappa}$ is nearly perpendicular to the plane of the 2D layer. The in-plane component of $\vec{\kappa}$, which we call $\vec{k}$, has typical values of $|\vec{k}|=k<10^4 cm^{-1}$ [5,6].

The kinematics of inelastic light scattering processes follow from conservation of momentum. Translational invariance, which is expected in high mobility 2D electron systems with weak residual-disorder, is associated with conservation of wavevector in inelastic light scattering processes. In single-layer 2D electron systems, the ones considered here, the excitations have only in-plane components of $\vec{q}$. Thus conservation of wavevector is written as

$$\vec{q} = \vec{k} \tag{2}$$

The range of wavevectors which is relevant in studies of dispersive collective excitations is determined by $1/\ell_o$. Since typical values are $\ell_o \lesssim 100\mathring{A}$, the experiments that make use of the scattering geometry shown in Fig. 2, where $k<10^4 cm^{-1}$, have $k\ell_o<<1$. Wavevector conservation, as given by Eq. (2), implies that only long wavelength dispersive excitations are active in such inelastic light scattering experiments.

Inelastic light scattering experiments in the quantum Hall regimes, particularly those at $\nu=1/3$, take advantage of sharp and strong resonant enhancements of the cross-sections[9-12]. An example of spectra measured with sharp resonant enhancement profile is displayed in Fig. 3, which shows measurements of inter-Landau-level excitations at the cyclotron energy ω_c[10]. The data in Fig. 3 uncovered the astonishing fact that intensities of inelastic light scattering by collective excitations of the dilute 2D electron gas could be comparable to peak intensities of the strong photoluminescence of the GaAs quantum well.

The spectra in Fig. 3 demonstrate an extremely sharp resonant enhancement profile of width close to 0.2meV. Such sharp resonances are associated with excitonic optical transitions in the light scattering mechanisms and cross-sections[5,11,29]. Resonant inelastic light scattering mechanisms that invoke *virtual* intermediate excitonic states are given by third- or higher-order time-dependent perturbation theory[5,11,29,30]. The matrix elements of such processes have resonances when incident and scattered photon energies overlap the energies of exciton states. The one at $\hbar\omega_L$ is the *incoming resonance*. The one at the scattered photon energy $\hbar\omega_S$ is the *outgoing resonance*. Such resonance enhancement doublets are characteristic signatures of third- and higher-order light scattering mechanisms.

Resonant inelastic light scattering experiments in the quantum Hall regimes display incoming and outgoing resonances[10,11,31,32]. A remarkable example of an outgoing resonance is displayed in Fig. 3. Here the outgoing resonance occurs as the enhancement maximum in the spectrum that almost exactly overlaps the L_o luminescence due to optical recombination at the fundamental gap of GaAs quantum well.

The observations of inter-Landau-level excitations at ω_c, including those displayed in Fig. 3, offer direct evidence that under strong resonance conditions the light scattering cross-sections for $q\rightarrow 0$ excitations may not be related to conventional dynamical structure factors of the 2D electron gas. This follows from the fact that for excitations at ω_c the dynamical structure factor has a a q^2-dependence at long wavelengths[11]. The impact of mixing between valence Landau levels on the resonant light scattering cross-sections has been suggested as an explanation for the light scattering observations of $q\rightarrow 0$ inter-Landau-level excitations[10,11,32]. In GaAs quantum structures nonparabolicity of valence bands and spin-orbit coupling result in extensive, B-dependent, mixing of eigenstates of orbital and spin angular momentum. Such effects should allow, in the dipole-approximation, the two optical transitions required for light scattering by excitations at ω_c.

The access to excitations that are not predicted by conventional response functions of the electron gas, such as the dynamical structure factors $S(q,\omega)$, offers significant opportunities to study long wavelength excitations in the quantum Hall regimes. As stated above, these collective modes often have $S(0,\omega)\rightarrow 0$.

Breakdown of wavevector conservation, due to loss of translational invariance, gives access to light scattering by large wavevector excitations. Breakdown of wavevector conservation activates resonant inelastic light scattering processes with $q\neq k$. Such processes have been invoked to explain the observation of large wavevector excitations in the quantum Hall regimes[31-34]. These experiments enabled the determinations of magnetoroton minima and other critical points that occur in the dispersions of the collective excitations at wavevectors $q\sim 1/\ell_o$. The loss of translational symmetry is attributed here to the well-known effects of localization and residual-disorder that prevail in the insulating states of the quantum Hall regimes[12,31-34].

3. Studies of Gap Excitations of the FQHE state at $\nu=1/3$

The sharp peak at 1.18meV in the spectra of Fig. 1 is assigned to a collective gap mode of the electron quantum fluid of the FQHE state with $\nu=1/3$[9]. The sharpness of the peak suggests that wavevector is conserved in these measurements, and that the observed excitations have $q=0$. The assignment is supported by the striking temperature and magnetic field dependences displayed in the figure. In these results the sharp inelastic light scattering peak is observed for temperatures $T<1.3K$ and in the small magnetic field interval $\Delta B\simeq 0.5T$ near the field of $\nu=1/3$. These are the temperature and magnetic field dependences seen in magnetotransport measurements of FQHE states.

The comparisons of the light scattering results with theoretical predictions for long wavelength gap excitations of the FQHE state need to take into consideration the effect of the finite width of the 2D layer along the perpendicular direction. Such finite width effect has been assessed within the single-mode approximation[26] and by numerical evaluations of finite systems with small numbers of electrons[35,36]. Both calculations are in excellent agreement with the light scattering results.

The strong luminescence seen in the spectra of Fig. 1 occurs because in those resonant light scattering experiments the photon energies overlap the fundamental optical transitions of the GaAs quantum structure. It is clearly apparent in Fig. 1 that this luminescence signal creates serious difficulties for the identification of inelastic light scattering lineshapes. To overcome the negative impact of strong luminescence backgrounds on inelastic light scattering spectra we have searched for resonances above the fundamental optical gap of the GaAs quantum structures.

In this search we have been guided by conjectures on the light scattering mechanisms of collective gap excitations of the FQHE state at $\nu=1/3$. We have seen that resonant inelastic light scattering processes involve two intermediate optical transitions associated with virtual exciton states. Light scattering takes place here because

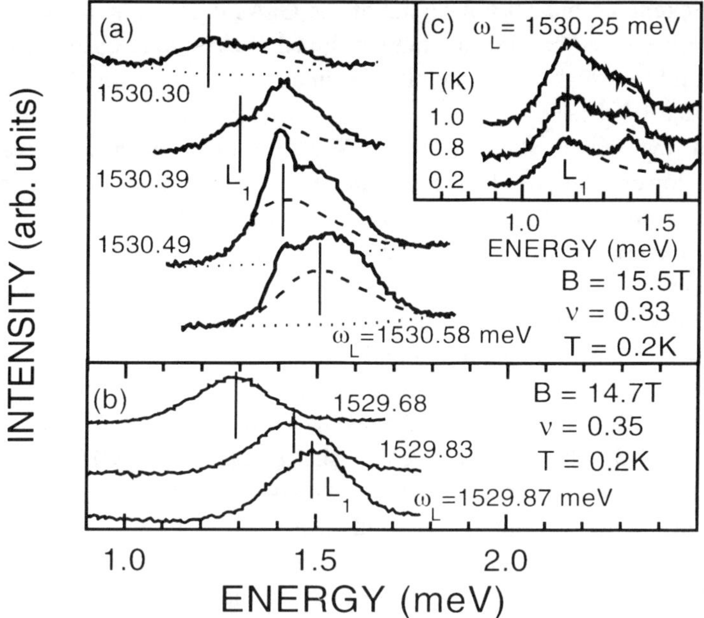

Figure 4. Resonant inelastic light scattering in the energy range of $q \rightarrow 0$ gap excitations of the FQHE state at $\nu=1/3$. The sample is a 250Å GaAs quantum well. The electron density is $n=1.25 \times 10^{11} cm^{-2}$ and the low temperature mobility is in excess of $5 \times 10^6 cm^2/Vs$. ω_L are the incident photon energies. The dotted lines are estimated backgrounds. The dashed lines are estimated non-equilibrium luminescence L_1. The vertical bars indicate the peak positions of L_1. (a) spectra at the magnetic field of $\nu=0.33$, showing a sharp peak at 1.38meV that is interpreted as the long wavelength gap mode of the FQHE state. (b) spectra obtained at $\nu=0.35$. (c) temperature dependence of spectra at $\nu=0.33$.(After Refs. [12] and [33]).

electron-electron interactions introduce processes in which the virtual excitons change state with emission of *real* collective excitations of the 2D system[11,12,29,30].

To be specific, consider the scenario in which the q=0 gap excitation of the $\nu=1/3$ FQHE is constructed from two-roton states. Within the framework of time-dependent perturbation theory light scattering by two-roton states are fourth-order processes in which the exciton-roton interaction acts twice[12]. Here exciton-roton interactions may arise from the large electric dipole associated with gap excitations near the magnetoroton minimum.

Given that fourth-order processes are associated with large enhancements at the outgoing resonance[30], we designed experiments in which ω_S overlaps the energies of higher-lying excitonic optical transitions of the GaAs quantum structure[12,33,37]. in this work we measured several modulation doped GaAs single quantum wells having width 250Å and very high electron mobilities in excess of $4x10^6 cm^2/Vs$.

The results of these recent experiments show the sharp peak of the $q \rightarrow 0$ gap excitation of the FQHE state at $\nu=1/3$. We also observe strongly resonant light scattering at lower energies. In the interpretation of the lower-lying band we have invoked breakdown of wavevector conservation associated with weak residual-disorder. The measured energy was identified with the critical point of the dispersive collective gap excitations of the $\nu=1/3$ FQHE[12,33]. This seems to have been the first experimental evidence of the magnetoroton minimum predicted by current theories of the incompressible electron quantum liquid.

Figure 4 displays the results obtained in the energy range of the long wavelength gap excitation. In Fig. 4(a) we show the results measured at B=15.5Tesla, the field of filling factor $\nu=1/3$. The spectra reveal a sharp peak(FWHM≃0.07meV) at energy 1.38meV, that overlaps a broader structure. The results shown in Fig. 4(b) indicate that the sharp peak is observed only at the fields near $\nu=1/3$, and Fig. 4(c) shows that this peak has the temperature dependence associated with the FQHE. The broader structure(FWHM≃0.25meV) is largely from luminescence due to recombination between Landau levels above the fundamental gap of the 250Å GaAs quantum well. This non-equilibrium luminescence, labeled L_1, is at energy 0.7meV above that of the fundamental recombination L_o.

The measurements shown in Fig. 4 reveal a sharp excitonic enhancement of the scattering cross section of the mode at 1.38meV when the peak overlaps the center of the L_1 luminescence. This is the signature of the outgoing resonance with the excitonic transition associated with L_1. The intensities of the L_1 luminescence are about 40 times weaker than the fundamental recombination L_o. Thus the light scattering intensities in Fig. 4(a) are comparable to those of non-equilibrium luminescence, and these spectra offer enhanced contrast of light scattering against luminescence.

Close examination of the spectra in Fig. 4(a) indicates that the broader structure that overlaps the sharp peak of the $q \rightarrow 0$ gap mode may also incorporate an inelastic light scattering signal. This is suggested by spectra measured with $\hbar\omega_L$ closest to the outgoing resonance(1530.49 and 1530.39meV). In this spectra we see evidence that light scattering by long wavelength excitations of the FQHE state at $\nu=1/3$ may consist of a sharp peak followed by a continuum at higher energy.

The existence of such continua of excitations at energies slightly above the lowest q=0 gap mode is predicted by numerical evaluations of systems with small numbers of electrons[24,35,36,38-41]. In recent work, based on evaluations of systems of up to ten electrons, Song He and P.M. Platzman re-examined the structure of collective intra-Landau-level excitations of the FQHE state at $\nu=1/3$[35,36,41]. These numerical results indicate the existence of a large two-roton component in the long wavelength

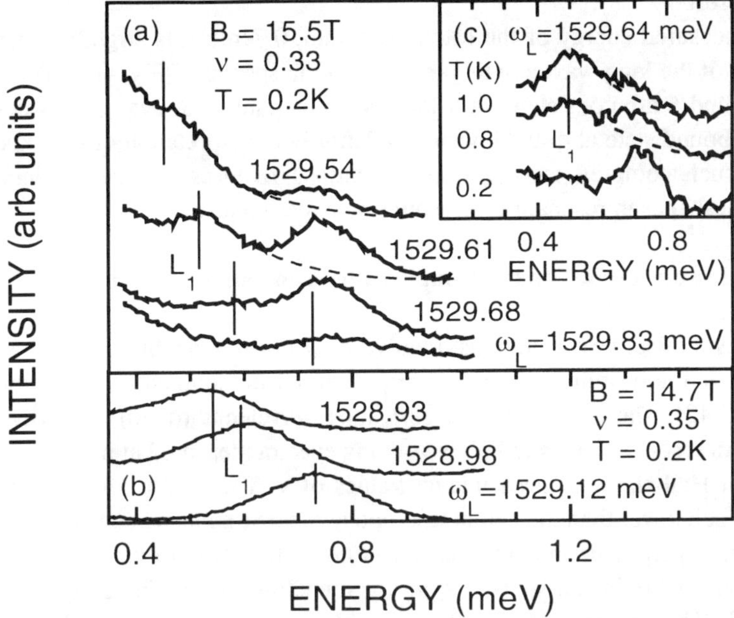

Figure 5. Resonant inelastic light scattering spectra in the magnetic field range of the FQHE state at ν=1/3. The sample is the 250Å GaAs quantum well of Fig. 4. (a) spectra at the field of ν=0.33, showing a narrow light scattering band at 0.74meV. This band is interpreted as the peak in the density of states at the magnetoroton minimum in the dispersion of collective gap excitations of the FQHE state. (b) spectra measured at ν=0.35. (c) temperature dependence of spectra at ν=0.33. (After Refs. [12] and [33]).

excitations, and find numerical evidence that the lowest q=0 mode consists of a weakly bound two-roton state at energy slightly below that of the two-roton continuum.

In Fig. 5 we display the results obtained in the energy range of the magnetoroton minimum of dispersive gap excitations. The spectra measured at the field of ν=0.33 reveal a narrow(FWHM=0.15meV) inelastic light scattering band at energy 0.74meV. This band, clearly seen as separate from the L_1 luminescence, has the magnetic field and temperature dependences that characterize the FQHE states(see Figs. 5(b) and 5(c)). We interpret this band as due to the peak in the density of states at the magnetoroton minimum of the collective mode dispersion. Since the wavevector of this critical point, $q \sim 1/\ell_o$, is much larger than k we assume there is a breakdown of wavevector conservation. Thus resonant inelastic light scattering by the modes at the magnetoroton minimum is activated by the loss of translational invariance due to localization and residual-disorder.

The measured energy of the roton minimum, 0.74meV, is slightly larger than half the energy of the long wavelength gap mode in the spectra of Fig. 4(a). Thus the results in Figs. 4 and 5 are consistent with the results of numerical evaluations that predict a two-roton bound state at q=0. However, a definitive interpretation of the experiments in terms of such complex quasiparticle excitations requires a better understanding of spectral lineshapes in resonant light scattering experiments.

4. Unstable Spin-excitations in Coupled Electron Double-layers

Low-energy collective excitations involving electron tunneling transitions between symmetric and antisymmetric states were observed in coupled GaAs double quantum wells[12,13,42]. The excitations are *soft*, long wavelength($q \rightarrow 0$), spin-density modes that occur in resonant inelastic light scattering spectra measured at magnetic fields close to quantum Hall states of *even* integer values of ν. These soft collective modes have energies much lower than the symmetric-antisymmetric subband spacings.

Recent experiments indicate that quantum Hall states at even-integer values of ν the electron double-layers support spin-flip tunneling excitations of vanishingly small energy[13]. The emergence of such *unstable* spin-excitations could drive quantum phase transitions to broken-symmetry states with strong spin correlations between the two layers[13,43].

The softening of $q \rightarrow 0$ spin-excitations are associated with excitonic bindings within the particle-hole pairs of the tunneling transitions[12,13,42]. These excitonic effects result from vertex-corrections due to exchange Coulomb interactions in the neutral quasiparticle-quasihole pairs of the excitations[13,14,44,45].

Similar soft, unstable, collective excitations play major roles in current theories of strongly correlated states of electrons in low-dimensional quantum structures. We recall that excitonic electron-electron interactions are responsible for the magnetoroton minima in the dispersions of charge-density excitations of quantum Hall states[19]. Roton minima are predicted for gap modes of the FQHE-liquid[24-26], and for tunneling excitations of double-layers at $\nu=1$[46-50]. These minima, at wavevectors $q \sim 1/\ell_o$, are believed to be responsible for the excitonic instabilities associated with the FQHE-liquid to Wigner-solid transition[26,51].

Vertex-correction-driven collective mode instabilities are also expected to be responsible for the suppression of the quantum Hall state in electron bi-layers at $\nu=1$[46-50,52]. In fact, electron double-layers at $\nu=1$ exhibit rich phase diagrams[53-55], in which collective tunneling excitations may play major roles[56].

The direct observation of unstable collective modes has not been reported in experiments carried out in the integer and fractional quantum Hall regimes. Thus, the observations of extremely soft spin-density tunneling excitations of electron double-layers in the quantum Hall regime offers insights that reach beyond previous experimental studies of instabilities in the 2D electron gas. By uncovering such evidence of unstable collective modes, these light scattering experiments indicate new venues for studies of the intriguing physics in low-dimensional electron systems.

The soft spin-density excitations are measured in coupled electron double-layers that have small symmetric-antisymmetric tunneling gaps Δ_{SAS}. In a qualitative discussion of vertex-corrections we consider the following approximate expression for the energy of spin-excitations at B=0[44]

$$\omega_{SE}^2 = \Delta_{SAS}^2 - 2(n_S - n_{AS})\beta\Delta_{SAS} \tag{3}$$

where n_S and n_{AS} are the occupations of the symmetric and antisymmetric states, and β parametrizes the strength of vertex-corrections. When the vertex-correction term approaches the value of Δ_{SAS}^2 the spin-density mode may become unstable[57]. However, the instability does not occur at B=0 in coupled GaAs double quantum wells[45].

In the electron double-layers of the studies considered here $\Delta_{SAS} \lesssim 1$meV, and at B=0 electrons populate symmetric as well as antisymmetric states because Δ_{SAS} is smaller than the Fermi energy. Thus, the magnitude of the vertex-corrections term in Eq. (3) is limited by the available phase-space for the excitations (n_S-n_{AS}).

Enhancement of vertex-corrections occurs when B≠0. In a perpendicular magnetic field the effective electron density participating in the tunneling excitations can be manipulated and increased by changes in Landau level filling associated with changes in B. This is readily seen in the expression[13]

$$n_S - n_{AS} = n(\nu_S - \nu_{AS})/\nu \tag{4}$$

where $n = n_S + n_{AS}$. ν_S and ν_{AS} are the filling factors of the symmetric and antisymmetric states ($\nu = \nu_S + \nu_{AS}$). Equation (4) shows that a maximum in available in phase-space for the tunneling excitations occurs at $\nu=2$, when all the electrons participate in the excitations($\nu_S=2$ and $\nu_{AS}=0$).

The tunneling excitations can be classified according to the angular momenta of the neutral quasiparticle-quasihole pair states that enter in the collective modes[13]. In coupled GaAs double quantum wells the conduction subband states of electrons have spin 1/2 and orbital angular momentum is zero.

We consider the cases of even Landau level filling factor, when the two spin states of the populated Landau levels are fully occupied by the electrons. Spin-excitations here are triplet states in which the change in total angular momentum is $\delta S=1$[13,19]. These states are classified by changes in spin angular momenta along the magnetic field. Spin-density excitations have $\delta S_z=0$ and energies ω_{SDE}. Spin-flip excitations have $\delta S_z=\pm1$, and energies $\omega_{SF}^{\pm}$. Within the Hartree-Fock approximation the energies of spin-flip tunneling excitations at even values of ν are given by[13,19]

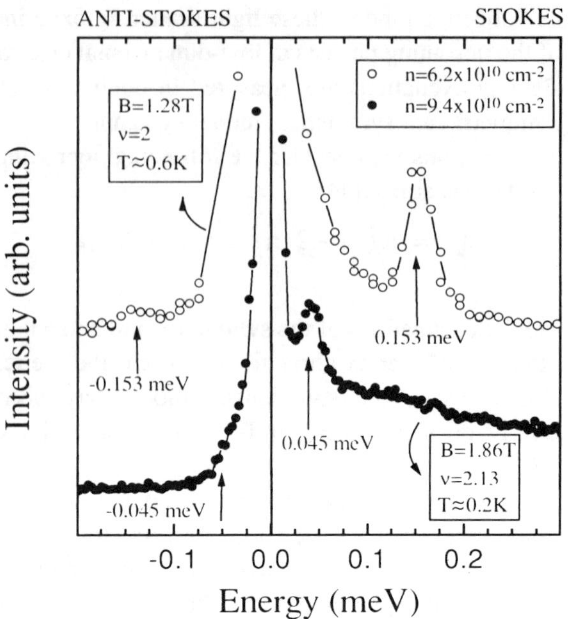

Figure 6. Resonant inelastic light scattering spectra from electron double-layers at filling factors close to ν=2. The spectra were measured with orthogonal linearly polarized incident and scattered photons. The sharp peaks(FWHM≲0.03meV) correspond to long wavelength(k=q~0) spin-density excitations. The results for two electron densities are shown.(After Ref. [13]).

$$\omega_{SF}^{\pm} = \omega_{SDE} - E_z \delta S_z \tag{5}$$

where $E_z = g\mu_B B$ is the Zeeman energy. μ_B is the Bohr magneton and g is the g-factor for the GaAs double quantum wells($g \simeq -0.4$).

Light scattering polarization selection rules for the collective tunneling excitations derived for bulk semiconductors also apply to 2D electron layers in the quantum structures[4,13]. For the backscattering geometry displayed in Fig. 2 these selection rules predict that SDE modes are active in the spectra measured with orthogonal incident and scattered light polarizations[58]. For spin-flip transitions these selection rules require that one of the photons has to be polarized along the magnetic field. Thus spin-flip excitations are forbidden in the experimental configuration of Fig. 2.

We consider here the results from two modulation doped double quantum wells grown by molecular beam epitaxy. They consist of two nominally identical GaAs quantum wells of thickness 180Å, that are separated by a 79Å wide $Al_{0.1}Ga_{0.9}As$ undoped barrier. These structures have electron densities of $n=6.2x10^{10}cm^{-2}$ and $n=9.2x10^{10}cm^{-2}$. The low temperature electron mobilities are close to $10^6cm^2/Vs$. In resonant inelastic light scattering experiments at B=0 we determined that in these electron double-layers $\Delta_{SAS}\simeq0.7meV$ and $\omega_{SE}\simeq0.45meV$[13,45].

Figure 6 shows light scattering spectra of spin-density tunneling excitations measured at $\nu\simeq2$. These results demonstrate a large, density-dependent, softening of the collective mode energy. The spectrum measured in the higher electron density sample displays a very sharp peak at $\omega_{SDE}=0.045meV$. This energy is *one-tenth* of the value determined at B=0. This extremely low energy is very close to the $E_z=0.043meV$ calculated with g=-0.4.

The observed collapse of the spin-density tunneling mode to energies very close to the Zeeman splitting implies the existence of spin-flip excitations with $\delta S_z=1$ at vanishingly small energies, as indicated by Eq. (5). The direct observation of such ultra-low-energy modes is beyond the reach of the present light scattering experiments. There is, however, ample evidence of anomalous behaviors in the high density sample at filling factors close to $\nu=2$[13]. For example, light scattering spectra at $T\simeq0.2K$ display a marked decrease of the intensities measured at ω_{SDE} for fields such that $\nu\leq2.05$. Surprisingly the peak recovers at temperatures $T\gtrsim0.5K$.

The anomalous temperature dependences of light scattering intensities are unexpected. They suggest that striking changes occur in the bi-layers at $\nu\simeq2$ and $T\leq0.5K$. The anomalies are observed when ω_{SDE} collapses to an energy close to E_z. It is thus conceivable that the collapse of spin-flip tunneling excitations leads to the spontaneous generation of large numbers of spin-flip excitonic pairs. The generation of such bound particle-hole pairs does not change the insulating behavior of the quantum Hall state, but causes the emergence of a net spin-polarization along the magnetic field axis. While the anomalous temperature dependence of the light scattering intensities is presently not understood, their occurrence could be consistent with such a hypothesis.

The recent work of Zheng et al. suggests a different interpretation[43]. Here the collapse of the spin-flip tunneling excitation is associated with a quantum phase transition to a state with antiferromagnetic in-plane spin correlations between the layers. This state is expected to sustain new collective spin-excitations. The anomalous temperature dependence of the light scattering intensities could be regarded as evidence of a finite temperature phase transition that destroys the order in the antiferromagnetic phase[59]. Further light scattering experiments might offer evidence about the existence of such novel spin states.

5. Conclusion

This article has presented a succinct review of recent resonant inelastic light scattering research carried out in the regimes of the integer and fractional quantum Hall effects. Such studies exemplify the class of fundamental physics that is accessible by resonant inelastic light scattering of low-dimensional electron systems in semiconductor quantum structures.

We anticipate increasing interest in studies of collective modes of electron liquid states in the magnetic field regime of the FQHE. Great progress has been achieved in studies of the state with $\nu=1/3$. Much work remains to be done to measure gap excitations of other incompressible states such as those at filling factors 2/5 and 3/7. Such studies could uncover aspects of composite-fermion quasiparticles[20,60], and thus offer new significant insights into fundamental interactions associated with the electron quantum fluid. Experimental advances will require here imaginative methods to enhance the contrast between the resonant inelastic light scattering signals and the backgrounds due to luminescence. In the cases of weaker FQHE states with $\nu\neq1/3$, the capability to extract the light scattering intensities that overlap strong luminescence is essential for successful measurements of low-lying gap excitations. Of great current interest are also studies of the intriguing low-lying excitations of the compressible state at $\nu=1/2$[61].

Light scattering research may also play a key role in gaining an understanding of the impact of the spin degree of freedom and of spin-excitations in the incompressible and compressible states of the 2D electron gas in the extreme magnetic quantum limit($\nu<1$). The capabilities of the resonant light scattering method to access spin-excitations in the FQHE regime have been demonstrated[9,62,63]. Further experiments could be significant in the elucidation of fundamental interactions associated with spin in the FQHE liquid.

The discovery of unstable spin-excitations in the electron bi-layers of double quantum wells suggests novel applications of resonant light scattering methods in studies of quantum phase transitions. They include phase transitions at even and odd integer filling factors, and the transition from the electron liquid to a Wigner solid at small values of ν. While there is a large body of theory predicting that quantum phase transitions are triggered by unstable collective excitations[43,46-56], our recent results in electron double-layers are the first to offer spectroscopic evidence of the existence of such modes[13]. Thus these are areas in which light scattering spectroscopy of low-lying excitations could yield significant results.

Finally, we believe that the insights and know-how gained in low temperature light scattering studies of highly correlated states of 2D electron systems creates opportunities for the study of fundamental electron interactions in the lower dimensionality electron systems of semiconductor quantum wires and quantum dots. Because these interactions become apparent in low-disorder systems, progress in these areas will benefit from advances in semiconductor nanofabrication technologies.

6. Acknowledgements

We gratefully acknowledge valuable discussions with E. Burstein, S. Das Sarma, B.I. Halperin, Song He, P.B. Littlewood, A.H. MacDonald, P.M. Platzman and L. Zheng.

We are also grateful to K.W. Baldwin for magneto-transport measurements. ASP gratefully acknowledges financial support from the Leverhulme Trust, UK.

7. References

1. Prange, R.E. and Girvin, S.M. (eds.) (1990) *The Quantum Hall Effect*, 2nd Ed., Springer, New York.
2. Chakraborty, T. and Pietilainen, P. (1988) *The Fractional Quantum Hall Effect: Properties of an incompressible quantum fluid*, Springer-Verlag, Berlin.
3. Das Sarma, S. and Pinczuk, A. (eds.) (1996) *Perspectives in Quantum Hall Effects: Novel Quantum Liquids in Low Dimensional Semiconductor Structures*, J. Wiley & Sons, New York.
4. Burstein, E., Pinczuk, A., and Buchner, S. (1979) Resonance inelastic light scattering by charge carriers at semiconductor surfaces, in B.L.H. Wilson (ed.), *Physics of Semiconductors 1978*, The Institute of Physics, London, pp. 1231-1234.
5. Abstreiter, G., Cardona, M., and Pinczuk, A. (1984) Light scattering by free carrier excitations in semiconductors, in M. Cardona and G. Guentherodt (eds.), *Light Scattering in Solids IV*, Springer-Verlag, Berlin and Heidelberg, pp. 5-150.
6. Pinczuk, A. and Abstreiter, G. (1989) Spectroscopy of free carrier excitations in semiconductor quantum wells, in M. Cardona and G. Guentherodt (eds.), *Light Scatteering in Solids V*, Springer-Verlag, Berlin-Heidelberg, pp.153-211.
7. Pinczuk, A., Dennis, B.S., Pfeiffer, L.N., West, K.W., and Burstein, E. (1994) Inelastic light scattering by the two-dimensional electron gas: fractional quantum Hall regime and beyond, *Phil. Mag.* B**70**, 429-442.
8. Pinczuk, A. (1992) Inelastic light scattering by the two-dimensional electron gas, *Festkoerperprobleme(Advances in Solid State Physics)* **32**, 45-60.
9. Pinczuk, A., Dennis, B.S., Pfeiffer, L.N., and West, K.W., (1993) Observation of collective excitations in the fractional quantum Hall effect, *Phys. Rev. Lett.* **70**, 3983-3986.
10. Pinczuk, A., Dennis, B.S., Pfeiffer, L.N., and West, K.W., (1994) Inelastic light scattering in the regimes of the integer and fractional quantum Hall effects, *Semicond. Sci. Technol.* **9**, 1865-1870.
11. Pinczuk, A., (1996) Resonant inelastic light scattering from quantum Hall systems, in *Ref. [3]*, pp. 307-341.
12. Pinczuk, A., Dennis, B.S., Plaut, A.S., Pfeiffer, L.N., and West, K.W. (1997) Light scattering by low-energy collective excitations of quantum Hall states, in *Procs. of the Int. Conf. on the Application of High Magnetic Fields in Semiconductor Physics*, to be published.
13. Pellegrini, V., Pinczuk, A., Dennis, B.S., Plaut, A.S., Pfeiffer, L.N., and West, K.W., (1997) Collapse of spin-excitations in quantum Hall states of coupled electron double-layers, *Phys. Rev. Lett.* **78**, 310-313.
14. Ando, T., Fowler, A.B., and Stern, F. (1982) Electronic properties of two-dimensional systems, *Revs. Mod. Phys.* **54**, 437-672.

15. Tsui, D.C., Stormer, H.L., and Gossard, A.C. (1982) Two-dimensional magnetotransport in the extreme magnetic quantum limit *Phys. Rev. Lett.* **48**, 1559-1562.
16. Laughlin, R.B. (1990) Elementary theory: the incompressible quantum fluid, in *Ref. [1]*, pp. 233-301.
17. Chiu, K.W. and Quinn, J.J. (1974) Plasma oscillations of a two-dimensional electron gas in a strong magnetic field, *Phys. Rev.* B**9**, 4724-4732.
18. Lerner, I.V. and Lozovik, Yu.E. (1980) Mott exciton in a quasi-two-dimensional semiconductor in a strong magnetic field, *Sov. Phys. JETP* **51**, 588-592.
19. Kallin, C and Halperin, B.I. (1984) Excitations from a filled Landau level in the two-dimensional electron gas, *Phys. Rev.*B **30**, 5655-5668.
20. Stormer, H.L. and Tsui, D.C. (1996) Composite fermions in the fractional quantum Hall effect, in *Ref. [3]*, pp. 385-421.
21. Kukushkin, I.V., Pulsford, N.J., Klitzing, K.v., Ploog, K., Ploog., K., Haug, R.N., Koch, S., and Timofeev, V.B. (1992) Energy gaps of the fractional quantum Hall effect measured by magneto-optics, *Europhys. Lett.* **18**, 63-68.
22. Arovas, D., Schrieffer, J.R., and Wilczek, F. (1984) Fractional statistics and the quantum Hall effect, *Phys. Rev. Lett.* **53**, 722-723.
23. Halperin, B.I. (1984) Statistics of quasiparticles and the hierarchy of fractional quantized Hall states, *Phys. Rev. Lett.* **52**, 1583-1586.
24. Haldane, F.D.M. and Rezayi, E.H. (1985) Finite-size studies of the incompressible state of the fractionally quantized Hall effect and its excitations, *Phys. Rev. Lett.* **54**, 237-240.
25. Girvin, S.M., MacDonald, A.H., and Platzman, P.M. (1985) Collective excitation gap in the fractional quantum Hall effect, *Phys. Rev. Lett.* **54**, 581-583.
26. Girvin, S.M., MacDonald, A.H., and Platzman, P.M. (1986) Magneto-roton theory of collective excitations in the fractional quantum Hall effect, *Phys. Rev.* B**33**, 2481-2494.
27. Heiman, D., Goldberg, B.B., Pinczuk, A., Tu, C.W., Gossard, A.C., and English, J.H. (1988) Optical anomalies of the two-dimensional electron gas in the extreme magnetic quantum limit, *Phys. Rev. Lett.* **61**, 605-608; Goldberg, B.B., Heiman, D., Pinczuk, A., Pfeiffer, L., and West, K. (1990) Optical investigations of the integer and fractional quantum Hall effects: energy plateaus, intensity minima, and line splitting in band-gap emission, *ibid* **65**, 641-644.
28. Lee, D.H. and Zhang, S.C. (1991) Collective excitations in the Guinzburg-Landau theory of the fractional quantum Hall effect, *Phys. Rev. Lett.* **66**, 1220-1223.
29. Danan, G., Pinczuk, A., Valladares, J.P., Pfeiffer, L.N., and West, K.W. (1989) Coupling of excitons with free electrons in light scattering from GaAs quantum wells, *Phys. Rev.* B**39**, 5512-5515.
30. Ganguly, A.K. and Birman, J.L. (1967) Theory of lattice Raman scattering in insulators, *Phys. Rev.* **162**, 806-816.
31. Pinczuk, A., Valladares, J.P., Heiman, D., Gossard, A.C., English, J.H., Tu, C.W., Pfeiffer, L., and West, K. (1988) Observation of roton density of states in two-dimensional Landau-level excitations, *Phys. Rev. Lett.* **61**, 2701-2704.

32. Pinczuk, A., Dennis, B.S., Heiman, D., Kallin, C., Brey, L., Tejedor, C., Schmitt-Rink, S., Pfeiffer, L.N., and West, K.W. (1992) Spectroscopic measurement of large exchange enhancement of a spin-polarized 2D electron gas, *Phys. Rev. Lett.* **68**, 3623-3626.
33. Pinczuk, A., Dennis, B.S., He, S., Sohn, L.L., Pfeiffer, L.N., and West, K.W. (1995) Light scattering by collective gap excitations of the fractional quantum Hall liquid, *Bull. Am. Phys. Soc.* **40**, 515.
34. Marmorkos, I.K. and Das Sarma, S. (1992) Magnetoplasmon excitation spectrum for integral filling factors in a two-dimensional electron system, *Phys. Rev.* **45**, 13396-13399.
35. He, S. (1996) Resonant Raman scattering from bound magnetorotons in the fractional quantum Hall regime, *Bull. Am. Phys. Soc.* **41**, 76.
36. He, S. and Platzman, P.M. (1996) Resonant Raman scattering and numerical study in the fractional quantum Hall regime, *Surf. Sci.* **361/362**, 87-90.
37. Pinczuk, A., Dennis, B.S., Pfeiffer, L.N., and West, K.W. (1995) unpublished.
38. He, S., Simon, H., and Halperin, B.I. (1994) Response function of the fractional quantized state on a sphere. II. Exact diagonalization, *Phys. Rev.* **B50**, 1823-1831. 1823.
39. Wu, X.G. and Jain, J.K. (1995) Excitation spectrum and collective modes of composite fermions, *Phys. Rev.* B**51**, 1752-1761.
40. Kamilla, R.K., Wu, X.G., and Jain, J.K. (1996) Composite fermion theory of collective excitations in fractional quantum Hall effect, *Phys. Rev. Lett.* **76**, 1332-1335.
41. He, S. (1996), private communication.
42. Pinczuk, A., Dennis, B.S, Plaut, A.S., Eisenstein, J.P., Pfeiffer, L.N., dnd West, K.W. (1996) Soft low energy spin excitations of electrons in bouble quantum wells, *Bull Am. Phys. Soc.* **41**, 482.
43. Zheng, L., Radtke, R.J., and Das Sarma, S. (1996), Spin-excitation- instability-induced quantum phase transitions in double-layer quantum Hall systems, Phys. Rev. Lett. , submitted.
44. Decca, R., Pinczuk, A., Das Sarma,S., Dennis, B.S., Pfeiffer, L.N., and West, K.W. (1994) Absence of spin-density excitations in quasi-two-dimensional electron systems, *Phys. Rev. Lett.* **72**, 1506-1509.
45. Plaut, A.S., Pinczuk, A., Tamborenea, P.I., Dennis, B.S., Pfeiffer, L.N., and West, K.W. (1997), Absence of unstable zero-field intersubband spin-excitations of dilute electron bi-layers, *Phys. Rev.* B, to be published.
46. Fertig, H.A. (1989) Energy spectrum of a layered system in a strong magnetic field, *Phys. Rev.* B**40**, 1087-1095.
47. MacDonald, A.H., Platzman, P.M., and Boebinger, G.S. (1990) Collapse of integer Hall gaps in a double-quantum-well system, *Phys. Rev. Lett.* **65**, 775-778.
48. Brey, L. (1990) Energy spectrum and charge-density-wave instability of a double quantum well in a magnetic field, *Phys. Rev. Lett.* **65**, 903-906.
49. Chen, X.M. and Quinn, J.J. (1992) Correlated charge-density-wave states of double-quantum-well systems in a strong magnetic field, *Phys. Rev.* B**45**, 11054-11066.

50. He, S., Das Sarma, S., and Xie, X.C (1993), Quantized Hall effect and quantum phase transitions in coupled two-layer electron systems, *Phys. Rev.*B**47**, 4394-4412.
51. Kamilla, R.K. and Jain, J.K. (1996), Excitonic instability and termination of fractional quantum Hall effect, *Phys. Rev. Lett.*, submitted.
52. Boebinger, G.S., Jiang, H.W., Pfeiffer, L.N., and West, K.W. (1990) Magnetic-field-driven destruction of quantum Hall states in a double quantum well, *Phys. Rev. Lett.* **64**, 1793-1796.
53. Murphy, S.Q., Eisenstein, J.P., Boebinger, G.S., Pfeiffer, and West, K.W. (1994) Many-body integer quantum Hal effect: evidence for new phase transitions, *Phys. Rev. Lett.* **72**, 728-731.
54. Yang, K., Moon, K., Zheng, L., MacDonald, A.H., Girvin, S.M., Yoshioka, D., and Zhang, S.C. (1994) Quantum ferromagnetism and phase transitions in double-layer quantum Hall systems, *Phys. Rev. Lett.* **72**, 732-735.
55. Lay, T.S., Suen, Y.W., Manoharan, H.C., Ying, X., Santos, M.B., and Shayegan, M. (1994) Anomalous temperature dependence of the correlated $\nu=1$ quantum Hall effect in bilayer electron systems, *Phys. Rev.* B**50**, 17725-17728.
56. Girvin, S.M. and MacDonald, A.H. (1996), Multicomponent quantum Hall systems: the sum of their parts and more, in *Ref. [3]*, pp. 161-224.
57. Das Sarma, S., and Tamborenea, P.I. (1994) Vertex-correction-driven intersubband spin-density excitonic instability in double quantum well structures, *Phys. Rev. Lett.* **73**, 1971-1974.
58. Yafet, Y. (1966) Raman scattering by carriers in Landau levels, *Phys. Rev.* **152**, 858-862.
59. Das Sarma, S. (1996), private communication.
60. Jain, J.K. (1996), Composite fermions, in *Ref. [3]*, pp.265-305.
61. Halperin, B.I. (1996), Fermion Chern-Simons theory and the unquantized quantum Hall effect, in *Ref. [3]*, pp. 225-263.
62. Turberfield, A.J., Davies, H.D.M., Harris, J.C., Brockbank, R.L., Snelling, M.J., and Ryan, J.F. (1997), Raman scattering from a Laughlin liquid at $\nu=1/3$, in *Ref. [12]*, to be published.
63. Pinczuk, A., Dennis, B.S., Pfeiffer, L.N., and West, K.W.(1996), work in progress.

ELECTRIC-FIELD DOMAINS, POCKELS EFFECT AND COHERENT ACOUSTIC PHONONS IN SUPERLATTICES

R. MERLIN
Department of Physics
University of Michigan
Ann Arbor, MI 48109-1120, *U. S. A.*

1. Introduction

These notes center on experiments [1-35] and the theory [36-46] of electric field domains in quantum-well (QW) structures. We also discuss the QW Pockels effect [48-53] and the generation of high frequency (terahertz) coherent sound in superlattices using light pulses [54-55]. Our intent is not so much to provide a comprehensive description, but a brief and didactic account of selected topics in these areas. However, the list of references is fairly complete. The reader will find answers to questions not considered here and alternative views on some of the issues in the many papers available from the literature (for a review of experiments on domains, see [47]).

2. Electric-Field Domains and Sequential Resonant Tunneling

Sequential resonant tunneling (SRT), combining conventional resonant tunneling with phonon-induced intersubband decay, is the dominant transport mode of *weakly-coupled* QW structures [56]. Let F be the electric field for which the potential drop across a superlattice period, d, exactly matches the energy separation between the two lowest eigenstates of the wells, i.e., $F = (E_2 - E_1)/ed \equiv \mathcal{E}_{12}$. In this situation, the first energy level of a given well and the second level of a neighboring well become degenerate and, thus, carriers can resonantly tunnel between the two sites. Since the tunneling current and, therefore, the net current flow (arising primarily from intersubband scattering due to carrier-phonon coupling) *decreases* for small departures from the resonant condition, we have that the current is maximum at $\mathcal{E}_{12}$ as shown schematically in Fig. 1(a). Collisions which randomize the momentum component parallel to the layers play an additional important role in the carrier transport. In particular, these processes lead to broadening of the resonance by destroying the tunneling coherence [56].

The first observation of QW-SRT was reported by Esaki and Chang in 1974 [1]. However, their work on unipolar GaAs/AlAs superlattices did not reveal the predicted *I-V* peak, but a ratchet-like behavior as in Fig. 1(b) with a voltage period approximately equal to $\mathcal{E}_{12}d$. Much later, in 1985, Capasso *et al.* reported data on a photoexcited *p-i-n* (Al,In)As/(Ga,In)As structure which exhibited the expected peak at the voltage corresponding to $F \approx \mathcal{E}_{12}$, and also a second one at the field associated with the alignment of the first and third subband: $F \approx (E_3 - E_1)/ed \equiv \mathcal{E}_{13}$[2]. As illustrated in Fig. 1(b), the results of Esaki and Chang [1] - which have since been reproduced by several groups for GaAs/(Al,Ga)As [4,5,7-35] and other heterostructure systems [2,3,6] - reflect the fact that QW structures

G. Abstreiter et al. (eds.), Optical Spectroscopy of Low Dimensional Semiconductors, 83–97.

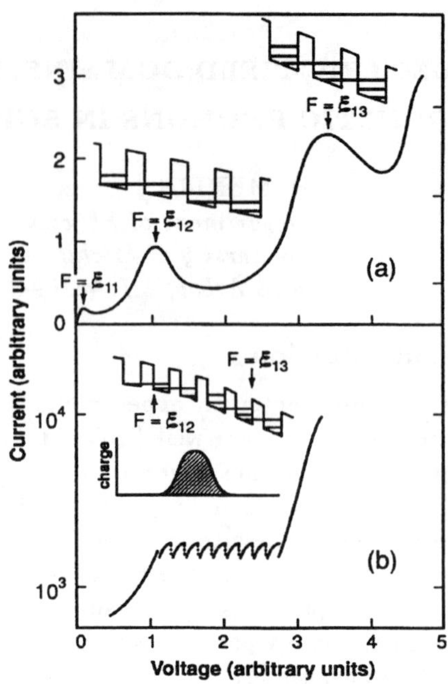

Figure 1. Schematic *I*- *V* traces corresponding to (a) uniform-field (low density), and (b) the domain regime (high density). Current maxima in (a) are due to tunneling involving the level alignment shown in the figure. In (b), the oscillatory structure is due to the motion of the wall by one superlattice period. Inset: *z*- dependence of the potential and charge in the vicinity of the domain boundary. After [21].

spontaneously break into regions referred to as electric-field *domains* so that the field is not uniform but piecewise constant across the sample. In each domain, the field adopts values close to those at the tunneling resonances; see [19]. Domains are but one of the possible manifestations of SRT-induced negative-differential conductance (NDC). In contrast, the behavior of the weakly-conducting sample of Capasso *et al.* [2] is such that the field is uniform (or nearly so) and NDC is shown explicitly. Here, we notice that domains have also been seen in nominally undoped QW structures under strong above-gap photoexcitation [9,21,26]. Photoinjection of carriers allows one to continuously tune the density so that both the low- and high-density regimes as well as the transition between the two can be studied in a single sample [9,21,26]. As reported in [25], domains can also be induced by THz electromagnetic radiation through photon-assisted tunneling.

The phenomena described in the previous paragraph illustrates the characteristic NDC-scenario where a single physical mechanism manifests itself in dramatically different ways depending on sample parameters and boundary conditions [57]. For instance, conventional NDC semiconductors - such as *n*-GaAs [58] and *p-Ge* [59] - may alternatively display Gunn oscillations (i.e., moving high-field domains [60]), static domains like those in quantum- wells or spatiotemporal chaos under alternating bias [59]. Factors determining the pattern selection include the nature of the contacts, the past history of the sample (hysteresis) and the specific profile's stability against fluctuations. As it is well known, the various patterns are but particular solutions to a class of non-linear transport equations exhibiting instabilities arising from NDC [57].

Other than the transport work, studies of photoluminescence (PL) [9,21,26],

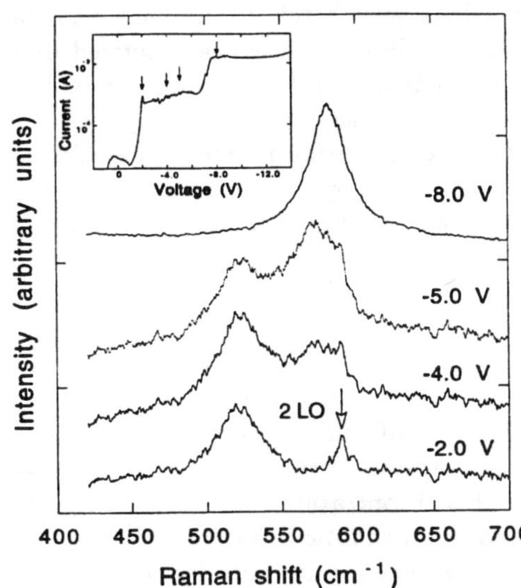

Figure 2. Raman spectra at 4 K showing intersubband excitations for a photoexcited structure consisting of 50 periods of 144-ÅGaAs/34−ÅAlAs. The peak labeled 2LO is due to scattering by two zone- center longitudinal optical phonons. Inset: current vs. voltage; arrows denote voltages at which spectra were obtained. After [31].

cathodoluminescence (CL) [23,32] and Raman scattering (RS) [12,17,31] have been shown to provide direct information on the existence of domains which support the accepted picture shown in Fig. 1(b). These methods rely on the Stark effect for heavy-hole excitons (PL and CL; this is referred to in the literature as the quantum-confined Stark effect [61]) or, in the case of RS, intersubband transitions [62] to characterize the field distribution in the samples. The Raman results in Fig. 2 show $e_1 \rightarrow e_2$ charge-density scattering, where $e_1(e_2)$ denotes the first (second) QW state [31]. The top and bottom spectra correspond to cases where the sample is a monodomain. At $V = -2\text{V}(-8\text{V})$, the first subband of a given well aligns with the second (third) subband of a neighboring well to promote tunneling. The presence of two intersubband peaks at intermediate voltages reflect the coexistence of the two domains.

Experimentally, the question of the formation time of domain walls and, more generally, that of the time-dependence of tunneling instabilities are addressed in [27,29,33] for *doped* structures, and in [21,26] for *undoped* superlattices under photoexcitation. These cases differ primarily in that photoexcited samples exhibit a regime of *damped* oscillatory behavior [21] while the related oscillations in doped systems are *self-sustained* [27]. The PL data in Fig. 3, obtained using step-like profiles, correspond to the former situation. Following the onset of photoexcitation at $t = 0$, the single feature splits into two peaks corresponding to two domains. For $t \geq 100$ns the intensities of peaks oscillate with time, but their positions do not vary much. At a given bias, the latter agree well with values from steady-state spectra, an indication that the domain wall forms within $50 - 100$ns after the laser is turned on. The V-dependence of the fundamental frequency is shown in the inset of Fig. 3. It exhibits several minima as well as discontinuities which seem to correlate with those of the steady-state photocurrent; the decay time ($\approx 0.2 - 0.5\mu$s) shows a similar trend. The *damped* oscillations, which reflect the back and forth motion of the domain boundary approaching equilibrium, are characterized by parameters depending both on V and the power density P [21]. At a fixed bias, the oscillations exist only in a narrow window of intensities, i.e., they disappear if P is either too high or too low. Measurements of the oscillation frequency as a function of laser power for a doped superlattice are shown in Fig. 4. We remark that, unlike photoexcited structures, doped systems exhibit a range of parameters for which steady-state solutions are always unstable so that the corresponding oscillations do not decay in time. As shown in Fig. 4, the time-dependence can be strongly modified by illumination. In particular, notice that the fundamental frequency decreases linearly with laser power above 150 mW and that it vanishes at $\approx$ 500mW. As recently reported [33], doped structures represent an ideal system with a large number of degrees of freedom for studies of nonlinear dynamics. QW samples driven with alternating sources exhibit spontaneous periodic and *chaotic* current oscillations as well as transitions between synchronization and chaos via pattern forming bifurcations (a different form of temporal chaotic behavior of domains was investigated theoretically in [42,45]). At large driving amplitudes, chaos is suppressed giving rise to oscillations involving subharmonics of the driving frequency [33].

Most theoretical studies are based on a discrete drift model which considers a set of weakly-interacting wells characterized by average values of the electric field, F_j, and carrier densities, n_j(electrons) and p_j(holes); the integer j labels the wells [38,39]. Such mean-field-like approach is motivated by the fact that the relevant time for oscillations ($\approx 0.1\mu$s) is considerably larger than those for tunneling and for reaching equilibrium with the lattice ($\approx$ 10ps; see [63]). For doped (n-type) structures, the one-dimensional transport equations read [38,39]:

$$F_j - F_{j-1} = \frac{4\pi e d}{\in}(n_j - \mathcal{N}_j) \quad , \tag{1}$$

$$\in \frac{dF_j}{dt} + ev(F_j)n_j = J \quad , \tag{2}$$

where (1) and (2) are, respectively, Poisson equation and Ampère's law. The dielectric constant is denoted by $\in$, $\mathcal{N}_j$ is the donor impurity concentration and, as before, d is the superlattice period. $v(F)$ is an effective electron velocity and J is the current density. The problem becomes well-defined by adding the bias equation, $\Sigma F_j = V/\ell$, and an additional (non-trivial) boundary condition. These expressions apply also to photoexcited samples if we ignore the contribution of the heavier holes to the current, replace n_j by $n_j - p_j$ and add the hole rate-equation

$$\frac{dp_j}{dt} = \gamma - rn_jp_j \tag{3}$$

where γ is the photogeneration rate (proportional to the laser power-density) and r is the recombination constant [39].

Much of the physics of the model is contained in $v(F)$ which can be either calculated [38] or taken as a datum [39] under the assumption that $v(F)$ does not

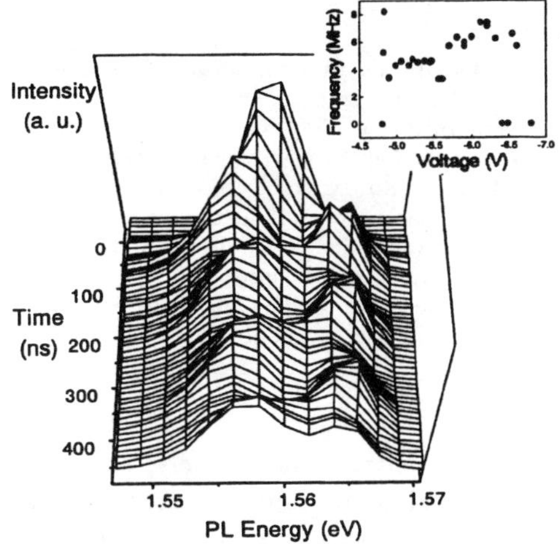

Figure 3. Time-resolved PL data for an ***undoped*** structure (40 periods of 90-ÅGaAs/40-ÅAlAs) after step-like photoexcitation showing ***damped*** oscillatory behavior. The bias voltage $V \approx -6.2V$ is in the domain range. Spectra are given at intervals of 10 ns. Oscillating peaks at ≈ 1.557 and $\approx 1.565 eV$ correspond to the high- and low-field domain. The V-dependence of the real part of the fundamental frequency is shown in the inset. After [21].

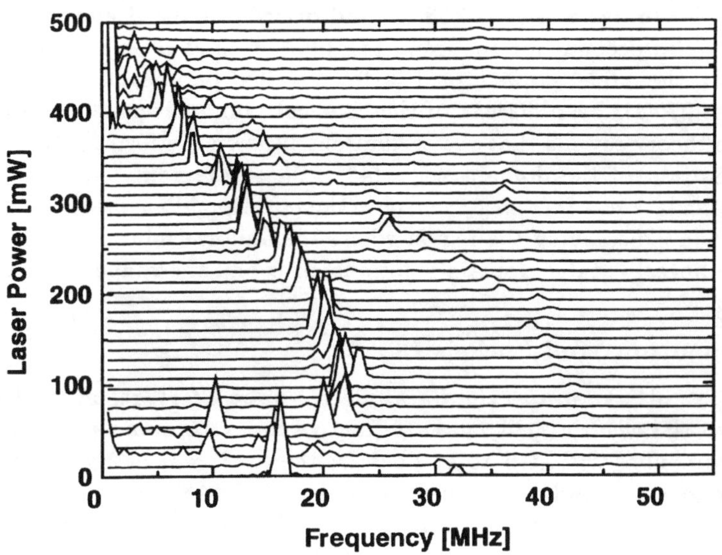

Figure 4. Doped superlattice. Fourier transform of self-sustained (***undamped***) current oscillations as a function of laser power. Data are for a structure consisting of 40 periods of 90-ÅGaAs/40-ÅAlAs at $V = 7$V and $T = 18$K. After [27].

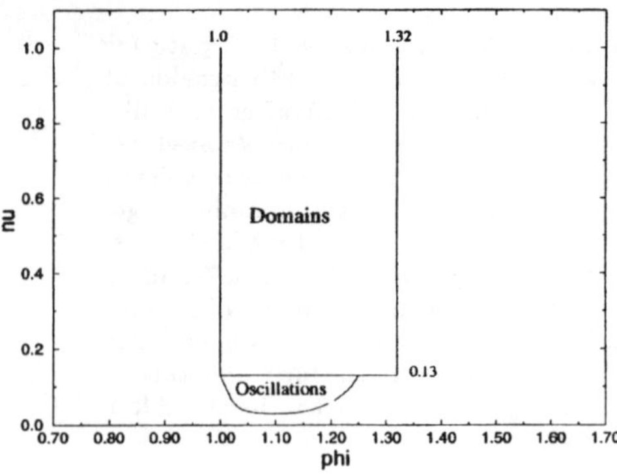

Figure 5. Theoretical phase diagram of a ***doped*** structure. The doping ν and the voltage bias ϕ are given in units of $\in F_M/ed$ and $F_M\ell$, respectively. F_M is the field at the relevant maximum of the *I-V* trace and ℓ is the total length of the sample. The range for self-sustained ***oscillations*** and that for which the time-independent ***domain*** solution is stable [see Fig. 1(b)] are shown. Outside these regions, the field is uniform across the superlattice [Fig. 1(a)]. After Bonilla [43].

vary with density. Qualitatively, the results do not depend on the precise shape of $v(F)$ provided the function exhibits maxima at the resonant fields [39].

Steady-state solutions follow a discrete mapping which, depending on the parameters, has either one or more fixed points. The latter is a necessary condition for the existence of domains [38,39]. In general, there are two or more solutions at a given V with a similar field profile, but with the wall displaced by one or more wells. Consistent with experiments, this gives rise to flat, ratchet-like *I-V* traces similar to that in Fig. 1(b). The discrete drift model has been very successful in describing stationary domains and, in particular, phenomena such as multistability (giving rise to hysteresis) [16,38,39] and disorder-induced effects [28,35]. The calculations also account for the importan photoexcited *p-i-n* samples [21,39,43]. Interestingly, the model shows for the latter case that photoexcitation is the source of damping [27,41]. Reminiscent of the classical Gunn-effect [60], the continuum limit of the theory reveals that the oscillations bear on the formation, annihilation and regeneration of the domain walls [43]. Phase diagrams can also be gained from the calculations. In Fig. 5, we show the corresponding diagram for a doped superlattice [43] displaying the region where the uniform solution is stable, the region where linearly stable domain solutions exist and the range for self-sustained oscillations.

3. The Quantum-Well Pockels Effect

Pockels effect, also known as the *linear* electro-optic effect, refers to changes proportional to the electric field in the refractive index components of a solid. Although commonly attributed to F. Pockels, the effect was actually discovered by W. C. Röntgen and, independently, by A. Kundt in 1883 [64]. The reason why the phenomenon bears Pockels' name traces back to a series of illuminating papers which were published in the period 1889-1893 [65]. In these studies, Pockels uses symmetry arguments to make the crucial steps of establishing the proportionality between the observed changes in optical constants and the electric polarization, and identifying the phenomenon as being unique to piezoelectric materials. Experimentally, Pockels also manages to separate intrinsic terms due to the field from those associated with piezoelectric distortions representing, respectively, a measure of the *electronic* and *ionic* contributions to the nonlinear susceptibility $\chi^{(2)}(\omega,\omega,0)$. It is interesting to point out that the discovery of the *quadratic* electro-optic effect preceded that of Pockels' by nearly one decade; Kerr found it in glasses in 1875. Unlike its linear counterpart, Kerr's effect is not limited to piezoelectric compounds and covers as well optically isotropic media.

Linear electro-optic (or Pockels) coefficients are usually defined by the third-rank tensor r_{ijk} measuring changes of the *impermeability* tensor $\eta_{ij}(\equiv 1/\epsilon_{ij}, \epsilon_{ij}$ is the dielectric response) which are proportional to the electric field components F_k[66]:

$$\eta_{ij} = \eta_{ij}(0) + \sum_k r_{ijk} F_k + \dots \tag{4}$$

For a lossless medium, the principal values of η_{ij} are $1/n_s^2 (s = 1, 2, 3)$ where n_s denotes refractive indices associated with the optical axes. The latter depend on the direction of the electric field. We recall that the piezoelectric response is also characterized by a third-rank tensor. Accordingly and as mentioned earlier, only piezoelectric materials can exhibit linear electro-optic behavior. Clearly, compounds having inversion symmetry are immediately excluded.

Pockels' effect affords a very convenient means of manipulating the phase or the intensity of electromagnetic radiation. Because of its high speed (modulation frequencies are well above 10^{10}Hz), broad bandwidth and the fact that it does not require moving parts, the so-called Pockels cell forms the basis of numerous applications widely used in light communication systems [66]. Important as Pockels effect is for applications, it is somehow unfortunate that its magnitude for most substances remains very small up to the breakdown field [66]. Relevant to our discussion is the case of GaAs where the only non-vanishing electro- optic coefficient is $r_{123} = r_{312} = r_{231} (\equiv r_{41}$ in the contracted- index notation) [67]. Using Eq. (4), it can be shown that a field oriented, say, along the [001] crystal-axis breaks the optical degeneracy between [110] and [1$\bar{1}$0] (i.e., the sample becomes biaxial; note, however, that one cannot optically separate [100] from [010]). The field-induced anisotropy is $\Delta n \equiv n_{[110]} - n_{[1\bar{1}0]} = n^3 r_{41} F$, where F is the z-component of the field. Experiments give $r_{41} \sim 1.5 \times 10^{-10}\text{V}^{-1}\text{cm}$ or $\Delta n/F \sim 10^{-8}\text{V}^{-1}\text{cm}$, with no significant wavelength dependence [68]. Changes in the imaginary part of the refractive index - which matter for frequencies above the band gap - are of the same order [68].

Effects due to electric field in quantum-wells, particularly the system GaAs-$Al_xGa_{1-x}As$, have been extensively investigated in the past several years [61,69]. However, little in the literature has dealt with Pockels' [48-53]. Unlike the latter, we notice that effects such as the quantum-confined Stark effect [61] and Wannier-Stark localization and ladders [69] are, to lowest order, *quadratic* in the field, i.e.,

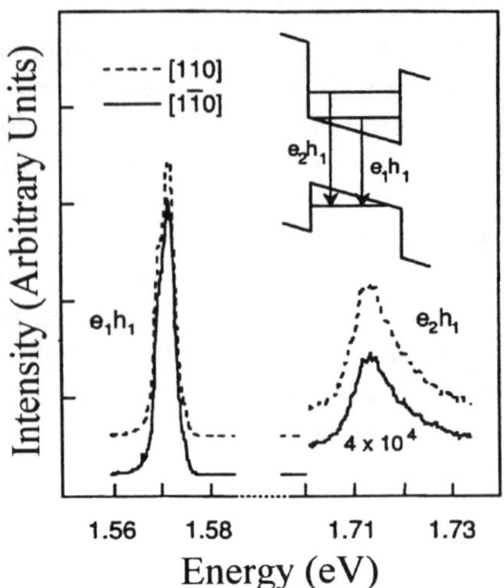

Figure 6. Polarized PL spectra of the quantum-well structure at $V = 4$ V. The relevant electron and hole states are shown in the inset. Adapted from [49].

the response does not change sign when the field is reversed and, moreover, the films remain uniaxial.

With the numbers of bulk-GaAs in mind, it has to come as a surprise that Pockels-like behavior can be easily uncovered in a QW and, more so, that one can observe it in PL experiments [48]. Results on a 90-ÅGaAs/40-ÅAlAs structure giving a summary of these findings are reproduced in Figs. 6-7. The two peaks in Fig. 6 are due to the recombination of electrons in the first (e_1) and second (e_2) levels of the well with empty states at the ground hole-level (h_1). The anisotropy is pronounced for e_2h_1, a transition that is *nearly* forbidden at zero bias [70], although not for the dominant e_1h_1, which is allowed. A measure of the anisotropy is given by the polarization ratio

$$\rho \equiv \frac{\mathcal{I}_{[110]} - \mathcal{I}_{[1\bar{1}0]}}{\mathcal{I}_{[110]} + \mathcal{I}_{[1\bar{1}0]}} \tag{5}$$

which is plotted in Fig. 7 for e_2h_1; $\mathcal{I}_{[110]}$ and $\mathcal{I}_{[1\bar{1}0]}$ are PL- intensities for light polarized parallel to the [110] and [1$\bar{1}$0] axis, respectively. The remarkable fact that ρ *decreases* with F indicates that the magnitude of the field is well beyond the range of linear theory. Symmetry considerations dictate the following form [49]

$$\mathcal{I}(\phi, F) = I_g(F) + \cos(2\phi)I_u(F) \tag{6}$$

for the dependence of the intensity $\mathcal{I}$ on F and the angle ϕ between the polarization of the emitted light and the [110] GaAs crystal axis. I_g and I_u are, respectively, even and odd functions of the field which cannot be determined solely from the symmetry analysis. As shown in the inset of Fig. 7, the expected ϕ-dependence is well obeyed in practice. It follows from Eq. (5) that $\rho = I_u(F)/I_g(F)$. The solid line in Fig. 7 is a fit using $\rho = (F/F_A)/[1 + (F/F_B)^2]$ where F_A and F_B are

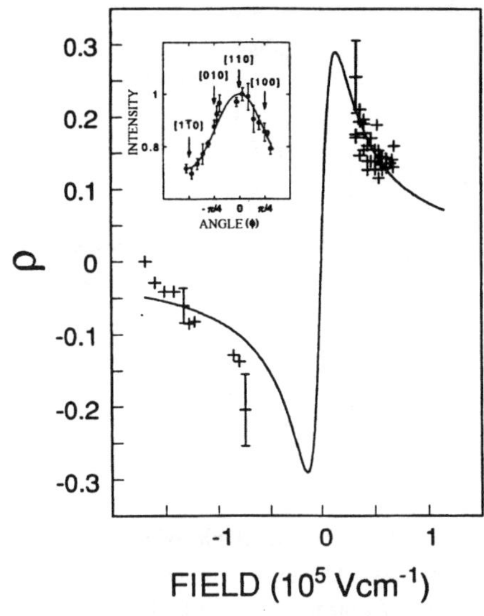

Figure 7. Field dependence of the polarization ratio ρ for e_2h_1PL. The fit uses the expression described in the text. Inset: Dependence of the e_2h_1 recombination intensity on the angle ϕ [Eq. (6)]; solid line is $\cos(2\phi)$. Adapted from [49].

constants. This expression includes only the lowest-order contributions to I_u and I_g.

The current understanding of the quantum-well Pockels effect is based on perturbative analyses of the field-dependence of the matrix elements determining the optical behavior [48,51,53]. Here, there are two contributions involving, on one hand, the field-induced coupling between the $\Gamma_{7,8}$ and Γ_1 conduction bands and, on the other hand, the weaker lattice term [51]. Pockels- type anisotropy results primarily from the interference between different hole spin-components which, due to heavy-light hole mixing, is significantly enhanced with respect to that in bulk materials [51].

4. Generation of Coherent Acoustic Phonons in Superlattices

Following recent advances in femtosecond laser technology, several groups demonstrated that the propagation of light pulses in solids is accompanied by intense THz lattice vibrations showing a high degree of spatial and temporal coherence [71-76]. The availability of *coherent* optical phonons at such frequencies has led to a variety of suggestions for applications and experiments involving, in particular, time-domain spectroscopy of phonons [71-74], conversion of mechanical into coherent electromagnetic energy [75], and intriguing proposals bearing on photon control of the ionic motion [77]. The most common experiment on coherent phonons involves two laser pulses obtained by splitting a single fs-pulse. As shown in Fig. 8, the stronger *pump* pulse creates a vibrational wave which perturbs the weaker *probe* pulse that follows behind. Here, the signal of interest is the transmitted or reflected intensity of the probe beam as a function of the time- delay, as measured by the relative distance between the two pulses.

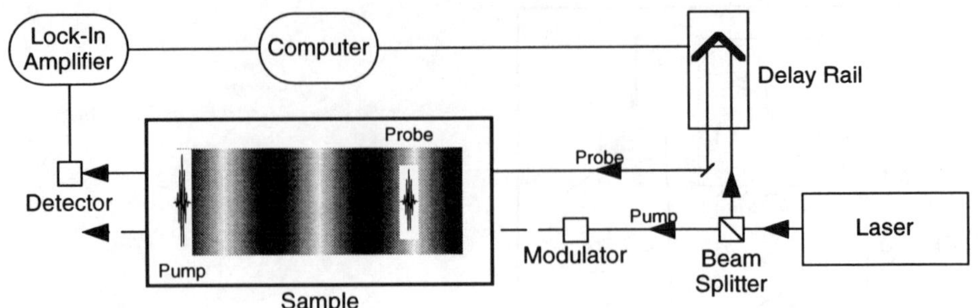

Figure 8. Schematic snapshot diagram of a pump-probe transmission geometry experiment. The gray scale represents the phonon displacement field which moves with and is generated by the pump.

In its simplest form, the generation of coherent modes and their effect on the probe beam rely on the modulation of the refractive index n by the ion motion [76]. This is referred to in the early literature as transient stimulated RS (TSRS) [78]. Consider the *field* $\mathcal{Q}$ associated with a particular optical phonon branch. To first order in $\mathcal{Q}$, we have that $\delta n(\mathbf{r},t) = 2\pi\delta\chi/n \approx 2\pi[\partial\chi/\partial\mathcal{Q}]\mathcal{Q}(\mathbf{r},t)/n$ (χ is the electronic susceptibility). The resulting change in electromagnetic energy density is

$$\delta U = \frac{1}{2}\delta\chi \mid E_0(\mathbf{r},t) \mid^2 = \frac{1}{2}\left(\frac{\partial\chi}{\partial\mathcal{Q}}\right)\mathcal{Q}(\mathbf{r},t) \mid E_0(\mathbf{r},t) \mid^2 \qquad (7)$$

where E_0 is the magnitude of the pump electric field. Since $\delta U \propto \mathcal{Q}$, $Eq.(7)$ gives a force density f acting on $\mathcal{Q}$ which is proportional to the electric field intensity. If we ignore phonon dispersion, the equation of motion for the lattice field becomes identical to that of a driven harmonic oscillator, i.e.,

$$\frac{d^2\mathcal{Q}}{dt^2} + \Omega^2\mathcal{Q} = \frac{1}{2}\left(\frac{\partial\chi}{\partial\mathcal{Q}}\right) \mid E_0(\mathbf{r},t) \mid^2 = f(\mathbf{r},t) \quad . \qquad (8)$$

The acoustic counterpart to TSRS is transient stimulated Brillouin scattering where two overlapping pump pulses give rise to sound through the photoelastic modulation of the refractive index [79]. From phase matching conditions, the coherent acoustic frequency is $\Omega_A \approx 2(\omega_0 nv/c)\sin(\theta)$ where v is the appropriate sound velocity and θ is the angle between the pump beams (since $v \ll c$, the sound wave propagates in a direction nearly perpendicular to that of the light). In transparent bulk media, attainable acoustic frequencies are in the range 1–100 GHz and, thus, ps- and fs- sources give comparable results. Periodic superlattices provide us with an alternative method which, in principle, offers some advantages for coupling to THz-modes. In this case, the wavevectors of the coherent phonons are defined not so much by the optical properties of the material, but by the period of the structure [54-55]. Using picosecond sources, this approach was used on ~500-

Å-thick GaAs-$Al_xGa_{1-x}As$ multilayers to generate frequencies up to ~300 GHz [54]. More recently, fs-data on acoustic modes were reported for (001) GaAs-AlAs structures with ~50-Å-thick layers using laser energies in the vicinity of the fundamental gap [55]. The results reproduced in Fig. 9 show coherent oscillations at $\Omega \approx 0.6$THz due to a longitudinal-acoustic (LA) mode of wavevector $q \approx 2\pi/d$ propagating along the growth axis; d is the period of the superlattice. This value corresponds to the center of the *folded* Brillouin zone. The fact that the power spectrum reveals a single frequency is noteworthy; this is unlike spontaneous backscattering, which shows characteristic folded-phonon *doublets*, but like forward-scattering [80]. Below, we discuss a simple model based on the photoelastic mechanism which captures the essential features of folded-phonon generation. The theory treats the layers in the continuum approximation and, thus, it applies primarily to "thick" structures such as those in [54]. A more microscopic description involving QW states and their interaction with acoustic modes is required to fully analyze the resonant results in Fig. 9, particularly the phase shift by π shown in the figure.

Consider a superlattice where the layers are perpendicular to the $\hat{z}$-axis. The structure extends from $z = 0$ to $z = \ell$. The equation of motion for the acoustic displacement of components u_i is [81]

$$\sigma \ddot{u}_i = \sum_{\mathrm{klm}} \frac{\partial}{\partial x_k}\left(\lambda_{iklm}\frac{\partial u_l}{\partial x_m} - p_{iklm}\frac{E_l E_m}{4\pi}\right) \quad . \tag{9}$$

$\sigma(z)$ is the piecewise-constant mass density and E_l denotes a component of the pump field. The elastic stiffness and photoelastic coefficients, depending also on z, are denoted by λ_{iklm} and p_{iklm}, respectively. Typical GaAs-AlAs structures have $\ell \sim 1\mu$m which is a few times smaller than the size of a ~ 100 fs-pulse in GaAs ($n \approx 3.6$). Since the relevant acoustic period is considerably larger that the time it takes for the optical pulse to cross the superlattice, we approximate $c \to \infty$ and, thus, the boundary conditions become $u_i = 0$ for $t < 0$ and

$$(4\pi\sigma)\frac{\partial u_i}{\partial t}\bigg|_{\mathrm{t=0}} = -\sum_{\mathrm{klm}}\left(\frac{\partial p_{iklm}}{\partial x_k} \times \int_{-\infty}^{0^+} E_l E_m dt\right) \quad . \tag{10}$$

For LA modes propagating normal to the layers in (001) structures, Eq. (9) reduces to

$$\sigma(z)\ddot{u}_{LA} = \frac{\partial}{\partial z}\left(C_{44}(z)\frac{\partial u_{LA}}{\partial z} - p_{12}(z) \mid E_0(z - ct/n) \mid^2 /(4\pi)\right) \tag{11}$$

with $u_{LA} = u_z, p_{12} = p_{xxzz}$ and $C_{44} = \lambda_{zzzz}$. Let $\{\mathcal{U}_q\}$ represent the set of eigenfunctions of Eq. (11) at $E_0 = 0$. Then, the solution satisfying the condition (10) is

$$u_{LA} = \sum_q r_q \mathcal{U}_q(z)\sin(\Omega_q t) \tag{12}$$

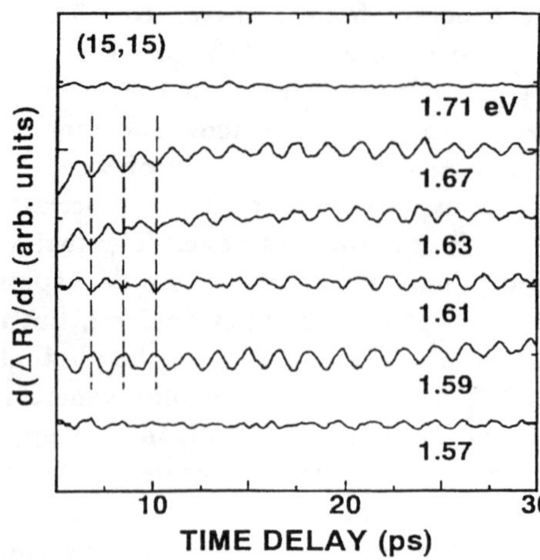

Figure 9. Time derivative of the reflectivity modulation for a (001) superlattice. The period is 42.4-ÅGaAs/42.4-ÅAlAs. The central photon energy of the pulse is indicated. Oscillations are due to acoustic modes at the center of the folded Brillouin zone. Notice the phase shift by π in the vicinity of the heavy-hole exciton energy at ≈ 1.6eV. After Yamamoto *et al.* [55].

where Ω_q are the eigenfrequencies and

$$r_q = -\frac{\int E_0^2 dt}{4\pi\Omega_q} \times \left(\int_0^\ell \mathcal{U}_q(z) \frac{\partial p_{12}(z)}{\partial z} dz \right) \quad . \tag{13}$$

Here, we used the fact that $\int \sigma \mathcal{U}_p \mathcal{U}_q dz = \delta_{pq}$. If the number of periods is large, the index q becomes the Bloch-wavevector and $\mathcal{U}_q$ can be approximated by a combination of Bloch-like functions at $\pm q$. Because $p_{12}(z)$ is a periodic function, $r_q = 0$ unless $q = q_m = 2\pi m/d$ (m is an integer). Hence, the excitation affects only modes which are at the center of the folded zone (this is an exact result at $c = \infty$). In the vicinity of q_m, Bragg-mixing due to the $\pm q_m/2$-components of the modulation conduces to gaps involving, for a given m, two modes of symmetries A_1 and B_2[80]. The reason why forward-RS spectra show singlets is because only the A_1-mode is Raman active [80]. This also accounts for the results of Fig. 9 showing a single frequency and, accordingly, the oscillations are due to the A_1-phonon. In closing, we note that the reflection of the pump beam at the interface between the sample and the substrate leads to an additional contribution which is equivalent to backscattering. As in the latter case, we expect this term to produce *doublets* and, hence, beating in the time domain since the modes at the corresponding wavevectors contain a large admixture of A_1 and B_2 symmetries [80].

Acknowledgments - This work was supported by the U. S. Army Research Office under Contract Number DAAH04-96-1-0183, and by the NSF through the Center for Ultrafast Optical Science under STC PHY 8920108.

5. References

1. L. Esaki and L. L. Chang, *Phys. Rev. Lett.* **33**, 495 (1974).
2. F. Capasso, K. Mohammed and A. Y. Cho, *Appl. Phys. Lett.* **48**, 478 (1986).
3. Y. Kawamura, K. Wakita and K. Oe, *Jpn. J. Appl. Phys.* **26**, L1603 (1987).
4. K. K. Choi *et al.*, *Phys. Rev. B* **35**, 4172 (1987).
5. K. K. Choi *et al.*, *Phys. Rev. B* **38**, 12362 (1988).
6. T. H. H. Vuong, D. C. Tsui and W. T. Tsang, *Appl. Phys. Lett.* **52**, 981 (1988).
7. M. Helm *et al.*, *Phys. Rev. Lett.* **63**, 74 (1989); M. Helm, J. E. Golub and E. Colas, *Appl. Phys. Lett.* **56**, 1356 (1990).
8. H. T. Grahn, H. Schneider and K. v. Klitzing, *Appl. Phys. Lett.* **54**, 1757 (1989).
9. H. T. Grahn , H. Schneider and K. von Klitzing, *Phys. Rev. B* **41**, 2890 (1990).
10. H. T. Grahn, R. J. Haug, W. Müller and K. Ploog, *Phys. Rev. Lett.* **67**, 1618 (1991).
11. P. Helgesen, T. G. Finstad and K. Johannessen, *J. Appl. Phys.* **69**, 2689 (1991).
12. S. H. Kwok, E. Liarokapis, R. Merlin and K. Ploog, in *Light Scattering in Semiconductor Structures and Superlattices*, ed. by D. J. Lockwood and J. F. Young, NATO ASI Ser. B **273** (Plenum, New York, 1991), p. 491.
13. H. T. Grahn *et al.*, *Surf. Sci.* **267**, 579 (1992).
14. A. Shakouri *et al.*, *Appl. Phys. Lett.* **63**, 1101 (1993).
15. Y. Zhang *et al.*, *Appl. Phys. Lett.* **65**, 1148 (1994).
16. J. Kastrup *et al.*, *Appl. Phys. Lett.* **65**, 1808 (1994).
17. S. Murugkar *et al.*, *Phys. Rev. B* **49**, 16849 (1994).
18. S. H. Kwok, R. Merlin, H. T. Grahn and K. Ploog, *Phys. Rev. B* **50**, 2007 (1994).
19. S. Kwok *et al.*, *Phys. Rev. B* **51**, 9943 (1995).
20. S. A. Stoklitskiĭ*et al.*, *Pis'ma Zh. Éksp. Teor. Fiz.* **61**, 399 (1995) [*JETP Lett.* **61**, 405 (1995)].
21. S. H. Kwok *et al.*, *Phys. Rev. B* **51**, 10171 (1995).
22. Z. Y. Han, S. F. Yoon, K. Radhakrishnan and D. H. Zhang, *Appl. Phys. Lett.* **66**, 1120 (1995).
23. S. H. Kwok *et al.*, *Appl. Phys. Lett.* **66**, 2113 (1995).
24. Y. Xu, A. Shakouri and A. Yariv, *Appl. Phys. Lett.* **66**, 3307 (1995).
25. B. J. Keay *et al.*, *Phys. Rev. Lett.* **75**, 4098 (1995); **75**, 4102 (1995).
26. R. Klann, S. H. Kwok, H. T. Grahn and K. Ploog, *Phys. Rev. B* **52**, 8680 (1995).
27. J. Kastrup *et al.*, *Phys. Rev. B* **52**, 13761 (1995).
28. A. Wacker *et al.*, *Phys. Rev. B* **52**, 13788 (1995).
29. J. Kastrup *et al.*, *Phys. Rev. B* **53**, 1502 (1996).
30. S. H. Kwok *et al.*, *Phys. Rev. B* **53**, 7634 (1996).
31. S. Murugkar *et al.*, *Solid State Electron.* **40**, 153 (1996).
32. S. H. Kwok *et al.*, *Solid State Electron.* **40**, 527 (1996).
33. Y. Zhang *et al.*, *Phys. Rev. Lett.* **77**, 3001 (1996).
34. Yu. A. Mityagin and V. N. Murzin, *Pis'ma Zh. Éksp. Teor. Fiz.* **64**, 146 (1996) [*JETP Lett.* **64**, 155 (1996)].
35. G. Schwarz *et al.*, *Appl. Phys. Lett.***69**, 626 (1996).
36. B. Laikhtman, *Phys. Rev. B* **44**, 11260 (1991).
37. B. Laikhtman and D. Miller, *Phys. Rev. B* **48**, 5395 (1993).
38. F. Prengel, A. Wacker and E. Schöll, *Phys. Rev. B* **50**, 1705 (1994).
39. L. L. Bonilla *et al.*, *Phys. Rev. B* **50**, 8644 (1994).
40. D. Miller and B. Laikhtman, *Phys. Rev. B* **50**, 18426 (1994).
41. R. Döttling and E. Schöll, *Solid State Electron.* **37**, 685 (1994).
42. O. M. Bulashenko and L. L. Bonilla, *Phys. Rev. B* **52**, 7849 (1995).
43. L. L. Bonilla, in *Nonlinear Dynamics and Pattern Formation in Semiconductors*, ed. by F. J. Niedernostheide (Springer, Berlin, 1995), Chap. 1.
44. L. L. Bonilla *et al.*, *Solid State Electron.* **40**, 161 (1996).
45. O. M. Bulashenko, M. J. García and L. L. Bonilla, *Phys. Rev. B* **53**, 10008 (1996).
46. G. Schwarz and E. Schöll, *Phys. Stat. Sol.* (b) **194**, 351 (1996).

47. H. T. Grahn, in *Semiconductor Superlattices: Growth and Electronic Properties*, ed. by H. T. Grahn (World Scientific, Singapore, 1995), Chap. 5.
48. S. H. Kwok, H. T. Grahn, K. Ploog and R. Merlin, *Phys. Rev. Lett.* **69**, 973 (1992).
49. S. H. Kwok, H. T. Grahn, K. Ploog and R. Merlin, in *Proc. 21th Int. Conf. Phys. Semiconductors*, ed. by P. Jiang and H.-Z. Zheng (World Scientific, Singapore, 1992), p. 1180.
50. G. Armelles, J. Meléndez and P. Castrillo, *Phys. Rev. B* **49**, 17444 (1994).
51. B. Zhu and Y-C. Chang, *Phys. Rev. B* **50**, 11932 (1994).
52. J. A. Prieto, J. Findeisen and G. Armelles, *Solid State Electron.* **40**, 767 (1996).
53. O. Krebs and P. Voisin, *Phys. Rev. Lett.* **77**, 1829 (1996).
54. P. Basséras, S. M. Gracewski, G. W. Wicks and R. J. D. Miller, *J. Appl. Phys.* **75**, 2761 (1994).
55. A. Yamamoto, T. Mishina, Y. Masumoto and M. Nakayama, *Phys. Rev. Lett.* **73**, 740 (1994).
56. R. F. Kazarinov and R. A. Suris, *Fiz. Tekh. Poluprov.* **6**, 148 (1972) [*Sov. Phys. Semicond.* **6**, 120 (1972)].
57 See, e.g., *Negative Differential Resistance and Instabilities in 2-D Semiconductors*, ed. by N. Balkan, B. K. Ridley and A. J. Vickers, NATO ASI Ser. B **307** (Plenum, New York, 1992).
58. H. Kroemer, *IEEE Trans.* **ED-15**, 819 (1968).
59. A. M. Kahn, D. J. Mar and R. M. Westervelt, *Phys. Rev. Lett.* **46**, 369 (1992).
60. J. B. Gunn, *IBM J. Res. Dev.* **8**, 141 (1964).
61. See, e.g., D. A. B. Miller *et al.*, *Phys. Rev. B* **32**, 1043 (1985); L. Viña *et al.*, *Phys. Rev. Lett.* **58**, 832 (1987).
62. K. Bajema *et al.*, *Phys. Rev. B* **36**, 1300 (1987).
63. See, e.g., J. Shah, *Hot Carriers in Semiconductor Nanostructures: Physics and Applications* (Academic, Boston, 1992), p. 279.
64. W. C. Röntgen, *Wied. Ann.* **18**, 213; *ibid.* **18**, 534 (1883); *ibid.* **19**, 319 (1883); A. Kundt, *ibid.* **18**, 228 (1883).
65. F. Pockels, *Anhandl. der Mathematisch-physikal. Klasse der k. Ges. d. Wiss. zu Göttingen* **39**, 1 (1893), and references therein.
66. A. Yariv and P. Yeh, *Optical Waves in Crystals* (Wiley, New York, 1984), Chap. 7-8.
67. See, e.g., J. F. Nye, *Physical Properties of Crystals* (Clarendon Press, Oxford, 1985), Chap. 13.
68. See, e.g., S. Adachi, *J. Appl. Phys.* **72**, 3702 (1992).
69. E. E. Méndez, F. Agulló-Rueda and J. M. Hong, *Phys. Rev. Lett.* **60**, 2426 (1988); P. Voisin, J. Bleuse, C. Bouche, S. Gaillard, C. Alibert and A. Regreny, *Phys. Rev. Lett.* **61**, 1639 (1988).
70. The transition is strictly forbidden at $q_{\|} = 0$, but becomes allowed for $q_{\|} \neq 0$ due to heavy-light hole mixing; $q_{\|}$ is the wave-vector component parallel to the layers. See: W. T. Masselink *et al.* *Phys. Rev. B* **32**, 8027 (1985).
71. G. C. Cho, W. Kütt and H. Kurz, *Phys. Rev. Lett.* **65**, 764 (1990).
72. H. J. Zeiger *et al.*, *Phys. Rev. B* **45**, 768 (1992).
73. T. Pfeifer, W. Kütt, H. Kurz and R. Scholz, *Phys. Rev. Lett.* **69**, 3248 (1992).
74. G. A. Garrett, T. F. Albrecht, J. F. Whitaker and R. Merlin, *Phys. Rev. Lett.* **77**, 3661 (1996).
75. T. Dekorsy *et al.*, *Phys. Rev. Lett.* **74**, 738 (1995).
76. For reviews, see: Y.-X. Yan and K. A. Nelson, *J. Chem. Phys.* **87**, 6240 and 6257 (1987); R. Merlin, *Solid State Commun.*, to be published (1996).
77. S. Fahy and R. Merlin, *Phys. Rev. Lett.* **73**, 1122 (1994).
78. See: Y. R. Shen, *The Principles of Nonlinear Optics* (Wiley, New York, 1984), and references therein.
79. See, e.g., L. T. Cheng and K. A. Nelson, *Phys. Rev. B* **37**, 3603 (1988), and references therein.

80. C. Colvard, R. Merlin, M. V. Klein and A. C. Gossard, *Phys. Rev. Lett.* **43**, 298 (1980). For a review, see: B. Jusserand and M. Cardona, in *Light Scattering in Solids V*, Topics in Applied Physics Vol. 66, ed. by M. Cardona and G. Güntherodt (Berlin, Springer, 1989), Chap. 3.
81. See, e.g., N. M. Kroll, *J. Appl. Phys.* **36**, 34 (1965).

STRUCTURE AND OPTICAL PROPERTIES OF SELF-ORDERED V-GROOVE QUANTUM WIRES AND QUANTUM WELLS

Eli KAPON

Department of Physics
Swiss Federal Institute of Technology
1015 Lausanne, Switzerland

ABSTRACT. Recent progress in the realization of low-dimensional quantum structures formed by self-ordering on nonplanar substrates is reviewed. In particular, the structure and the optical properties of GaAs-based vertical quantum wells (VQWs) and quantum wires (QWRs) grown by organometallic chemical vapor deposition on V-grooved substrates are described. Distinct nano-faceting at the growth front results in well defined QWR and VQW structures with lateral dimensions in the sub-10 nm range. Clear subband structure and polarization anisotropy related to the low-dimensionality of the confined carriers are revealed in photoluminescence excitation spectra of these nanostructures. Diode lasers incorporating such self-ordered QWRs and VQWs are also discussed.

1. Introduction

Low-dimensional semiconductor nanostructures, particularly quasi-one dimensional (1D) quantum wires (QWRs) and quasi-0D quantum dots (QDs), are expected to show new optical and electronic properties associated with multi-dimensional quantum confinement and reduced dimensionality. Examples of such properties include new subband structure and valence band mixing effects [1,2], modified Coulomb correlations [3,4] and Luttinger liquid behavior of 1D carriers at low temperatures [5]. These novel features make such low-dimensional structures of interest both from the fundamental viewpoint as well as for future applications in novel electronic and optoelectronic devices.

However, the extremely small lateral size (of the order of several 10 nm or less), high uniformity (on the atomic level) and defect free interfaces required in useful QWRs and QDs present considerable challenges concerning their fabrication technology [6]. The small lateral dimensions are necessary for ensuring observable quantum size effects at reasonably high temperatures (particularly at room temperature). High uniformity in terms of size and composition are required for avoiding excessive inhomogeneous broadening effects which tend to smear out the characteristic singularities in the density of states (DOS) functions of ideal 1D or 0D structures. Defect free interfaces are particularly important for QWRs and

G. Abstreiter et al. (eds.), Optical Spectroscopy of Low Dimensional Semiconductors, 99–125.

QDs used in optical studies and applications, in order to minimize nonradiative recombination effects.

Conventional, lithography-based techniques are of little use in this context due to limitations in achievable minimum size and uniformity. For example, electron beam lithography techniques are usually limited to size features of several 10 nm or more and result in typical width fluctuations of at least several nm. Furthermore, definition of the lateral quantum structures with subsequent etching or implantation techniques usually results in the introduction of defects into the etched surfaces. For optical studies and applications, lateral confinement of both electrons and holes is often desirable, which renders certain techniques relying on electrostatic confinement, e.g., gated 2D electron gas systems [7], less useful for this purpose.

Self-ordering of nanostructures is an attractive approach for the realization of low-dimensional quantum systems, which can in principle overcome the limitations mentioned above. *Spontaneous* self-ordering of semiconductor heterostructures, however, usually leads to a broad distribution of the size of the quantum structures and does not offer a direct control of their positioning. An important example of such spontaneous self-ordering processes is the strain-induced formation of wire and dot structures prepared, e.g., in the InGaAs/GaAs or InGaAs/InP strained semiconductor systems [8,9]. The *in situ* formation of the interfaces in these self-ordered structures avoids the introduction of process-induced interface defects. Whereas some lateral and vertical ordering effects, originating in the strain fields, have been observed in these structures [10,11], the achievable ordering is not complete. Most importantly, their size distributions exhibit relative widths of at least 10-20%, much higher than the above-mentioned uniformity requirements.

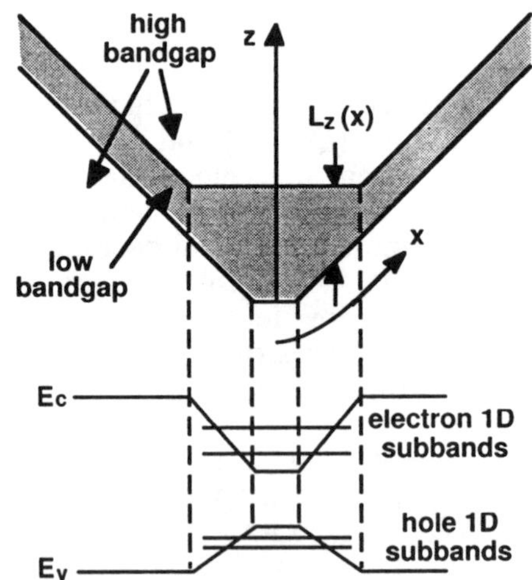

Figure 1. Lateral quantum confinement in a tapered QW heterostructure grown on a nonplanar substrate. The lateral QW thickness variations translate into lateral effective bandgap variations. Charge carriers can be laterally trapped in the thicker parts of the tapered QW. 1D conduction and valence subbands are formed for sufficiently narrow channels.

An alternative approach, which combines lithography for fixing the sites of the quantum structures with self-ordering processes in which the structures are formed "spontaneously", is growth on nonplanar substrates (for a review, see [12]). This approach involves two simple concepts. The first one concerns the achievement of lateral quantum confinement in laterally-tapered quantum well (QW) structures (see Fig. 1) [13]. Such structures can be fabricated, e.g., by molecular beam epitaxy (MBE) of an otherwise conventional QW heterostructure on nonplanar substrates. In this case, the lateral QW thickness variations can be obtained due to lateral variations in the molecular beam fluxes on the inclined surfaces of a groove or a ridge structure. The lateral thickness variations readily translate into effective bandgap variations via the transverse quantum size effect, i.e., the strong variation of the carrier confinement energy with the QW thickness. MBE growth in a groove or on a ridge can thus yield a lateral potential well (both for electrons and for holes, in a type I heterostructure) at the thicker part of the tapered QW. Lateral *quantum* confinement can be achieved provided this thicker QW section is sufficiently narrow (several 10 nm or less). Such structures could be obtained, in principle, by growth on extremely narrow channels or ridges fabricated, e.g., using electron beam lithography and etching. However, the required dimensions, and particularly the desired groove uniformity, are too demanding for such lithography techniques at present.

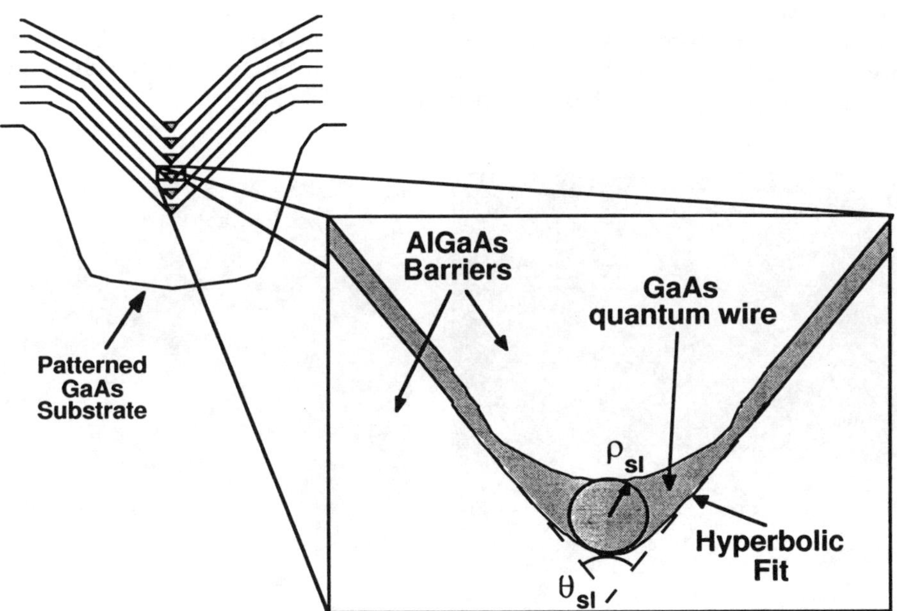

Figure 2. Schematic illustration of the self ordering of a V-groove profile grown by OMCVD on a nonplanar substrate.

Self ordering of extremely narrow channels or ridges, formed during epitaxy on nonplanar substrates under certain growth conditions, offers the second concept necessary for realizing sufficiently narrow, tapered QWs. In particular, it has been shown that OMCVD growth of AlGaAs in a groove oriented along the $[01\bar{1}]$ direction on (100) GaAs substrates leads to a self-limiting formation of an extremely narrow groove, whose profile is determined solely by the growth parameters and the Al mole fraction (see Fig. 2) [14]. Starting from an arbitrarily shaped channel, facets are formed along several crystal directions in which the growth rate is extremal. After a sufficiently thick grown layer, the growth becomes self-limiting in the sense that the surface profile near the bottom of the groove remains unchanged (until planarization takes place). Figure 3 shows an atomic force microscope (AFM) cross-section of a GaAs/AlGaAs heterostructure grown in two such grooves, each with a different initial profile. After a minimum growth thickness (several 100 nm for the square-shaped groove, several 10 nm for the V-groove), the groove assumes a characteristic V-shape profile defined by near-{111}A planes. Starting with an initial groove profile that better resembles the self-limiting profile minimizes the thickness required for achieving the self-limiting growth front. A closer look at the bottom of the V-shape reveals a detailed surface profile characterized by specific nano-facets whose size depends on the layer composition and the growth conditions, as further discussed in the next section. Subsequent growth of a lower bandgap (lower Al mole fraction) AlGaAs layer in such a self-limiting groove results in the formation of a crescent shaped QWR, in which the tapered QW structure is well controlled and extends over a lateral distance of only a few 10 nm. Similar self-ordering of extremely narrow ridges grown by MBE on [011] mesas has also been reported [15].

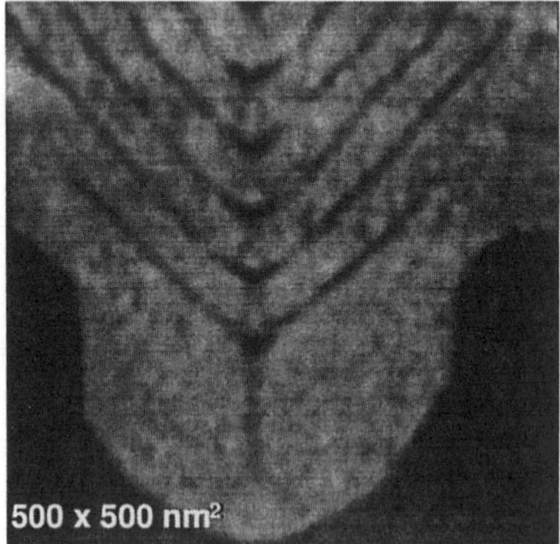

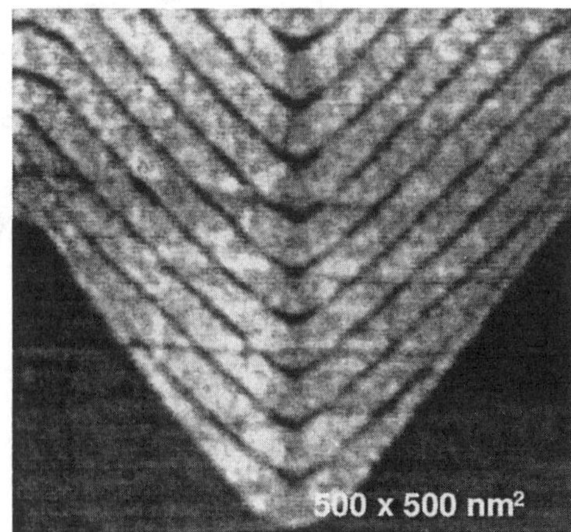

Figure 3. Atomic force microscope cross sections of GaAs/AlGaAs heterostructures showing the growth front during OMCVD on two $[01\bar{1}]$ channels of different profiles etched in a (100) GaAs substrate.

The self ordering of the extremely narrow groove allows the realization of narrow QWRs whose interfaces are formed *in situ* and are thus (potentially) defect free. Furthermore, since the shape of the groove is independent of the initial, fabricated groove profile, defects in the lithographic patterns used for defining the channels have a smaller effect on the resulting wires, as compared with techniques which exclusively rely on lithography. In fact, inadvertent *width* fluctuations in the initial channels give rise to *height* variations in the wires, whose cross sectional profile remains largely unchanged. On the other hand, the use of lithography in defining the channels allows the determination of the site of each wire: the crescent-shaped QWRs are formed directly above the bottom of the V-grooves. In this sense, the grooves act as *seeds* for initiating the self-ordering of the nanostructures.

The remaining part of this review will present the structural and optical properties of low-dimensional nanostructures obtained by this *seeded self ordering* process. Emphasis will be on recent progress in studies of GaAs/AlGaAs QWRs and AlGaAs vertical QWs (VQWs) grown by low-pressure (LP-)OMCVD grown on nonplanar GaAs substrates. Similar GaAs/AlGaAs, InGaAs/AlGaAs, InGaAsP/InP, SiGe/Si and GaAs/AlGaAs-on-Si self-ordered QWRs grown by OMCVD or MBE, on nonplanar substrates formed by etching or by growth on patterned substrates, have also been demonstrated and investigated [16-24].

2. Structure of Self-Ordered QWRs and VQWS

2.1. SELF-ORDERING OF QWR STRUCTURES

The self-limiting profile of the AlGaAs surface grown by atmospheric pressure (AP-) OMCVD in $[01\bar{1}]$ grooves can be well fitted with hyperbolas, as shown in Fig. 2 [25]. It is thus uniquely defined by two parameters, which we choose as the radius of curvature ρ_{sl}, measured at the bottom of the groove, and the angle θ_{sl} between the asymptotes of the hyperbola. The dependence of these parameters on the composition of the AlGaAs layer and the growth conditions has been investigated using the structures shown in the transmission electron microscope (TEM) cross sections of Fig. 4 [25]. Pairs of GaAs QW layers were grown with $Al_{0.5}Ga_{0.5}As$ barriers at 650°C; the pairs were separated by sufficiently thick barriers so as to achieve the characteristic self-limiting profile underneath the first GaAs layer of each pair. The separation of the QWs within each pair were varied in order to study the evolution of the growth front. It can be seen that the self-limiting AlGaAs profile is perturbed by the growth of the GaAs layer, which leads to a larger radius of curvature. During the growth of a thin AlGaAs layer above the GaAs crescent, the self-limiting profile partially recovers, leading to a smaller radius of curvature again; the self-limiting profile fully recovers upon growth of a sufficiently thick AlGaAs barrier.

This surface curvature recovery is described quantitatively in Fig. 5. The increase in the radius of curvature from its self-limiting value ρ_{sl} during growth of the GaAs layer follows the linear relation where t_{GaAs} is the thick-

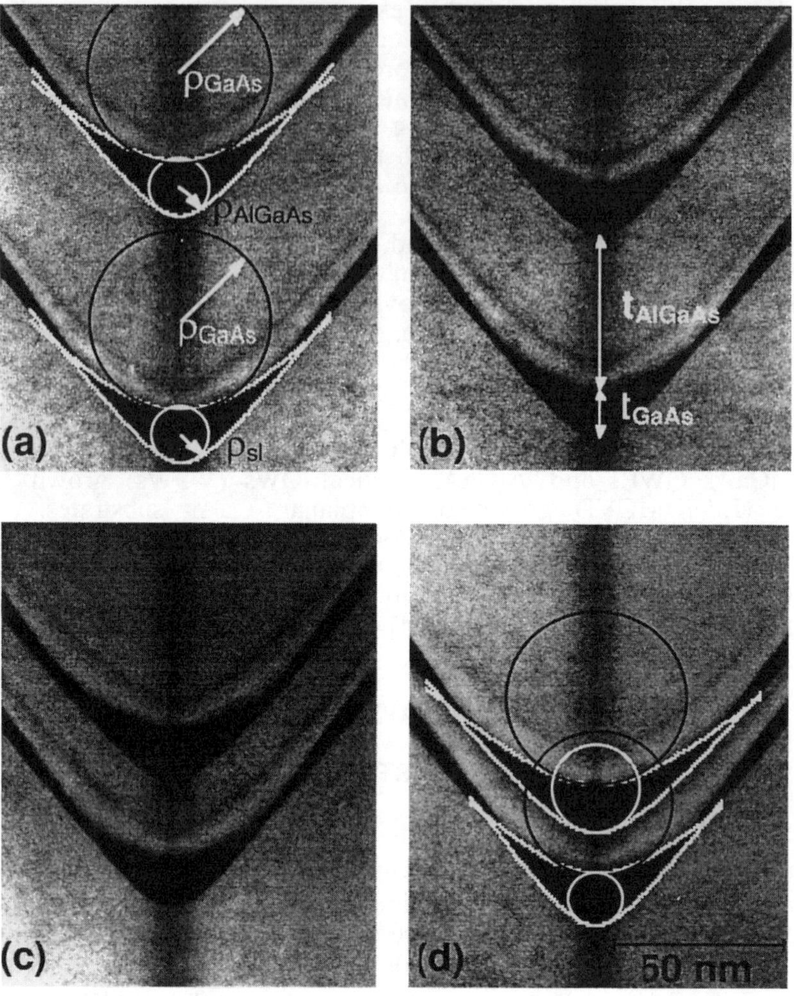

Figure 4. TEM cross sections of pairs of GaAs/$Al_{0.5}Ga_{0.5}As$ QWRs grown in a V-groove, showing the evolution of the surface at the bottom of a V-groove after growth of GaAs (darker areas) and $Al_{0.5}Ga_{0.5}As$ (brighter areas) layers [25].

$$\rho\left(t_{GaAs}\right) = \rho_{sl} + \alpha t_{GaAs}, \tag{1}$$

ness at the center of the crescent. The radius of curvature recovers *exponentially* during growth of the AlGaAs layer,

$$\rho\left(t_{AlGaAs}\right) = \rho_{sl} + \left(\rho_{GaAs} - \rho_{sl}\right)\exp\left(-t_{AlGaAs} / \tau\right). \tag{2}$$

This exponential recovery provides a strong driving force for the self ordering of the surface profile following a perturbation. The characteristic thickness τ is a measure of the minimum thickness required for complete relaxation of the curvature to its self-limiting value after the perturbation. For the structures described in Fig. 5, τ ranges between 4.8 and 10.3 nm for t_{GaAs} between 2.2 and 26.1 nm, respectively; the self limiting radii are fully recovered after growth of 25 and 60 nm, respectively. This full recovery of the surface profile allows vertical stacking of many, identical crescent-shaped QWRs, provided their separation exceeds the minimum thickness required for curvature recovery.

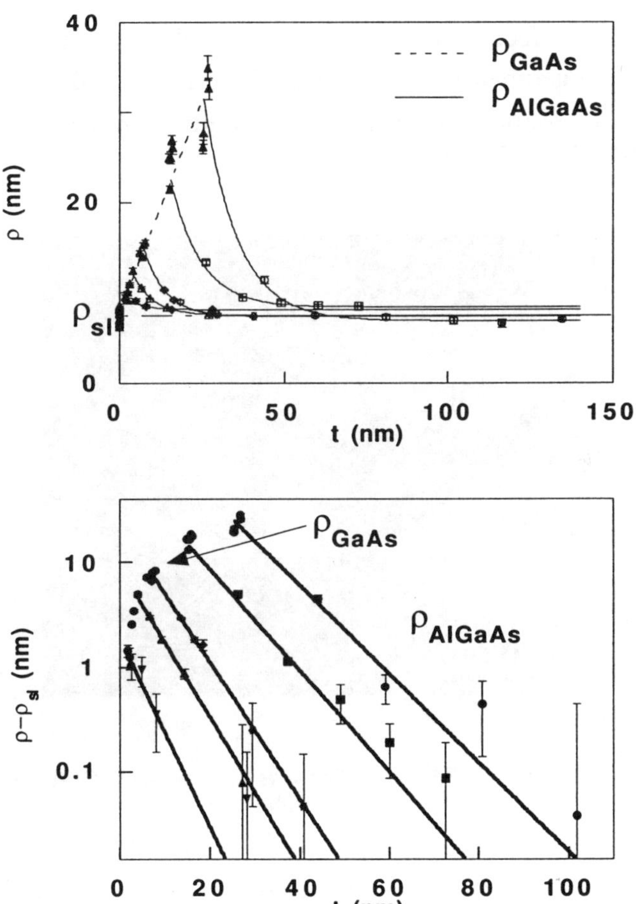

Figure 5. Variation of the radius of curvature versus the thickness of the GaAs and AlGaAs layers, both measured at the bottom of the groove. Upper and lower parts show the variations on a linear and a logarithmic scale, respectively.

The self-limiting radius of curvature increases with increasing growth temperature and decreasing Al mole fraction in the barriers. In the structures of Fig. 4, ρ_{sl} =7.7 nm. Thus, the achievable widths of such self-ordered grooves are sufficiently small to obtain significant lateral quantum confinement in crescent-shaped, tapered QWs formed on these highly curved surfaces. A model, assuming infinite potential barriers and hyperbolic crescent boundaries, yields an approximation for the energy separation of the 1D subbands in these wires [26]:

$$\Delta E \cong \frac{\hbar^2 \pi \alpha^{1/2}}{m^* t \bar{\rho}} \tag{3}$$

where m^* is the effective carrier mass, t is the crescent thickness, and $\bar{\rho} = \sqrt{\rho_l \rho_u}$ is the geometric mean radius of the lower and upper radii of the crescent. This approximation provides design guidelines for maximizing the lateral confinement energy in such wires.

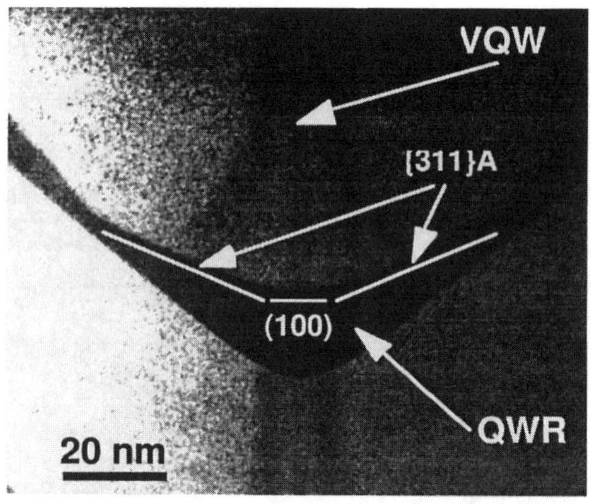

Figure 6. Dark field TEM image of a GaAs/AlGaAs single-QWR structure grown by low-pressure OMCVD.

A closer look at the self-limiting surface profile at the bottom of the grown V-grooves reveals that they consist of characteristic nano-facets. These facets are more distinct in similar wires grown by *low-pressure* (LP-) OMCVD (20 mbar), as evident in the structure of Fig. 6. The LP growth results in a self-limiting surface profile characterized by a center (100) facet flanked by two {311}A lateral facets. The extent of these facets depends on the Al content and the growth temperature. A high resolution TEM cross section of a V-groove QWR structure grown at LP is shown in Fig. 7 [27]. The wire interfaces are abrupt within one to two monolayers. Atomic force microscopy of similar (surface) GaAs QWRs shows that such monolayer-defined interfaces extend along sections of at least a few 100 nm along the wire axis [28]. The better interface quality of the QWRs grown at low pressure is attributed to the kinetically-limited nature of OMCVD under these conditions. In the case of AP-OMCVD, similar facets are obtained, however the less well defined interfaces (several monolayers) result in apparently smooth surface profiles. The hyperbolic fits to the surface profiles obtained by AP-OMCVD represent the approximate profile of the smoothed-out, faceted surfaces which result in that case.

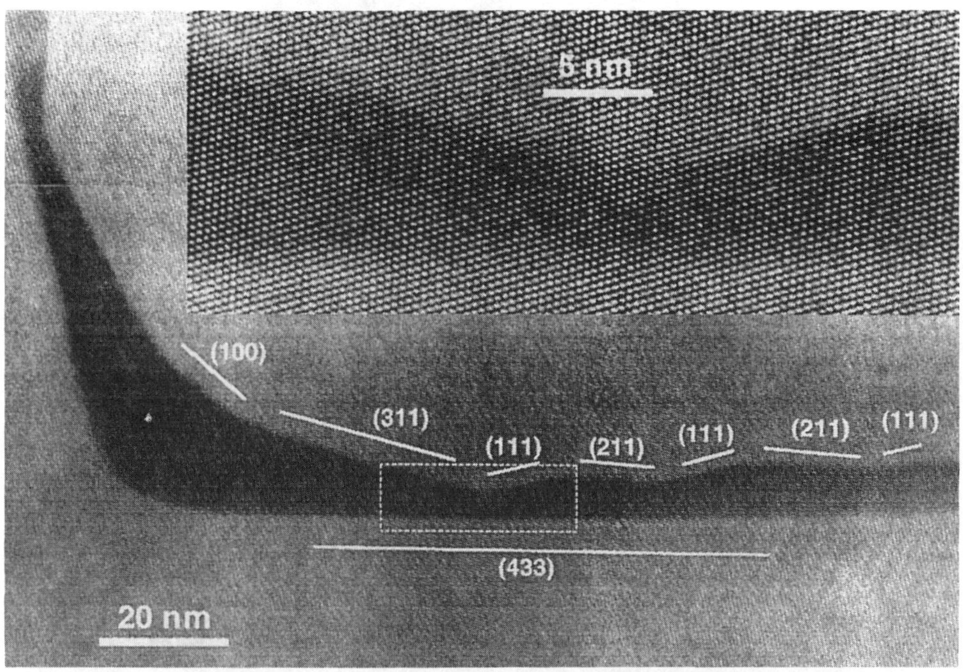

Figure 7. High resolution TEM cross section of a crescent-shaped QWR grown by low-pressure OMCVD. The inset shows a magnified view of a detail of the wire [27].

An example of a vertically-stacked array of GaAs/AlGaAs QWRs grown by LP-OMCVD is shown in Fig. 8 [27]. As for AP growth, the minimum vertical separation of such self-limiting, identical wires is determined by the surface curvature recovery. The crescent-shaped wires can also be arranged in dense lateral arrays, as shown in the structure of Fig. 9. In this case, the wires were grown on a 0.5 μm pitch grating made by holographic photolithography and wet chemical etching. The minimum lateral separation of such wires is limited by lithography. Vertically-stacked wire structures grown on such sub-μm pitch gratings can also be obtained for maximizing their packing density.

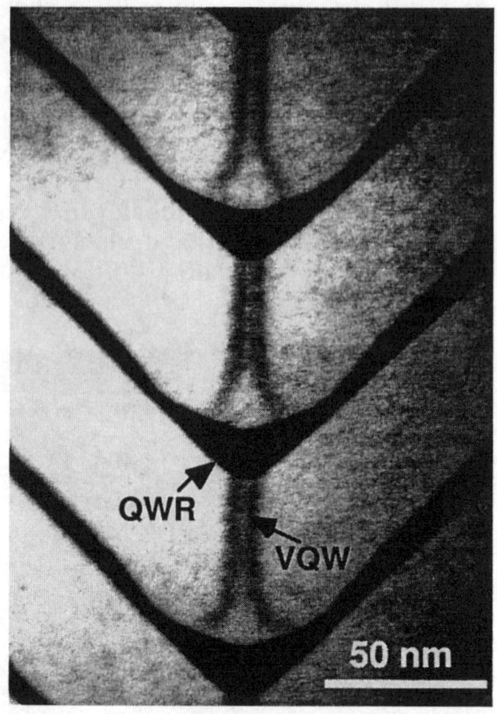

Figure 8. Dark Field TEM cross-section of a vertically-stacked array of GaAs/AlGaAs QWRs grown by low-pressure OMCVD [27].

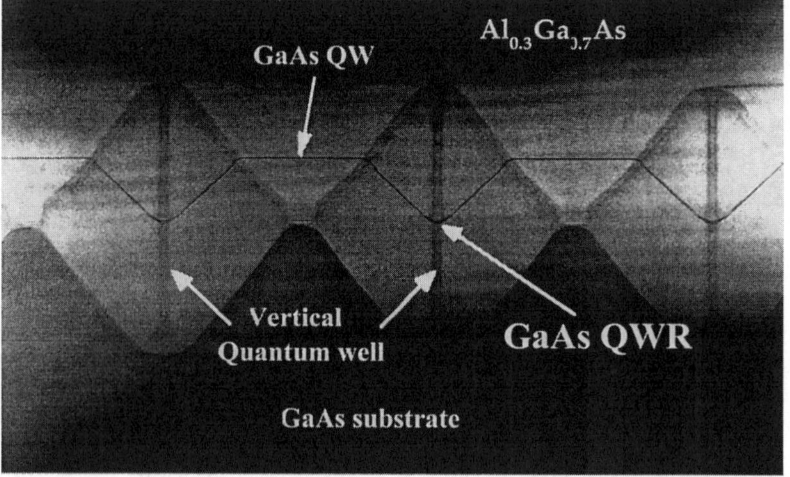

Figure 9. Dark Field TEM cross-section of a lateral array of GaAs/AlGaAs single-QWRs grown by low-pressure OMCVD on a 0.5μm pitch grating.

2.2 SELF-ORDERED VERTICAL QUANTUM WELLS

An interesting feature that is apparent in the TEM images of the GaAs/$Al_xGa_{1-x}As$ QWRs is the dark line that runs through the center of the groove. This darker contrast results from a Ga-rich region, which forms at the bottom of the groove due to group III segregation [29,30]. In the case of AP-OMCVD grown structures, this Ga-rich region is about 20 nm wide for x=0.5 and a growth temperature of 750°C, and it consists of a central, much narrower region surrounded by a diffused, wider Ga-rich region (see Fig. 4). Since the higher Ga mole fraction leads to a lower bandgap, these regions constitute thin potential wells running through the center of the grooves, and hence were termed *vertical* QWs (VQWs).

In similar structures grown by LP-OMCVD, the QW structures split into three characteristic branches (see Figs. 6, 8 and 9). The center branch originates from the center (100) facet at the bottom of the groove, whereas the two outermost ones stem from the {311}A facets. For sufficiently thick AlGaAs layers, the VQW structure reaches a self-limiting configuration with characteristic branch widths and separations [31]. In a transient growth, before reaching this self-limiting state (e.g., following a perturbation caused by growth of a layer with a different Al mole fraction), the outer branches change their separation until they reach the separation corresponding to the self-limiting state. This recovery of the self-limiting branch separation is another signature of the recovery of the surface profile previously discussed.

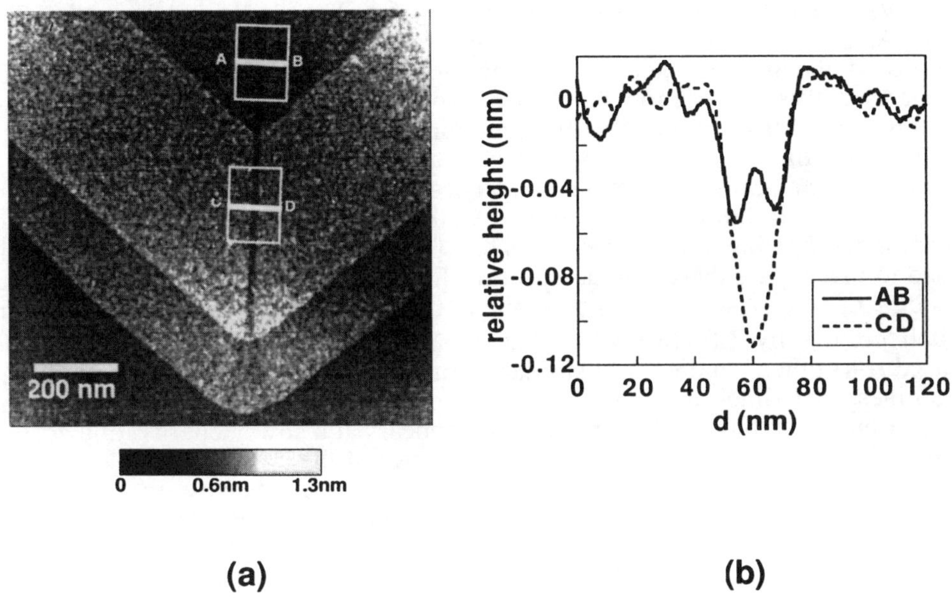

Figure 10. (a) Cross sectional AFM image of an AlGaAs heterostructure grown by low pressure OMCVD in a V-groove. The height contrast corresponds to variations in the oxide height associated with regions of different Al mole fraction. (b) Line scans across the VQWs, showing regions with Al mole fraction lower than that of the background [31].

The fact that the darker contrast observed at the VQW region in the TEM images is indeed associated with a Ga-rich region was established indirectly by low-temperature photoluminescence (PL) and cathodoluminescence measurements [30,32]. More directly, the increase in the Ga mole fraction was measured using electron energy loss spectroscopy techniques [33] and by a cross sectional AFM imaging method [31]. An example of the latter is shown in Fig. 10. The cleaved cross section of the AlGaAs heterostructure grown by LP-OMCVD in a V-groove was exposed to air, which results in the formation of an oxide layer on the cleaved surface. The oxide height depends on the Al mole fraction, being slightly (a nm or less) higher for higher Al mole fractions. Careful calibration of the oxide height as a function of the Al mole fraction and the exposure time to air permits the evaluation of the composition of the AlGaAs layers in such cross sections from the height image [34]. This can be done with the nm resolution provided by the AFM technique. The corresponding image of the V-groove heterostructure reveals the VQW structure, which indeed appears as a region of lower Al content.

3. Optical Properties

3.1. LUMINESCENCE AND ABSORPTION SPECTRA OF V-GROOVE QWRS

The V-groove QWRs are surrounded by several types of barriers: the 3D AlGaAs barrier material, the 2D side-wall and top GaAs QWs grown on the slopes of the grooves and on top of the ridges, respectively, and the 2D AlGaAs VQWs. Excitation of electron hole pairs in these barriers is usually followed by transport and finally recombination either at one of these barrier regions or in the wires. As a result, the luminescence spectra of these structures can be fairly complex and necessitate careful identification of the origin of each luminescence line. Cathodoluminescence (CL) [32] and scanning tunneling luminescence (STL) [35] imaging techniques have been used to investigate these optical spectra. Figure 11 shows low temperature CL spectra of a GaAs/$Al_{0.3}Ga_{0.7}As$ single-QWR structure grown on 0.5 μm pitch grating by LP-OMCVD. The identification of the different lines is based on other studies of similar structures with a wider pitch [36]. Luminescence lines originating from the different barriers are evident at higher energies; the QWR luminescence appears at a lower energy due to the larger thickness of the GaAs layer at the center of the crescents.

At higher probe currents, the higher carrier density in the QWRs results in band filling which allows observation of excited 1D states in the luminescence spectra (see Fig. 11). The observed subband separation in this sample is about 22 meV. Notice that in spite of the increase in the probe current by more than three orders of magnitude, the position of the ground state is unchanged. This possibly indicates the absence of bandgap renormalization in these 1D structures [22].

The low temperature photoluminescence (PL) spectrum of a GaAs/$Al_{0.3}Ga_{0.7}As$ V-groove QWR structure grown on a 0.5 μm-pitch

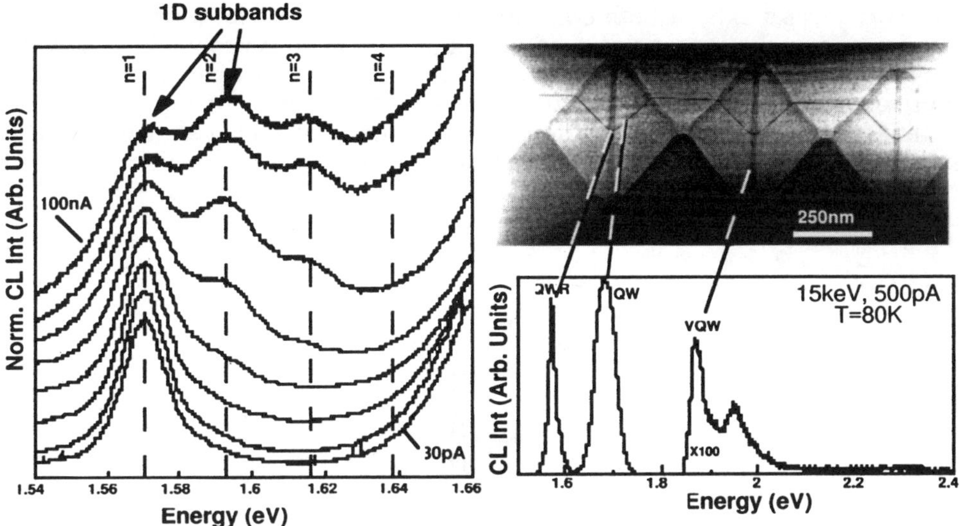

Figure 11. Low temperature CL spectrum of a $GaAs/Al_{0.3}Ga_{0.7}As$ QWR structure grown on 0.5 μm pitch grating (right). The origin of the different luminescence lines is identified in the dark-field TEM cross section at the top right. Exciting the same sample with higher probe currents results in filling of higher energy 1D subbands (left) [36].

grating by LP-OMCVD is shown in Fig. 12 (Ar^+-laser pumped) [37]. These PL spectra are characterized by an excitonic QWR line at lower energies, and several higher energy lines associated with the QW and bulk barriers surrounding the wires. Linewidths are typically narrower than the ones measured in the corresponding low temperature CL spectra, ranging from 5.5 to 7 meV for crescent (center) thicknesses of 14 to 4 nm, respectively. These linewidths are considerably narrower than those measured for similar QWR structures grown at atmospheric pressure. Moreover, they are comparable to the PL linewidths of GaAs/AlGaAs QWs of similar thickness, grown by LP-OMCVD on (100)-GaAs planar substrates.

Low temperature PL excitation (PLE) spectra of these QWRs, measured with a tunable Ti:sapphire laser, show well resolved peaks due to transitions between 1D conduction and valence subbands [37] (see Fig. 13). Detailed investigations of a series of such wires, with different QWR sizes and Al contents in their barriers, indicate good agreement between the observed subband separations and the calculated e-hh 1D transition energies based on a 2D model of the wire potential well. The PLE spectra, measured with an exciting beam linearly polarized parallel or perpendicular to the wires, show distinct polarization anisotropy associated with valence band mixing [38].

The PLE spectra exhibit both e-hh -like as well as e-lh -like transitions, which permits a quantitative analysis of the polarization anisotropy. It is worth mentioning that polarization anisotropy in such structures can also arise due to effects which are not related to the one-dimensionality of the structure, in particular, electromagnetic grating effects. These effects were ruled out in the samples discussed here by comparing V-groove QWR structures in which the upper surface had been planarized, with similar structures where a surface corrugation had been left above the wires [38].

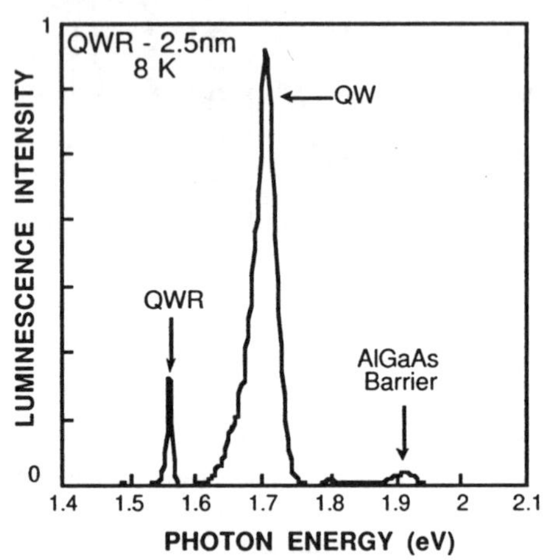

Figure 12. Low temperature (8K) photoluminescence spectrum of a $GaAs/Al_{0.3}Ga_{0.7}As$ V-groove QWR structure (9 nm crescent thickness) [37].

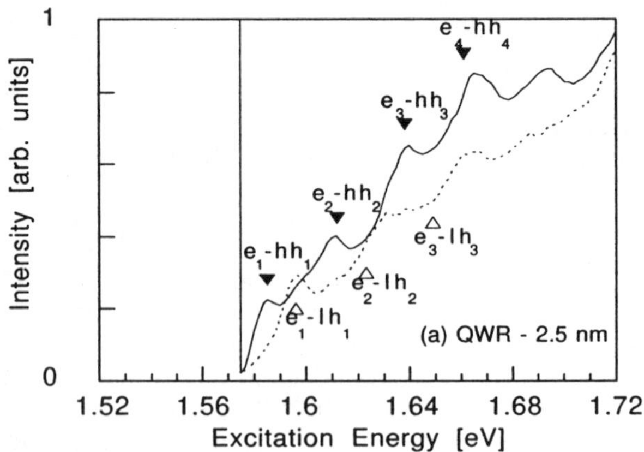

Figure 13. Low-temperature PLE spectra of a V-groove QWR structure (9 nm crescent thickness), shown for excitation polarization oriented parallel (solid line) and perpendicular (dashed line) to the wire axis. Results of model calculations of the e-hh and e-lh -like transition energies are also indicated [37].

The observed 1D subband separation in these V-groove QWRs generally increases with increasing Al mole fraction in the barriers and with decreasing growth temperature (for fixed GaAs crescent thickness). Both effects lead to a decrease in the self-limiting radius of curvature ρ_{sl} underneath the crescent and hence to a narrower lateral potential well. Figure 14 shows the polarized PLE spectra for the smallest wires investigated so far (grown by AP-OMCVD), for which the measured e-hh subband separation is about 45 meV [26]. Although the linewidth and Stokes shift of these narrower wires are larger than the ones described earlier, the 1D subbands are well resolved and the polarization anisotropy associated with the e-hh and e-lh -like transitions is evident.

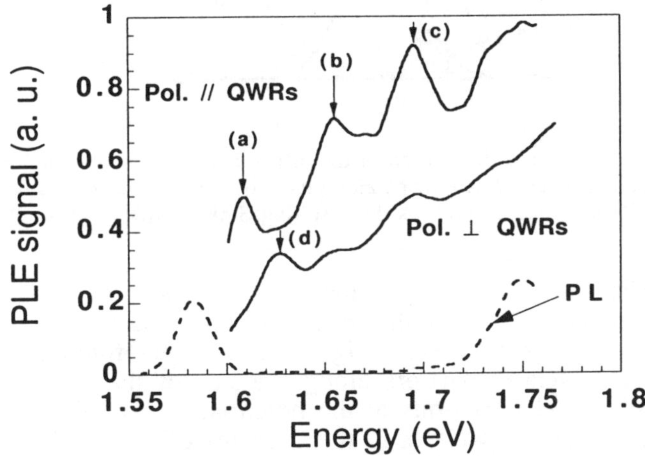

Figure 14. Low temperature (10K) PL and PLE spectra of a vertically-stacked GaAs/AlGaAs QWR structure with a large (45 meV) one-dimensional e-hh -like subband separation. Peaks (a), (b) and (c) denote e-hh -like transitions whereas (d) indicates an e-lh -like transition [26].

The cross section of the V-groove QWRs varies along their axis, both due to fluctuations in the widths of the lithography mask as well as because of nonuniformities in the growth of the AlGaAs and the GaAs layers. The latter effect alone is responsible for the formation of terraces separated by monolayers steps, which exist even in high quality QW structures that do not rely on any lithography. The lithography imperfections give rise to moderate height variations of the crescent along the axis of the wire. As mentioned above, AFM studies of V-groove surface QWRs show evidence for coherent wire domains, up to one μm long, in which the height variations are only several monolayers [28]. These coherent domains are separated by moderate slopes of roughly 5° inclinations and total height difference of up to about 10 nm, arising from the lithography imperfections. On the other

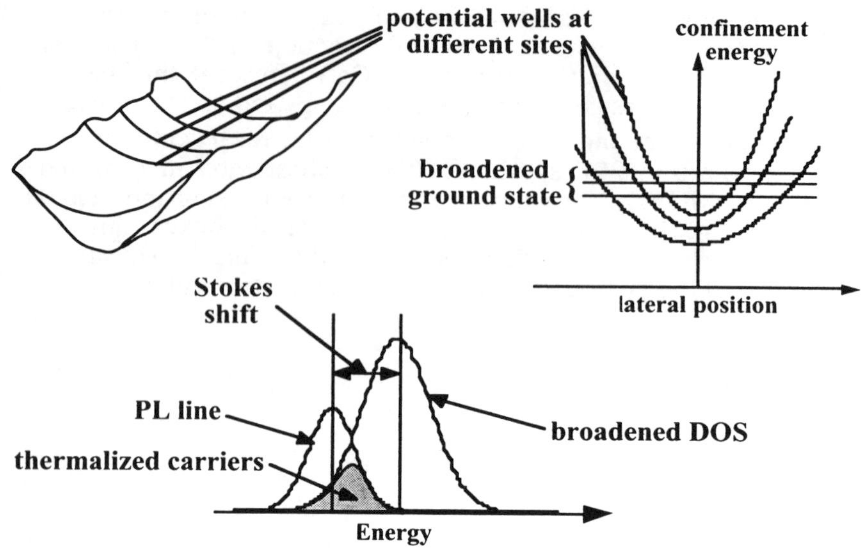

Figure 15. Schematic illustration of the effect of longitudinal variations in the V-groove QWR on the potential landscape along its axis. The expected luminescence and absorption lineshapes, as well as the resulting Stokes shift, are also shown.

hand, TEM studies suggest that the actual shape and size of the wires is quite uniform along its length. In addition to the longitudinal variations in the QWR interface positions, there might also exist fluctuations in the composition of the AlGaAs barriers along the axis of the wire. Both effects give rise to longitudinal variations in the potential well of the wire, as "seen" by the confined carriers, as well as in the energies of the confined 1D states; this *adiabatic* picture is schematically illustrated in Fig. 15. Whereas the luminescence spectrum is expected to peak near the lower energy part of the resulting distribution of the DOS, the interband absorption peaks near the maximum of the DOS distribution. This should give rise to a Stokes shift between the luminescence and the absorption spectra.

The low temperature (10K) Stokes shifts of the V-groove QWRs grown by LP-OMCVD range between 8 and 4 meV for crescent (center) thickness of 4 to 14 nm, respectively. The potential fluctuations which give rise to these Stokes shifts are responsible also for exciton localization in the potential minima at sufficiently low temperatures. As the temperature is increased, however, the excitons can be thermally activated to overcome these potential barriers and become delocalized. This effect manifests itself in the vanishing of the Stokes shift above a characteristic delocalization temperatures T_{del}, which is comparable to the height of potential fluctuations [37]. Figure 16 shows the variation of the Stokes shift with temperature for GaAs/AlGaAs QWRs grown by LP-OMCVD (4 nm center thickness). The Stokes shift vanishes at about 100K, indicating localizing potential barriers of height of about 8 meV, which is comparable to the Stokes shift at the lowest temperature.

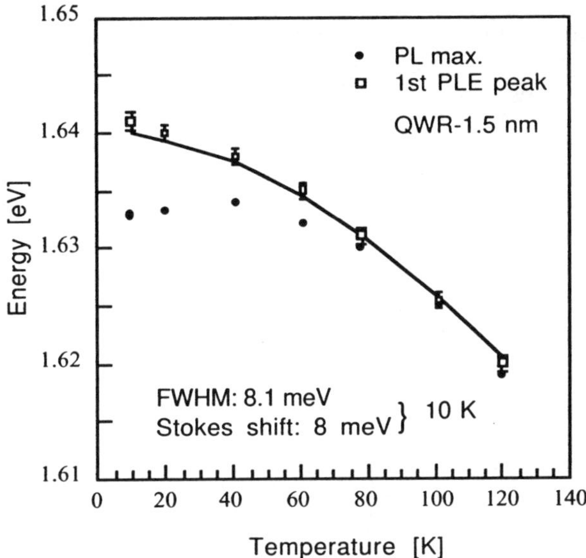

Figure 16. Stokes shift versus temperature for GaAs/AlGaAs V-groove QWRs grown by low-pressure OMCVD [37].

3.2 PHOTOLUMINESCENCE OF VQWS

Vertical QW heterostructures of well defined recombination regions have been fabricated for low temperature luminescence studies. These heterostructures consist of a core $Al_xGa_{1-x}As$ layer cladded by two $Al_yGa_{1-y}As$ layers with a higher Al mole fraction (y>x). Vertical QW structures are formed in both different AlGaAs regions, giving rise to a core VQW in the $Al_xGa_{1-x}As$ region, surrounded laterally by 3D $Al_xGa_{1-x}As$ barriers and vertically by the VQW regions formed at the $Al_yGa_{1-y}As$ cladding layers. Efficient carrier transfer from the cladding ($Al_yGa_{1-y}As$) bulk and VQW regions into the core regions result in luminescence spectra revealing lines associated with the various $Al_xGa_{1-x}As$ regions of the sample.

Figure 17 shows a low temperature PL (Ar^+- laser pumped) spectrum of such a VQW heterostructure grown by LP-OMCVD on a 3 μm pitch corrugated substrate (x=0.21) [33]. The various PL lines have been identified with the aid of low temperature CL imaging techniques. The VQW luminescence at 1.695 eV, with a linewidth of 6 meV meV. Lines due to recombination at the various facets of the AlGaAs core layers are identified at higher energies.

Lines associated with the recombination at similar VQW regions can also be observed in V-groove GaAs/AlGaAs QWR heterostructures, particularly at low temperatures [32]. At higher temperatures, these VQW lines disappear due to the efficient carrier transfer from the VQW regions to the connected QWR potential wells. Time resolved PL spectra of such QWR/VQW structures grown on sub-μm gratings demonstrated a 25 ps carrier transfer time from the VQWs to the QWRs [39].

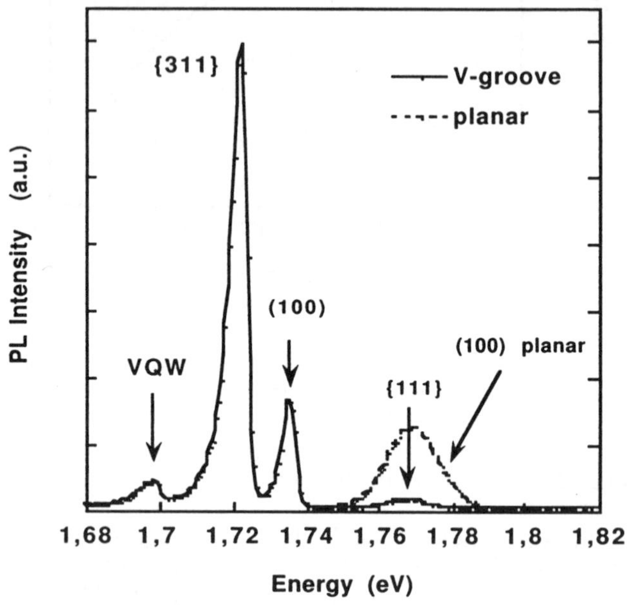

Figure 17. Low temperature photoluminescence spectrum of a VQW AlGaAs (x=0.21) structure grown on a 3 μm-pitch grating by low-pressure OMCVD [33].

4. Application in Diode Lasers

Low-dimensional QWR lasers have been predicted to show improved static and dynamic performance due to their small active volumes and peaked DOS profiles in 1D and 0D systems [40] (see also [41] for a review). In particular, the higher differential gain is expected to yield lower threshold currents, higher modulation speeds, narrower spectral linewidths and reduced chirp. The narrow spectral distribution of carriers is also expected to lead to a reduced temperature sensitivity of the threshold current of these devices. In addition to the high wire uniformity and near ideal interface quality needed for the realization of these 1D lasers, other device parameters have to be considered in order to utilize the inherent advantages of low-dimensional materials. Among these, most important are employing high density arrays of wires and dots and/or using very tight optical confinement (tight-confinement waveguides, micro cavities), in order to offset the inherently small overlap with the optical modes, reducing the optical cavity losses in order to cope with the low modal gain, and using efficient current confinement and carrier capture schemes to guide the carriers from the 3D contacts to the 1D or 0D active regions.

Various approaches have been used to realize 1D semiconductor lasers, including application of high magnetic fields, etching and regrowth and growth on vicinal substrates [41]. More recently, QWR lasers made by cleaved-edge overgrowth and [42] QD lasers fabricated by Stranski-Krastanow self-organized growth [43] have been demonstrated and studied. Here we review the structure and device characteristics of self-ordered QWR and QW diode lasers made by growth in V-grooves.

4.1. V-GROOVE QWR LASERS

The V-groove approach has been applied to the fabrication and studies of single- and multiple-QWR semiconductor lasers [44]. Longitudinal GaAs/AlGaAs QWR diode laser structures were realized by growing the crescent-shaped wires at the center of a 2D, tight optical waveguide, embedded in a p-n junction for providing carrier injection [14]. An example of a vertically stacked, multiple QWR laser structure is shown in Fig. 18. In this configuration, the wires run along the propagation direction of the optical waveguide mode, and the current is confined to the vicinity of the wire by means of proton implantation. The stacking of several wires is useful for increasing the otherwise extremely small optical confinement factor. Furthermore, models of the V-shaped waveguides reveal a very tight fundamental mode, with beam diameter of less than 0.5 μm, centered above the geometrical center of the waveguide [6]. This indicates that a higher

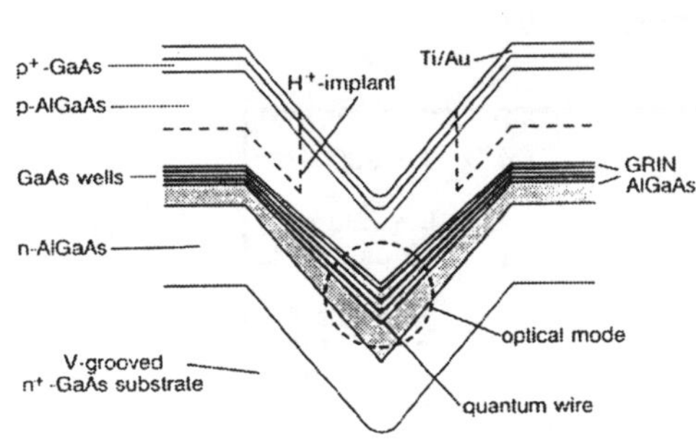

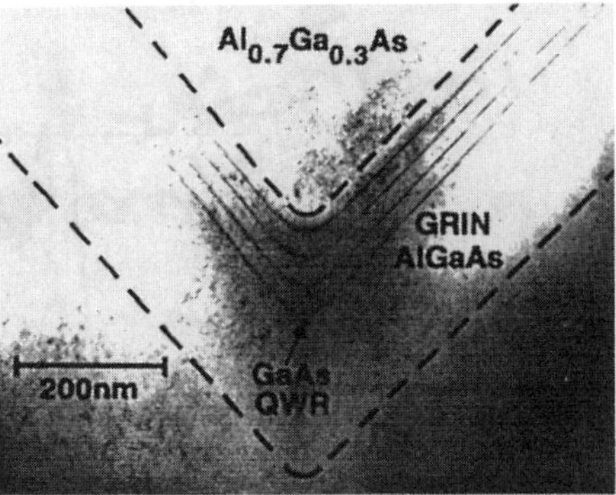

Figure 18. Schematic and TEM cross sections of a vertically-stacked GaAs/AlGaAs multiple QWR laser grown by AP-OMCVD.

optical confinement factor can be obtained with the asymmetric waveguide structure shown in Fig. 18. Diode lasers incorporating three GaAs/AlGaAs V-groove QWRs grown by atmospheric pressure OMCVD operated at room temperature with threshold currents as low as 0.6 mA (high-reflection coated devices) [45]. Strained InGaAs/GaAs V-groove QWR lasers grown by MBE exhibited threshold currents as low as 100 μA (uncoated mirrors) at room temperature [46].

The higher QWR material gain expected at the e-hh transition for light polarized parallel to the wire implies that *transverse* QWR laser configurations (i.e., structures in which the optical beam propagates perpendicular to the wire axis) should have lower threshold currents. Such strained InGaAs/GaAs V-groove QWR lasers grown on sub-μm gratings have also been demonstrated [47]. However, it should be noted that the threshold current is determined by the *modal* QWR gain, and hence the maximum optical confinement that can be obtained with a given configuration is also important. For this reason, the longitudinal QWR laser structures may have an advantage due to their very small mode spot size.

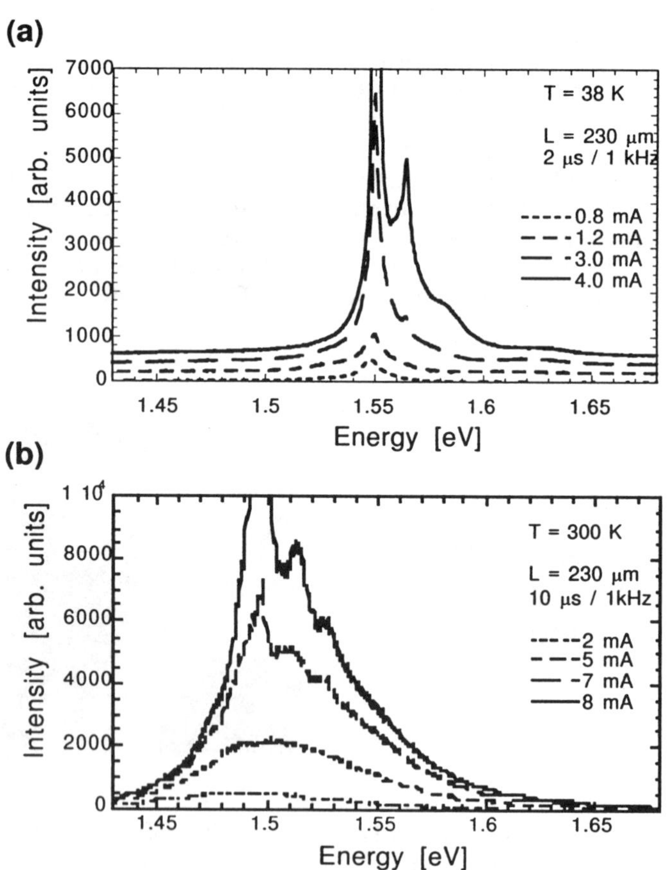

Figure 19. Low temperature (a) and room temperature (b) lasing spectra of a single-QWR V-groove diode laser [48].

Even with multiple-QWR active regions, the V-groove QWR lasers oscillate at room temperature only at relatively high carrier densities. This leads to band filling, and hence lasing due to transition between excited 1D electron and hole states, rather than the ground states. However, at lower temperatures, lasing at the ground states is possible owing to the increase in material gain. Figure 19 shows the amplified spontaneous emission and lasing spectra of single-QWR GaAs/AlGaAs diode lasers measured at room and at low temperatures. Peaks due to the enhanced DOS at the 1D subbands (separated by 10-15 meV) are evident in both cases. However, lasing starts at the ground state at lower temperatures (typically below 70K), as can be shown by an analysis of the subband structure and the temperature dependence of the GaAs band edge [48].

4.2. VQW V-GROOVE DIODE LASERS

The VQW heterostructures are interesting for laser and other optoelectronic device applications both since they assist in carrier capture into adjacent QWRs [29], and because they can serve as an active region with non-conventional features. Figure 20 shows TEM cross sections of an AlGaAs VQW diode laser structure grown by AP-OMCVD [49]. The device structure is similar to that of the longitudinal QWR laser structures discussed earlier, except that the GaAs QWR layer is skipped. Notice that the VQW active region is bounded vertically by two higher bandgap VQW structures, which can provide a carrier path from the outer AlGaAs barriers into the active VQW region.

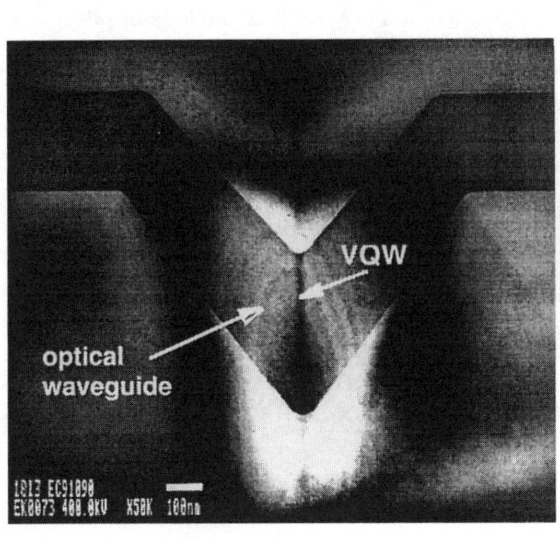

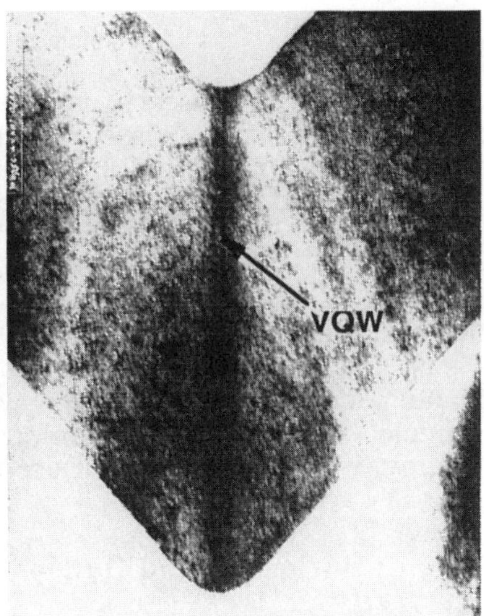

Figure 20. TEM cross sections of a VQW AlGaAs V-groove laser grown by AP-OMCVD [49].

In order to identify the different emission lines of these lasers, CL spectroscopy was performed at room temperature in spot mode. Two such CL spectra are shown in Fig. 21, one obtained with the electron beam positioned at the VQW active region, the other with the beam placed on the AlGaAs optical waveguide layer on the (100) planes next to the groove. Two distinct lines are emitted from each of these regions, respectively, with the well-resolved lower energy line emanating from the VQW active region. More detailed studies show that the composition, and hence the emission wavelength, of the AlGaAs waveguide layer directly next to the VQW region is similar to that of the AlGaAs region on the (100) sections.

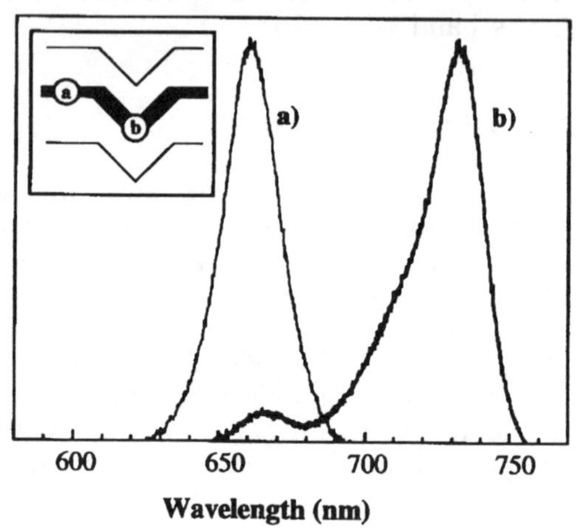

Figure 21. Cathodoluminescence spectra of a VQW AlGaAs diode laser [49].

The VQW lasers were operated at room temperature under pulsed conditions. Polarized laser spectra and light-current characteristics of the device are shown in Fig. 22. Comparison of the emission spectra below and above threshold with the CL spectra of Fig. 21 clearly show that lasing occurs due to recombination at the VQW, rather than in the AlGaAs "background" barrier on the slopes of the groove. The laser beam is polarized in the plane of the VQW (TM-like polarization), consistent with the expected stronger e-hh transition for light polarized in the plane of such a 2D structure.

Such VQW-based devices are also of interest since they offer a quantum confinement direction which is parallel to the wafer plane. As an example, intersubband transitions in such doped VQWs were observed recently, which suggests their use in top-illumination infrared devices [50].

5. Conclusions and Outlook

Considerable progress has been achieved recently in the realization of self-ordered QWR and VQW structures grown by low-pressure OMCVD on nonplanar substrates. The low-pressure growth appears to be particularly

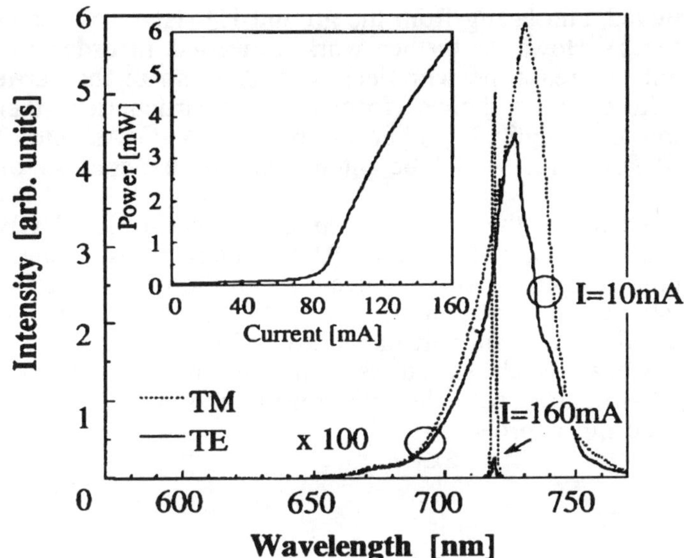

Figure 22. Polarized lasing spectra of a VQW AlGaAs laser (pulsed operation). Inset shows the light versus current characteristic of the device [49].

useful in providing kinetically-limited growth conditions which result in abrupt interfaces and distinct faceting at the nonplanar growth front. This leads to the formation of self-ordered QWR and VQW structures with interfaces defined on an atomic level. The currently achievable uniformity of these wire and well structures permits the clear observation of 1D and 2D quantum confinement effects, and thus makes these nanostructures useful in investigating the physics of low-dimensional systems. For example, the detailed absorption features observed in low temperature PLE spectra of these QWRs allow the investigation of polarization anisotropy effects arising from valence band mixing related to the 2D quantum confinement. The peculiar vertical configuration of the VQW heterostructures offers the possibility of constructing laterally coupled QW structures as well as coupled QW-QWR structures for studies of carrier transfer and tunneling between systems of different dimensionality.

The wire and well disorder in the currently achievable structures could be possibly further reduced by optimizing the lithography and the growth conditions. However, the moderate disorder existing in the present QWR structure is also interesting for studying weak localization effects of free carriers and excitons in 1D systems, which are more susceptible to such localization than their 2D counterparts. First observations established the transition between localized and delocalized excitons at a critical temperature in the V-groove QWRs.

Diode lasers incorporating these self-ordered quantum structures, operating at room temperature, have been demonstrated. Room temperature operation of V-groove QWR lasers with threshold currents as low as 100 μA has been achieved, and lasing from the ground 1D states has been observed at low temperatures. However, further work is needed in order to reduce the leakage currents, increase the wire density, and optimize the carrier injection paths in order to optimize their performance. In particular, further studies of the fabrication and optical properties of InGaAs/GaAs and InGaAs/InP strained QWR structures would be interesting in the context of improved QWR lasers.

An extension of the nonplanar epitaxy of self-ordered QW and QWR structures for fabricating self-ordered QD heterostructures is one of the most interesting future directions of this area. As in the case of self-ordered QW and QWR structures, one should identify specific self-limiting facet configurations which will support the controlled growth of three-dimensional QD potential wells. Such structures will combine the advantage of self-organized QD structures with the site control and better uniformity of the seeded self-ordering approach.

Acknowledgments

I would like to thank my collaborators at the Nanostructures Laboratory in EPFL, particularly G. Biasiol, M.-A. Dupertuis, B. Dwir, A. Gustafsson, P. Ils, E. Martinet, D. Oberli, F. Reinhardt, and F. Vouilloz for their contributions to the work described in this review. Special thanks are due to Giorgio Biasiol for his help in the preparation of the manuscript. This work was supported by the Fonds National Suisse de la Recherche Scientifique.

References

1. Bockelmann, U. and Bastard, G. (1991) Interband optical transitions in semiconductor quantum wires: selection rules and absorption spectra, Europhys. Lett. **15**, 215.
2. Sercel, P. and Vahala, K. (1991) Polarization dependence of optical absorption and emission in quantum wires, Phys. Rev. **B 44**, 5681.
3. Ogawa, T. and Takagahara, T. (1991) Interband absorption spectra and Sommerfeld factors of a one-dimensional electron-hole system, Phys. Rev. **B 43**, 14325; Optical absorption and Sommerfeld factors of one-dimensional semiconductors: an exact treatment of excitonic effects *ibid* **B 44**, 8138.
4. Rossi, F. and Molinari, E. (1996) Coulomb-induced suppression of band-edge singularities in the optical-spectra of realistic quantum-wire structures, Phys. Rev. Lett. **76**, 3642.
5. Kane, C.L. and Fisher, M.P.A. (1992) Transmission through barriers and resonant tunneling in an interacting one-dimensional electron gas, Phys. Rev. **B 46**, 15233.
6. Kapon, E., Walther, M., Christen, J., Grundmann, M., Caneau,, C., Hwang, D.M., Colas, E., Bhat, R., Song, G.H., and Bimberg, D. (1992) Quantum wire heterostructures for optoelectronic applications, Superlattices and Microstructures **12**, 491.
7. Beenakker, C.W.J. and van Houten, H. (1991) Quantum transport in semiconductor nanostructures, in Solid State Physic, *Semiconductor Heterostructures and Nanostructures*, Herenreich, H. and Turnbull, D., Eds. (Academic, New York).
8. Marzin, J.-Y., Gerard, J.-M., Izrael, A., Barrier, D., and Bastard, G. (1994) Photoluminescence of single InAs quantum dots obtained by self-organized growth on GaAs, Phys. Rev. Lett. **73**, 716.
9. Chen, A.C., Moy, A.M., Pearah, P.J., Hsieh, K.C., and Chen, K.Y. (1993) $Ga_xIn_{1-x}P$

multiple-quantum-wire heterostructures prepared by the strain induced lateral layer ordering process, Appl. Phys. Lett. **62**, 1359.

10. Grundmann, M., Christen, J. Ledentsov, N.N., Bohrer, J., Bimberg, D., Ruminov, S.S., Werner, P., Richter, U., Gosele, U., Heydenreich, J., Ustinov, V.M., Egorov, A.Yu., Zhukov, A.E., Kop'ev, P.S., and Alferov (1994) Ultranarrow luminescence lines from single quantum dots, Phys. Rev. Lett. **74**, 4043.
11. Xie, Q., Madhukar, A., Chen, P., and Kobayashi, N.P. (1995) Vertically self-organized InAs quantum box islands on GaAs(100), Phys. Rev. Lett. **75**, 2542.
12. Kapon, E. (1994) Lateral patterning of quantum well heterostructures by growth on nonplanar substrates, in *Epitaxial Microstructures*, A.C. Gosssard, Ed., Semiconductors and Semimetals **40** (Academic Press, New York, 1994), pp. 259-336.
13. Kapon, E., Tamargo, M.C., and Hwang, D.M. (1987) Molecular beam epitaxy of GaAs/AlGaAs superlattice heterostructures on nonplanar substrates, Appl. Phys. Lett. **50**, 347.
14. Kapon, E., Hwang, D.M., and Bhat, R. (1989) Stimulated emission in semiconductor quantum wire heterostructures, Phys. Rev. Lett. **63**, 430.
15. Koshiba, S., Noge, H., Akiyama, H., Inoshita, T., Nakamura, Y., Shimizu, A., Nagamune, Y., Tsuchiya, M., Kano, H., and Sakaki, H. (1994) Formation of GaAs ridge quantum-wire structures by molecular-beam epitaxy on patterned substrates, Appl. Phys. Lett. **64**, 363.
16. Bhat, R., Kapon, E., Werner, J., Hwang, D.M., Stoffel, N.G., and Koza, M.A. (1990) Organometallic chemical vapor deposition of InP/InGaAsP on noplanar InP substrates: Application to multiple quantum well lasers, Appl. Phys. Lett. **56**, 863.
17. Tsukamoto, S., Nagamune, Y., Nishioka, M., and Arakawa, Y. (1992) Fabrication of GaAs quantum wires on epitaxially grown V grooves by metal-organic chemical-vapor deposition, Appl. Phys. Lett. **71**, 533.
18. Vermeire, G., Yu, Z.Q., Vermaerke, F., Buydens, L., Daele, P.V., and Demeester, P. (1992), J. Crystal Growth **124**, 521.
19. Hasegawa, Y., Egawa, T., Jimbo, T., and Umeno, M. (1993) Fabrication of low-threshold AlGaAs/GaAs patterned quantum well lasers grown on Si substrates, Jpn. J. Appl. Phys. **32**, L997.
20. Usami, N., Mine, T., Fukatsu, S., and Shiraki, Y. (1993) Realization of crescent-shaped SiGe quantum-wire structures on a V-groove patterned Si substrate by gas-source Si molecular-beam epitaxy, Appl. Phys. Lett. **63**, 2789.
21. Shen, X.Q., Tanaka, M., Wada, K., and Nishinaga, T. (1994) Molecular beam epitaxial growth of GaAs, AlAs and $Al_{0.45}Ga_{0.55}As$ on (111)A-(001) V-grooved substrates, J. Crystal Growth **135**, 85.
22. Grundmann, M., Christen, J., Bimberg, D., and Kapon, E. (1995) Electronic and optical properties of quasi-one dimensional carriers in quantum wires, J. Nonlinear Optical Physics and Materials **4**, 99.
23. Pfister, M., Johnson, M.B., Alvarado, S.F., Salemink, H.W.M., Marti, U., Martin, D., Morier-Genoud, F., and Reinhart, F.K. (1995) Indium distribution in InGaAs quantum wires observed with the scanning tunneling microscope, Appl. Phys. Lett. **67**, 1459.
24. Wang, X.L., Ogura. M., and Matsuhata, H. (1995) Flow-rate modulation epitaxy of AlGaAs/GaAs quantum wires on nonplanar substrates, Appl. Phys. Lett. **66**, 1506.
25. Kapon, E., Biasiol, G., Hwang, D.M., and Colas, E (1995) Seeded self-ordering of GaAs/AlGaAs quantum wires on non-planar substrates, Microelectronics J. **26**, 881.
26. Kapon, E., Biasiol, G., Hwang, D.M., Colas, E., and Walther, M. (1996) Self-ordering mechanism of quantum wires grown on nonplanar substrates, Solid State Electronics **40**, 815.
27. Gustafsson, A., Reinhardt, F., Biasiol, G. and Kapon, E. (1995) Low-pressure organometallic chemical vapor deposition of quantum wires on V-grooved substrates, Appl. Phys. Lett. **67**, 3673.
28. Reinhardt, F., Biasiol, G., Gustafsson, A., Dwir, B. and Kapon, E. (1996) Morphology of low pressure OMCVD grown surface quantum wires, studied by atomic force microscopy, Electronic Materials Conference, Santa Barbara, California.

29. Walther, M., Kapon, E., Christen, J., Hwang, D.M., and Bhat, R. (1992) Carrier capture and quantum confinement in GaAs/AlGaAs quantum wire lasers grown on V-grooved substrates, Appl. Phys. Lett. **60**, 521.
30. Vermeire, G., Yu, Z.Q., Vermaerke, F., Buydens, L., Daele, P.V., and Demeester, P. (1992) Anisotropic photoluminescence behaviour of vertical AlGaAs structures grown on gratings, J. Crystal Growth **124,** 513.
31. Biasiol, G., Reinhardt, F., Gustafsson, A., Martinet, E., and Kapon, E. (1996) Structure and formation mechanisms of AlGaAs V-groove vertical quantum wells grown by low pressure organometallic chemical vapor deposition, Appl. Phys. Lett **69**, 2710.
32. Christen, J., Kapon, E., Colas, E., Hwang, D.M., Schiavone, L.M., Grundmann, M., and Bimberg, D. (1992) Cathodoluminescence investigation of lateral carrier confinement in GaAs/AlGaAs quantum wires grown by OMCVD on nonplanar substrates, Surf. Sci. **267**, 257.
33. Martinet, E., Gustafsson, A., Biasiol, G., Reinhardt, F., Kapon, E., and Leifer, K. (1996) Carrier quantum confinement in self-ordered AlGaAs V-groove quantum wells, submitted for publication.
34. Reinhardt, F., Dwir, B, and Kapon, E. (1996) Oxidation of GaAs/AlGaAs heterostructures studied by atomic force microscopy in air, Appl. Phys. Lett. **68**, 3168.
35. Pfister, M., Johnson, M.B., Alvarado, S.F., Salemink, H.W.M., Marti, U., Martin, D., Morier-Genoud, F., and Reinhart, F.K. (1994) Atomic-structure and luminescence excitation of $GaAs/(AlAs)_N(GaAs)_M$ quantum wires with the scanning tunneling microscope, Appl. Phys. Lett. **65**, 1168.
36. Gustafsson, A. and Kapon, E. (1996) Cathodoluminescence in the scanning electron microscope: application to low-dimension semiconductor structures, to appear in Scanning Microscopy International.
37. Oberli, D., Vouilloz, F., Dupertuis, M.-A., Fall, C., and Kapon, E. (1995) Optical spectroscopy of semiconductor quantum wires, Il Nuovo Cimento **17D**, 1641.
38. Vouilloz, F., Oberli, D., and Kapon, E. (1996) Polarization anisotropy in V-groove quantum wires, presented at the 23rd International Conference on Physics of Semiconductors, Berlin.
39. Haacke, S., Hartig, M., Oberli, D.Y., Deveaud, B., Kapon, E., Marti, U. and Reinhart, F. (1996) Carrier transfer and capture into quantum-wire arrays grown on V-groove substrates, Solid State Electron. **40**, 299.
40. Arakawa, Y. and Sakaki, H. (1982) Multidimensional quantum well laser and temperature dependence of its threshold current, Appl. Phys. Lett. **40**, 939.
41. Kapon, E. (1992) Quantum wire lasers, Proc. IEEE **80**, 398.
42. Wegscheider, W., Pfeiffer, L.N., Dignam, M.M., Pinczuk, A., West, K.W., McCall, S.L., and Hull, R. (1993) Lasing from excitons in quantum wires, Phys. Rev. Lett. **71**, 4071.
43. Kristaedter, N., Ledentsov, N.N., Grundmann, M., Bimberg, D., Ustimov, V.M., Ruminov, S.S., Maximov, M.V., Kop'ev, P.S., Alferov, Zh.I., Richter, U., Werner, P., Gosele, U, and Heydenreich, J. (1994) Low-threshold, large T-O injection-laser emission from (InGa)As quantum dots, Electron. Let. **30**, 1416.
44. Kapon, E., (1993) Quantum wire lasers grown by OMCVD on nonplanar substrates, OPTOELECTRONICS- Devices and Technologies **8**, 429.
45. Simhony, S., Kapon, E., Colas, E., Hwang, D.M., Stoffel, N.G., and Worland, P. (1991) Vertically stacked multiple-quantum-wire semiconductor diode lasers, Appl. Phys. Lett. **59**, 2225.
46. Tiwari, S. (1995) Operation of strained multi-quantum wire lasers, in *Low Dimensional Structures prepared by Epitaxial Growth or Regrowth on Patterned Substrates*, K. Eberl et al. (eds.), (Kluwer Academic) p. 335.
47. Walther, M., Kapon, E., Caneau, C. Hwang, D.M. and Schiavone L.M. (1993) InGaAs/GaAs strained quantum wire lasers grown by organometallic chemical vapor deposition on nonplanar substrates, Appl. Phys. Lett. **62**, 2170.
48. Chiriotti, N. Temperature dependence of the emission characteristics of quantum wire lasers, Diploma Thesis, EPFL, February 1996.

49. Kapon, E., Dwir, B., Pier, H., Gustafsson, A., and Bonnard, J.M. Lasing in self-ordered vertical quantum well heterostructures, Quantum Electronics Laser Sciences Conference (QELS '95), Baltimore, Maryland, May 1995.
50. Berger, V., Vermeire, G., Demeester, P., and Weisbuch, C. (1995) Normal incidence intersubband adsorption in vertical quantum wells, Appl. Phys. Lett. **66**, 218.

CLEAVED EDGE OVERGROWTH AND 1D LASERS

W. WEGSCHEIDER
Walter Schottky Institut, Technische Universität München
Am Coulombwall, D-85748 Garching, Germany

L. N. PFEIFFER AND K. W. WEST
Bell Laboratories, Lucent Technologies
600 Mountain Ave. Murray Hill, New Jersey 07974, USA

A. A. KISELEV
Ioffe Physico-Technical Institute
Polytekhnicheskaya 26, St. Petersburg 194021, Russia

M. HAGN
Walter Schottky Institut, Technische Universität München
Am Coulombwall, D-85748 Garching, Germany

AND

R. E. LEIBENGUTH
Bell Laboratories, Lucent Technologies, SSTC
Breinigsville, Pennsylvania 18031, USA

Abstract.

The cleaved edge overgrowth method — a molecular beam epitaxy technique which incorporates two sequential growth steps along orthogonal crystal directions — is reviewed. The application of this method for the fabrication of nanostructures exhibiting quantum confinement in more than one dimension is discussed. The T-shaped intersection of two quantum wells formed during the two growth steps represents a quantum wire with a cross-section of less than 10 by 10 nm, entirely defined by the growth process. The optical properties of this kind of quantum wires are compared with theoretical calculations of single particle as well as of exciton states in such structures. Substantial enhancement of exciton binding due to the confinement of charge carriers to one dimension is found to be responsible for the observation of stimulated emission from exciton states in T-quantum wire lasers. This result is consistent with the observed strong dependence of the laser emission on an applied magnetic field.

G. Abstreiter et al. (eds.), Optical Spectroscopy of Low Dimensional Semiconductors, 127–155.

1. Introduction

Semiconductor lasers operating due to recombination of charge carriers confined to quasi one- (1D) and zero-dimensional (0D) quantum wires (QWRs) and dots (QDs) have been a challenge to the photonics field for more than a decade. Just as the quantum well (QW) lasers (2D) have replaced bulk semiconductor devices (3D) due to their superior performance, further improvements in threshold current, differential gain as well as reduced sensitivity of threshold current and emission wavelength to ambient temperature are expected from 1D and 0D lasers [1, 2, 3]. The reduction in threshold current is a direct consequence of the smaller volume of optically active material in lower dimensional lasers. In addition, the $1/\sqrt{E}$ as well as the δ-function singularities in the density of states (DOS) at the band edge of 1D and 0D systems, respectively, as opposed to the square root and step-like DOS profiles in 3D and 2D, should lead to enhanced optical nonlinearities [4], narrower optical gain spectra and higher differential gain [2]. Furthermore, increased exciton binding [5, 6], anomalously strong concentration of the oscillator strength at the lowest-energy exciton state [7], and exciton condensation [8] are predicted as a result of Coulomb interaction in such strongly confined electron-hole systems.

The breakthrough of 2D semiconductor systems for optical and transport applications like the intersubband detector or the high electron mobility transistor was triggered by the development of highly sophisticated growth techniques like molecular beam epitaxy (MBE). Although these techniques are well suited to produce planar multilayer systems with interfaces which are flat on an atomic scale, the fabrication of structures in which charge carriers are quantum confined in more than one dimension remains a challenging task. Methods suggested and attempted include lithographic definition combined with etching and regrowth [9], growth on nonplanar [10] and vicinal substrates [11, 12], selective area deposition [13], and local interdiffusion [14, 15]. The pattern transfer by lithographic techniques introduces size fluctuations which blur most of the effects caused by the reduced dimensionality. Self-organisation has proven to be a powerful technique for the preparation of QWRs which form after organometallic chemical vapor deposition (OMCVD) on V-grooved substrates [10]. However, the relatively large size of the crescent-shaped QWRs (80-100 by 10 nm) currently achievable by this technique results in the occupation of more than one 1D subband, and therefore in optical properties reminescent to those of 2D systems. Recently, a different approach relying on self-assembling of strain induced QDs has attracted considerable attention [16]. Although low-threshold InGaAs/GaAs QD lasers fabricated by this technique have been demonstrated [17] the photoluminescence (PL) response of an ensemble of

QDs is characterised by a quite large linewidth (full width at half maximum FWHM $\approx$ 40-60 meV) which directly reflects the size distribution. Since the PL linewidth of each individual QD is only 0.02meV major advances in terms of the QD laser properties are expected if more homogeneous QD arrays can be realised. In this article we concentrate on another method for the fabrication of QWRs and QDs which makes use of MBE growth on atomically flat cleavage planes of previously prepared multilayer structures. This so-called cleaved edge overgrowth (CEO) technique enables us to prepare quantum size structures with characteristic dimensions down to a few lattice constants. This turns out to be of particular importance since the peculiar features associated with 1D excitons predicted by theory [7, 8] can be only observed if the QWR dimensions are comparable or smaller than the bulk exciton Bohr radius. Otherwise only the center-of-mass motion of the excitons shows 1D character while the electron-hole relative motion remains unaffected. In additon, arrays of QWRs and QDs fabricated by CEO are very uniform in size comparable to the homogeneity achievable in multiple QW layers. This is advantageous for laser applications where a large number of QWRs or QDs are necessary but should not degrade device performance because of size fluctuation induced broadening of the gain spectra.

This article is organized as follows. After the introduction the CEO technique is described in detail. The difficulties in high quality (110) GaAs crystal growth are outlined. The possible types of low-dimensional structures achievable with the CEO method and their relevance for optical and transport investigations are then discussed. We will then focus on the optical properties of 1D states of T-shaped QWRs i.e. states originating from the T-shaped confining potential of two intersecting QWs. Theoretical predictions for single particle binding energies to the QWRs as well as for the enhancement of exciton binding are presented together with experimental data. The design of a laser structure based on these 1D states and the properties of such a device are subject of the following section. The main topic of the discussion is the striking spectral constancy of the emission with large changes in excitation power which is interpreted as an increased stability of the 1D excitons against formation of an electron-hole plasma. The last part of this section is devoted to the effects of a strong magnetic field on the emission properties of the QWR laser.

2. Cleaved Edge Overgrowth

The concept of CEO is illustrated in Fig. 1. The conceptually simple and straightforward approach proposed almost 20 years ago [18] consists of two or three growth steps separated by an *in situ* cleave. The fabrication of a

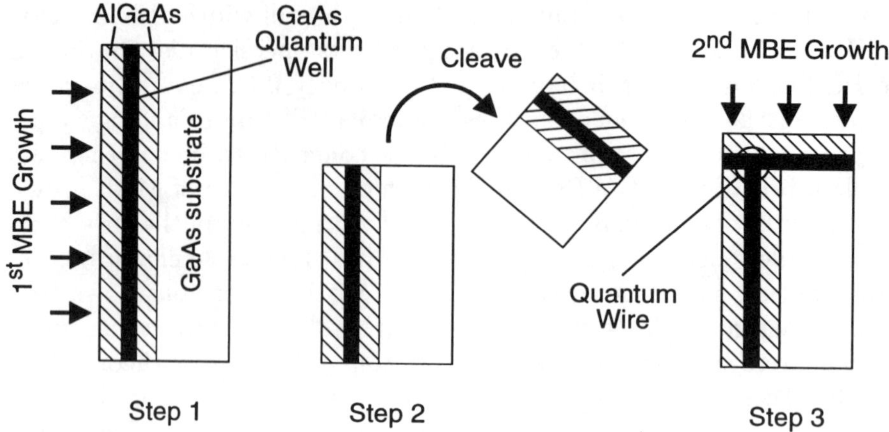

Figure 1. The concept of quantum wire fabrication by cleaved edge overgrowth.

complete CEO structure, successfully realized for the first time by Pfeiffer *et al.* [19] is summarized below. After the first MBE growth the wafer is removed from the machine and thinned from the backside using Br_2CH_3OH to a thickness of 100-150 μm. The thin wafer is then scribed and cleaved into rectangular pieces of about 5 by 10 mm. A scratch is made where the future *in situ* cleave is to occur. Up to eight of these pieces are mounted next to a GaAs(110) monitor wafer as shown in Fig. 2. In this way the (110) surface exposed after the cleave is parallel to the surface of the monitor wafer. The purpose of the monitor wafer is twofold. It provides a surface of proper emissivity for the measurement of the growth temperature, which turns out to be a critical parameter for GaAs(110) epitaxy, using a pyrometer. In addition it serves as a reference sample for the layer sequence grown after the cleave which can be more easily characterised than the CEO samples. Similar to conventional substrate introduction the CEO substrate holder assembly depicted in Fig. 2 is loaded into the growth chamber. After the machine is brought into running condition, the oxide from the (110) monitor wafer is desorbed so that a buffer layer can be grown. Then the cleave is performed. Within seconds MBE growth is initiated and proceeds both on the newly exposed cleave as well as on the adjacent monitor wafer.

2.1. DIFFERENCES IN (001) AND (110) GROWTH

For most applications the first growth step is carried out on GaAs(001), a substrate orientation on which QWs exhibiting narrow PL lines or high mobility two dimensional electron gases are readily achieved. The second and third growth step, however, has to be carried out on a (110) oriented GaAs

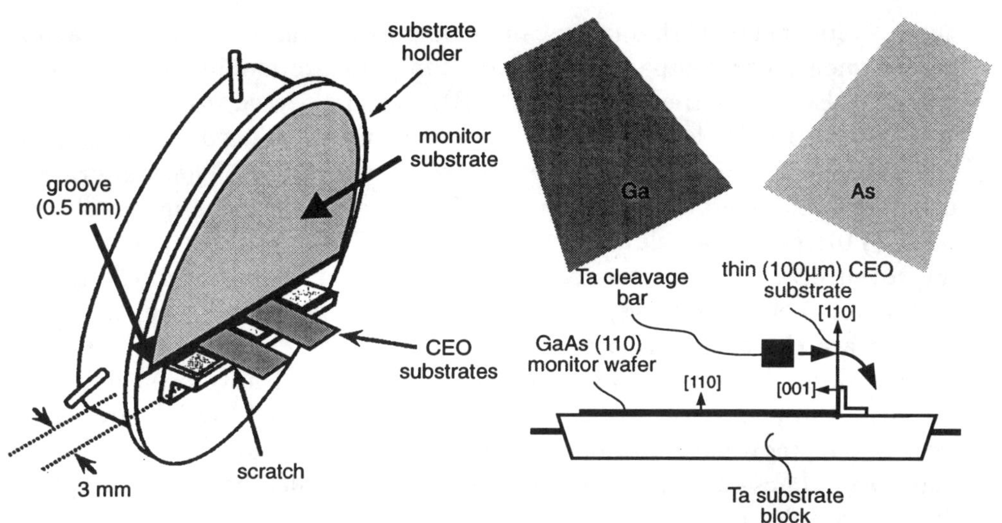

Figure 2. Experimental realisation of the cleaved edge overgrowth method.

or AlGaAs crystal face since these zinc-blende type semiconductors show a strong natural preference for cleavage on (110) lattice planes. Growth on these surfaces using conventional parameters for MBE along the [001] direction i.e. a substrate temperature in the range of 640 to 660 °C, a beam equivalent pressure (BEP) of about 10^{-5} Torr for As_4 and growth rates of 1 and 0.5 μm/hour for GaAs and AlAs, respectively, results in poor surface morphology and unacceptable optical and transport properties of QWs and modulation doped heterostructures. This is mainly due to the low sticking coefficient of the As tetramers on the non-polar (110) oriented surfaces compared to the (001) orientation, which leads to the formation of Ga droplets. One solution to this problem is the drastic reduction of the growth temperature below 500 °C, while the As_4 BEP is doubled and the growth rates are halved. In this way layers with mirrorlike featureless surfaces as inspected by Normarski microscopy can be prepared. The FWHM of PL lines originating from such (110) oriented QWs are still by a factor of two larger than that from corresponding (001) QWs. From reflection high energy electron diffraction (RHEED) investigations during growth of GaAs/AlAs multilayers grown under these conditions it is deduced that the interfaces break up into facets and a step structure forms on the GaAs(110) surface [20]. Nevertheless two dimensional electron gases (2DEGs) exhibiting low-temperature electron mobilities of $2 \cdot 10^6$ cm^2/Vs have been demonstrated in such layers [21]. However, it should be noted at this point, that since the mobility of modulation doped GaAs is extremely sensitive to the background impurity concentration which markedly increases with decreasing substrate temperature, layers of such high quality can be only grown in an ultra clean vacuum environment. For comparison, the low-temperature mobilties of (001) oriented 2DEGs prepared in such a MBE system exceed 10^7 cm^2/Vs [22]. Another approach to circumvent the problem of reduced As sticking on the (110) orientation is to use As_2 instead of As_4 [23]. Thermal cracking of the As tetramers to dimers indeed improves the surface morphology of GaAs(110) grown at elevated temperatures. This is directly reflected in improved optical properties of the layers [23, 24]. The hot cracking zone of the As_2 cell introduces, however, additional impurities, which tend to deteriorate the transport properties of modulation doped GaAs. Although improved cracker designs like the valved cracker cell have been developed recently, the peak electron mobilities of (110) oriented 2DEGs are, to the best of our knowledge, obtained for the growth conditions described above using As_4. These are also the growth conditions which have been employed for the samples described in the following.

2.2. TYPES OF LOW-DIMENSIONAL STRUCTURES ACHIEVABLE BY CLEAVED EDGE OVERGROWTH

The precise control of layer thicknesses and layer compositions currently achievable by MBE allows the design of very sophisticated band-gap engineered multilayer structures with applications ranging from light emitters and detectors based on tailored transitions between quantum confined states to high electron mobility transistors (HEMTs) where Coulomb scattering is suppressed because of spatial separation of the charge carriers from the ionized donors. The possibilty to overgrow a heterostructure by another set of layers just adds another degree of freedom in the design parameters. From the vast variety of structures which can be prepared by CEO [25] Fig. 3 illustrates three classes of QWRs. The intersection of two QWs as shown Fig. 3(a) confines both electrons and holes as explained in more detail in the next section and is therefore well suited for interband transition experiments requiring large overlap between the electron and hole wave functions. Overgrowth of a periodic sequence of δ-doped layers embedded in barrier material by a QW leads to a periodically modulated electron density in this QW as sketched in Fig. 3(b). The excitation spectrum of such a lateral superlattice is characterized by collective and single particle properties of the 1D quantum confined electron system. Since electrons as well as holes are again confined in this structure Raman scattering resonant to an interband transition in the QW can be utilized to study these excitations. The third class of QWRs which represent the 1D analogue to a high-mobility electron layer in 2D is shown in Fig. 3(c). In addition to the electrostatic confinement which arises as a result of charge transfer from the ionized donors over the barrier into the QW, the potential of the latter

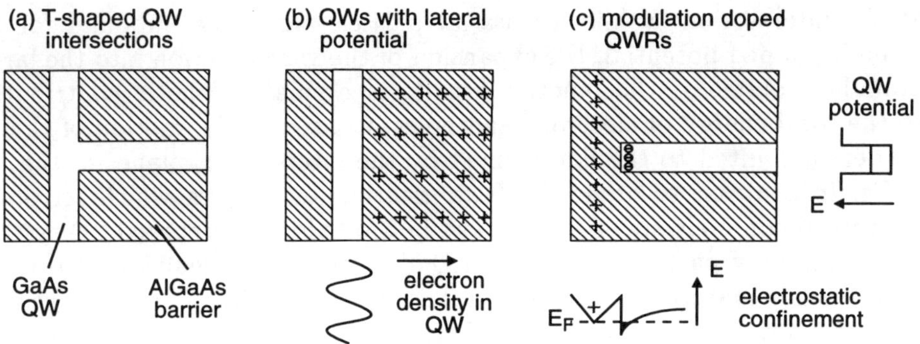

Figure 3. Classes of quantum wires which can be fabricated by cleaved edge overgrowth: (a) T-quantum wires, (b) quantum wells subject to a lateral potential and (c) modulation-doped quantum wires.

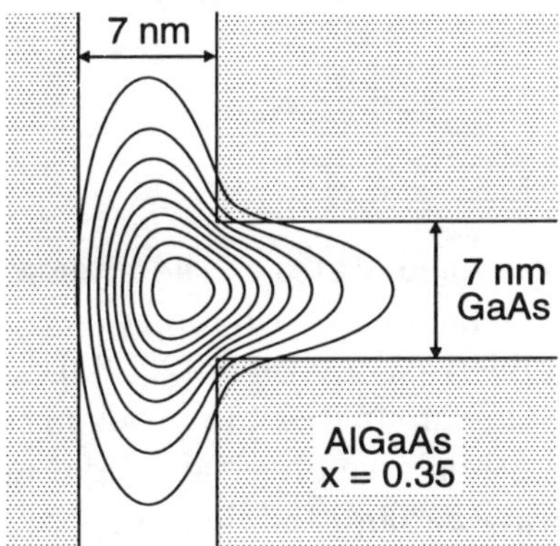

Figure 4. T-shaped intersection of two quantum wells in cross section. The contours are lines of constant probability ($|\psi|^2 = 0.1, 0.2, \ldots, 0.9$) for electrons confined in the T-shaped potential.

leads to an electron channel where carrier motion is limited to a line. The Fermi surface of such an electron system shrinks to two points at $\pm k_F$ in k space. As a consequence small angle scattering requiring $\Delta k \ll k_F$ is greatly suppressed and low-temperature electron mobilties of more than 10^8 cm^2/Vs are predicted [26]. Another peculiarity of ballistic transport in modulation-doped QWRs is the independence of the resistance from the wire length as recently demonstrated experimentally [27].

3. Optical properties of T-shaped quantum wires

Figure 4 illustrates the physics of QWR formation at the T-shaped intersection of two QWs. The figure shows the cross-section of such an intersection together with contours of constant probability for electrons. The origin of the quantum mechanical bound state is the relaxation of QW confinement at the intersection. While a classical particle would be unbound for the given T-shaped potential, the expansion of the wavefunction into the larger available volume at the junction results in a smaller kinetic contribution to the total energy of electrons and holes. Consequently, motion of the 1D carriers is limited to the line defined by the intersecting planes of the two QWs. In contrast to other QWR fabrication techniques which introduce additional confinement of the carriers in a previously prepared 2D system, for example by partial removal of a QW layer and subsequent overgrowth with barrier material, the QWR states in this structure are energetically located below the ground state QW transitions. The existence of such confinement relaxation QWR states was proposed already more than ten years ago by Störmer [28] and Chang *et al.* [29]. It should be noted at this point that the intersection of three QWs can be realized by another cleavage step followed by overgrowth. Analogous to the formation of 1D states at the intersection of two QWs carriers in such a structure are quantum confined in all 3 dimensions and a QD forms.

3.1. THEORY

The theoretical treatment of quantum confined states at the intersection of two QWs requires solving of the two-dimensional Schrödinger equation for the T-shaped potential. Although simplified calculations [29, 30] based on factorization of the structure potential, thus, reducing the problem to an effective one-dimensional Schrödinger equation, give physical insight and lead to analytical results, they are not accurate enough for this purpose. Figure 5(a) shows the electron probability density $|\psi|^2$ of the electron ground state for the structure depicted in Fig. 4, i.e. for two 7 nm wide GaAs QWs embedded in $Al_{0.35}Ga_{0.65}As$ barrier material obtained by the free-relaxation method where an anisotropic effective mass and nonparabolicity can be included [31]. Evidently, the wave function ψ differs strongly from a factorizable one. In order to optimize carrier binding to the 1D QWR state we have calculated the energy of the lowest wire-like state as a function of the width b of the first grown QW for a fixed width a of the overgrown QW (Fig. 6). It is intuitively clear that with the change of b the electron wave function rearranges between the two QWs. With $b \rightarrow 0$ the QWR state asymptotically approaches that of a 2D electron in the overgrown well (indicated by a dotted horizontal line). In the opposite case, with increasing b, the electron penetrates more and more into the well of width b. At finite b, when the QWR energy reaches that of the unbound 2D QW state given by the dotted line approximately proportional b^{-2} in Fig. 6, the lowest electron state loses its wire-like character and becomes purely 2D. The separation of the 2D-1D states, i.e. the binding energy of electrons to

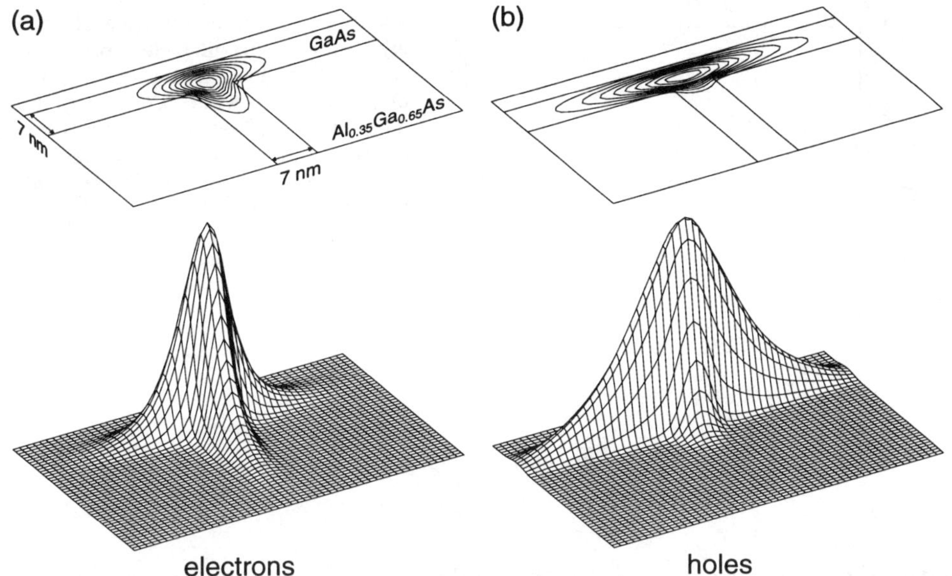

Figure 5. The electron (a) and hole probability density (b) for the wire-like state in the $GaAs/Al_{0.35}Ga_{0.65}As$ T-shaped structure with 7 nm wide wells.

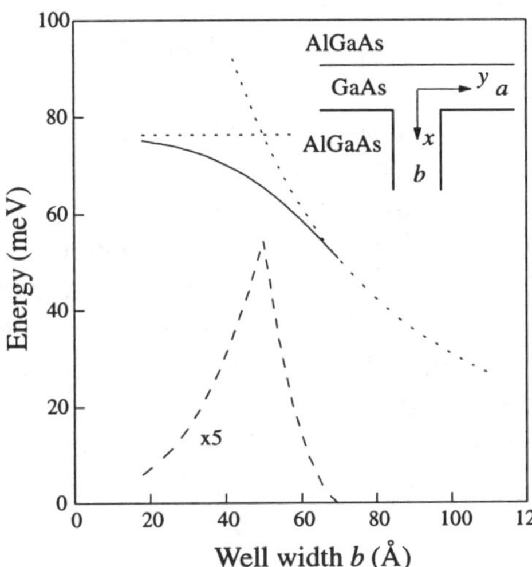

Figure 6. The dependence of the energy of the lowest electron state in the $GaAs/Al_{0.35}Ga_{0.65}As$ T-shaped structure on the width b of the (001) oriented QW (solid line) for fixed width of the overgrown QW of a = 5 nm. The energies of the lowest QW states are represented by the dotted lines. The 2D-1D energy separation multiplied by a factor of 5 is given by the dashed line.

the QWR is represented by the dashed line in Fig. 6. The QWR electron binding energy peaks sharply at the point $a = b$. This fact defines a criterion for an optimal choice of the parameters in the T-shaped structure: the equality of the energy values of the 2D states in the two QWs. This simple criterion holds also for more complex CEO geometries as discussed

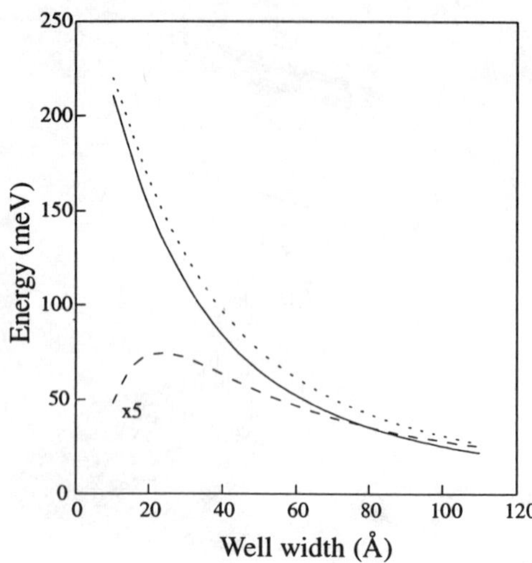

Figure 7. The dependence of the energy of the wire-like electron state in the T-shaped structure on the width of two equally wide QWs (solid line). The lowest QW states are represented by the dotted line. The dashed curve shows the 2D-1D energy separation magnified by a factor 5.

later. For the case $a = b$ the dependence of the QWR electron binding energy on the QW width is plotted in Fig. 7 (dashed line). In addition, the energetic positions of the QW and QWR states are indicated by the dotted and solid curves, respectively. The QWR binding energy decreases with decreasing well widths and reaches a maximum at $a = b \approx 2.5$ nm. This is a consequence of the finite conduction band offset which allows penetration of the electron wavefunction into the barrier material. For thinner QWs the electron is even more pushed out into the barrier regions and electron quantum confinement reduces. In order to predict transition energies of interband QWR transitions hole confinement and also excitonic effects have to be included. Although very sophisticated calculations for the valence band using a $\mathbf{k}\cdot\mathbf{p}$ model which includes six bands and, thus, mixing of the heavy-hole, light-hole, and split-off hole, have been carried out [32, 33], the small heavy-hole binding energies of less than 2 meV are very well reproduced by a simple one-band calculation with masses determined via the diagonal term in the Luttinger Hamiltonian with the angular momentum quantization axis parallel to the [110] overgrowth direction. As a result of the heavy-hole mass dependence on the crystallographic direction the eigenvalues for the two QWs are different and the holes are for equal well widths $a = b$ much more spread out into the overgrown QW as can be seen in Fig. 5(b). This modifies our criterion for maximum binding of carriers to the T-QWR and demands for a (110) QW width somewhat smaller than that of the (001) QW. Again, the maximum energetic difference between the 2D QW states and the 1D QWR state is obtained when the QW transition energies, including electron and hole confinement, are matched. The enlargement of exciton binding in such 1D structures can be estimated

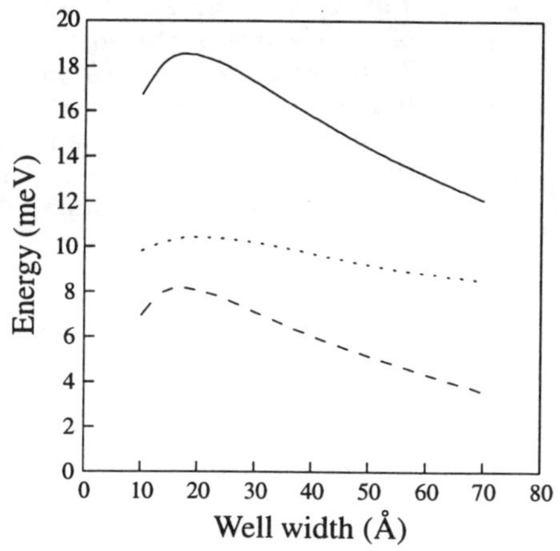

Figure 8. The binding energy of the 1D (solid line) and 2D excitons (dotted line) as a function of the well width $a = b$. Their difference is shown as the dashed curve.

applying a simple variational procedure [31]. Figure 8 shows the dependence of the exciton binding energy on the QW widths $a = b$ (solid line). For comparison the exciton binding energy in QWs obtained in the framework of the same rough model [34, 35] is also plotted in the figure (dotted line). As a consequence of the T-shaped potential the binding energy of the QWR exciton increases with decreasing well widths more rapidly than that of the 2D exciton. Their difference (dashed line) again has a maximum at $a = b \approx 2$ nm. The energy separation of the 2D and 1D exciton states is the sum of the confinement energy reduction of the free-electron-hole pair and the exciton binding energy enhancement. Combining the results depicted in Figs. 7 and 8 a maximum separation of about 33 meV is reached for the GaAs/$Al_{0.35}Ga_{0.65}As$ heteropair. This value slightly underestimates the maximal achievable QWR binding energy since the calculation is based on a structure with identical well widths which is only optimal if the effect of different hole confinement energies is neglected. For QW widths of $a = b = 7$ nm, a situation realized in the experiment described later, the 2D-1D exciton energy separation adds up to 14 meV.

Modifications of the conventional T-shaped structure can lead to an enhanced separation between the 2D and 1D single particle states. One possibility is the double T-shaped structure as indicated in the inset of Fig. 9 where two individual intersections are close enough to each other to permit coupling and repulsion of the 1D states. Due to the large extension of the ground state wave function into the wells at the T-junction (see Figs. 4 and 5) but negligible penetration into the barriers there is a range of barrier widths where coupling of 2D states is negligible and the interaction of the 1D states is large at the same time. These qualitative considerations have been verified by numerical calculations [32, 31]. The difference of the lowest

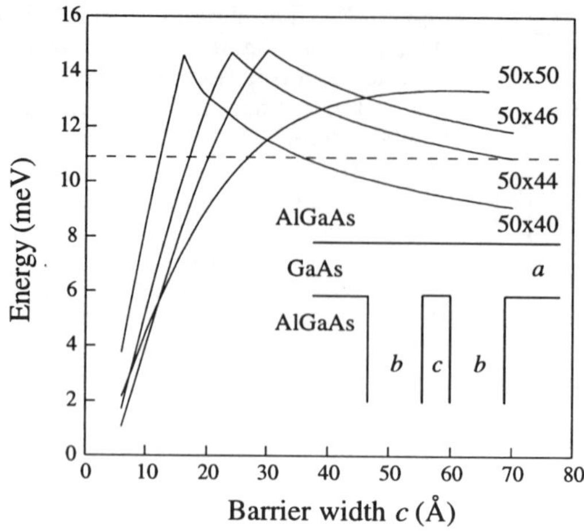

Figure 9. The 2D-1D separation of the electron states in the double T-shaped structure $a \times b$ (in Å) as a function of the barrier width c. The dashed line shows the same quantity for the single T-shaped intersection with $a = b = 5$ nm.

2D and 1D electron states as a function of the barrier width c in the double T-QWR structure is presented by the solid curves in Fig. 9. For comparison the corresponding energy of the single T-shaped intersection with $a = b = 5$ nm is also indicated in this figure (dashed line). When the QW widths are equal ($a = b = 5$ nm), the ground state in the double QW represents always the lowest 2D state in the system. For large barrier widths the T-QWRs decouple and their electron binding energy asymptotically goes to that of a single T-junction. For barrier widths in the range of 5–7 nm, however, the double QWR electron binding energy is indeed enhanced to more than 13 meV compared to about 11 meV for the single T-QWR. For even narrower barriers considerable lowering of the 2D state in the double QW with respect to the corresponding state in the overgrown QW occurs as a result of increased electron tunneling through the barrier. In this regime the double T-QWR structure represents an effective single T-junction of a wide QW with a narrow one. This explains the drastic reduction of electron binding below that in a single T-QWR for barrier widths $c < 2.5$ nm. The set of well widths $a = b$ considered so far, however, does not satisfy the criterion formulated above for maximum electron binding to the QWR. Consequently, by adjusting the width of the barrier separating two narrower QWs ($b < a$) so that the 2D state of the coupled QWs matches that of the overgrown QW, additional enhancement is seen.

Another possibility to enhance the separation between the 2D and 1D single particle states in T-QWRs closely related to this approach relies on overgrowth of a wide $Al_xGa_{1-x}As$ QW with a narrower GaAs QW. Since maximum electron binding to the T-QWR is again achieved when the transition energies in both QWs are equal, alignment of the 2D states determines the Al content in the (001) oriented QW. Using this concept of asymmetric T-shaped QWRs it has been shown theoretically that the QWR electron binding is doubled compared to symmetric T-QWRs with identical widths of the overgrown QW [33].

3.2. EXPERIMENTAL RESULTS

In 1990 Gershoni *et al.* [36] studied the optical properties of strain induced QWRs fabricated by CEO. The latter form due to modulation of the in-plane lattice constant of a GaAs/AlGaAs QW grown over a pseudomorphic InGaAs/GaAs strained-layer superlattice. The existence of confinement relaxation QWR states at the intersection of two QWs was experimentally demonstrated for the first time in 1992 [37]. Asymmetric T-QWRs fabricated in this work led to the identification of 1D bound states separated by more than 20 meV from the 2D QW states. The QWR transitions in these samples could be detected by optical emission and absorption spectroscopy up to room temperature. Temperature dependent measurements of the QWR PL intentsities yield activation energies for thermal dissociation of the 1D excitons of up to 30 meV. Although the observed PL response originating from the QWRs was relatively weak compared to the QW signal, the authors have already speculated about the possibility of stimulated emission in these structures. A different sample design which combines carrier confinement at the QW intersections with confinement of the optical mode by a T-shaped dielectric waveguide led to low-temperature T-QWR laser operation one year later by Wegscheider *et al.* [38]. The active region of the QWR laser sample consists of an array of symmetric T-QWRs with QW widths of 7 nm as described in detail in the next section. Polarization-resolved low-temperature (1.7K) PL spectra of such isolated QWRs, i.e. separated by a distance where coupling between the 1D states is negligible, recorded from an cleavage plane oriented perpendicular to the QWR axis are displayed in the upper panel of Fig. 10. At the low excitation power density of about 15 W/cm^2 used in the experiment exciton recombination in the QWRs and QWs is observed at about 1.563 and 1.58 eV as confirmed by spatially resolved optical spectroscopy by means of a scanning near-field optical microscope [39]. This indicates that the 1D state is about 17 meV deep with respect to the well, a value which compares well with the theoretical 2D-1D exciton energy separation of 14 meV for this structure and directly proofs the enhancement of exciton binding due to the T-shaped QWR confining potential. As presented earlier the total free-carrier QWR binding energy, composed of 9.9 and 0.7 meV for electrons and holes, respectively, cannot exceed 11 meV. The difference can, thus, only be due to an increase of the exciton binding energy by about 6 meV for the 1D structures. Assuming a 2D exciton binding energy in 7 nm wide QWs of

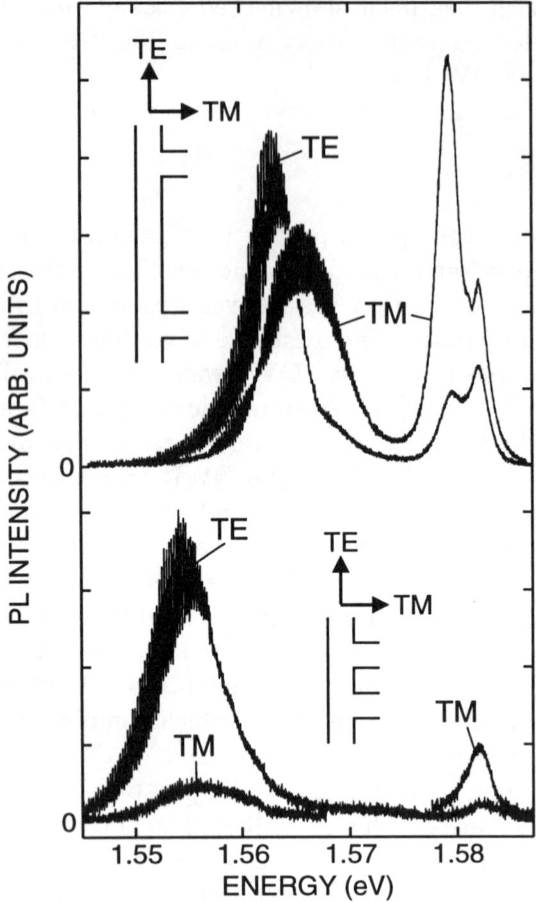

Figure 10. Polarization-resolved photoluminescence spectra of two T-QWR samples with QW widths of 7 nm measured at 1.7 K. The spectra displayed in the upper and lower panel were recorded from arrays of quantum wires with center-to-center distances of 38 and 14 nm, respectively. The direction of the **E** vector for the two polarization directions is given in the inset.

10-11 meV [40, 34] the absolute value of the 1D exciton binding energy in our structure adds up to about 17 meV, a value which exceeds that of the corresponding 2D structure by more than 50%. The doublet structure in the QW emission is believed to originate from slightly different confinement energies for the QWs grown along the [001] and [110] directions. This is consistent with the observed exchange of the relative peak intensities for the two polarization directions typical for a 2D system. In contrast, the TE- (**E** perpendicular to the [110] overgrowth direction) and TM-polarized (**E** parallel to the [110] overgrowth direction) luminescence are of nearly equal intensity, although of different shape and separated by about 3 meV. The TE-polarized QWR emission is characterized by a relatively narrow peak at 1.563 eV (FWHM $\approx$ 4 meV). Shifted by about 5 meV, we observe a shoulder on the high-energy side of this peak indicating an additional transition with oscillator strength in this polarization direction. The TM-polarized PL signal originating from the QWRs is much wider and exhibits no additional structure. The large width of this peak of about 8 meV strongly suggests the presence of several TM-active transitions in this spectral region, which we believe involve closely spaced hole states in the QWRs. The existence of excited states of the 1D exciton is unlikely in view of the specific symmetry of the T-QWR structure [41]. Qualitatively, we can interpret the reduced polarization anisotropy as a signature of carrier confinement to 1D since the emission from a QWR with circular cross-section would be isotropic, neglecting the anisotropy of the crystal structure. This argument can be verified by reducing the distance between T-QWRs. For a center-to-center distance of 14 nm, which corresponds to equal barrier and well widths, the 1D states at the T-junctions should be strongly coupled and indeed a QW-like polarization behaviour of the PL signal with a TE/TM intensity ratio of 7.7 is observed (lower panel of Fig. 10). In addition, the 2D-1D exciton energy separation increases to about 25 meV as predicted by theory.

The results of Wegscheider *et al.* described above on the enhancement of exciton binding and the polarization properties of T-shaped QWRs have been confirmed by Someya, Akiyama and Sakaki [42, 43, 44]. These authors prepared a variety of GaAs QW intersections with different well widths in the range of 5 nm. In this way the energetic position of the QWR state as a function of one QW width when the other one was kept fixed could be studied experimentally. This dependence is in qualitative agreement with the theoretical curve for the energetic positions of the electron QWR and QW states depicted in Fig. 6. Enhancement of the 1D excitons was either deduced from measurements of the diamagnetic shift of the QWR PL signal in a strong magnetic field or by comparing the observed QWR-QW PL energy separation with the calculated free-carrier QWR binding energy, a procedure identical to that outlined above. For T-QWRs embedded in pure AlAs barrier material a maximum 2D-1D exciton energy separation of 36 meV and an exciton binding energy of about 27 meV is found.

Recent experimental realizations of asymmetric T-intersections of wide AlGaAs QWs with narrower GaAs QWs by Gislason *et al.* [45] seem to confirm the theoretical predictions of an enhanced separation between the 2D and 1D single particle states compared to symmetric T-intersections of pure GaAs QWs [33]. The binding energy of excitons to such QWRs reaches a peak value of 54 meV for the intersection of a 12 nm wide $Al_{0.14}Ga_{0.86}As$ QW with a 2.5 nm wide GaAs QW embedded in $Al_{0.3}Ga_{0.7}As$. This large value significantly exceeds kT at room temperature and therefore shows the feasibility of light emitting devices based on confinement of electrons and holes to 1D in T-QWRs.

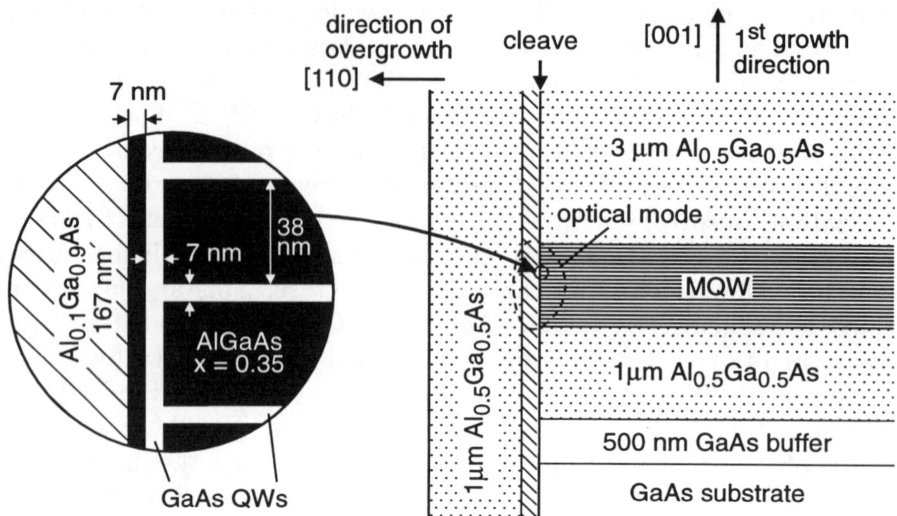

Figure 11. Schematic cross section of the quantum wire laser structure. The dashed line represents a contour plot of the optical mode at 10% of the maximum intensity.

4. T-QWR laser

The design of the first T-QWR laser [38] outlined below is based on the symmetric intersection of GaAs QWs with a width of 7 nm. This leads to a free carrier binding energy to the QWRs of almost 11 meV and a total binding energy for excitons to the 1D structures of 17 meV. The operation of such a device is accordingly limited to cryogenic temperatures. This is, however, not a principle limitation since the concepts and properties of T-QWR lasers described in this section should be also valid for modified T-structures with considerably larger binding energies.

4.1. INDEX GUIDED SEPARATE CONFINEMENT STRUCTURE

A schematic cross-section of the QWR laser structure is shown in Fig. 11. The first MBE growth formes the layer structure to the right of the arrow marked "cleave". It consists of a 1 μm $Al_{0.5}Ga_{0.5}As$ cladding layer followed by a 22-period $GaAs/Al_{0.35}Ga_{0.65}As$ multiple quantum well (MQW) structure with well and barrier thicknesses of 7 and 38 nm, respectively, as illustrated in the magnified area, followed by a 3 μm $Al_{0.5}Ga_{0.5}As$ cladding layer. After growth of the layers to the left of the arrow marked "cleave" 22 QWRs form at the T-intersections of the 7 nm wide QWs. The number of QWRs has been rather arbitrarily chosen so that maximum overlap of the optical mode with the active material is obtained while complete decoupling of the QWR states is maintained. The purpose of the $Al_{0.5}Ga_{0.5}As$ cladding layers is to guide the optical mode in a similar manner as realized in conventional separate confinement heterostructure laser designs. In addition, the high refractive index $Al_{0.1}Ga_{0.9}As$ layer, which is separated by a thin $Al_{0.35}Ga_{0.65}As$ barrier from the (110) oriented QW leads to confinement of the optical mode to the vicinity of the QWR array (see the dashed line in Fig. 11) as confirmed by waveguide calculations within the effective refractive index approximation as well as by a two-dimensional solution of the Maxwell equations for this structure [46]. In this way a completely index guided structure with an effective refractive index step of $\Delta n_{eff} = 0.029$ from the core of the T-shaped waveguide to the surrounding three-layer slab waveguides is obtained. The high degree of structural perfection attainable by the CEO method was demonstrated by transmission electron microscopy [38].

4.2. LASING CHARACTERISTIC OF OPTICALLY PUMPED DEVICES

In order to achieve lasing in the 1D structures, mirrors were cleaved perpendicular to the axis of the QWRs. The cleave mirrors were left uncoated so that each mirror had a reflectivity R of only about 0.3. Optical excitation with the samples immersed in superfluid He (1.7 K) from either the (001) or the (110) surface was performed by focusing the output of a dye laser tuned to $\lambda = 775$ nm to a stripe of about 700 μm in length and 5 μm in width oriented parallel to the QWRs. At this wavelength significant light absorption occurs only in the GaAs QW and QWR layers. However, because the QWR volume is so small, light absorption takes place mainly in the QWs.

Figure 12 compares emission spectra of a 600 μm long QWR laser below and above threshold for stimulated emission. Although the optically active volume of the QWRs is small compared to that of the QWs, the spectrum recorded at the lowest excitation power (0.25 mW) is dominated by the QWR signal. With increasing pump power the contrast in the Fabry-Pérot (FP) oscillations, which develop on the low energy side of the QWR peak, increases and at about 10 mW stimulated emission occurs. The peaks in the FP oscillations correspond to standing waves within the optical cavity formed by the two cleave mirrors. The development of these oscillations even at the lowest excitation powers where the QW signal is completely featureless reflects the high degree of transparency of the optical resonator within this wavelength region. This is due to the small fraction Γ of the optical intensity distribution, of about $3{\cdot}10^{-3}$, that overlaps with the QWRs. For comparison, the corresponding value for the QWs in our structure is $\Gamma \approx 0.15$. Further increase of the pump power leads to a significant narrowing of the QWR emission spectrum until at pump levels above about

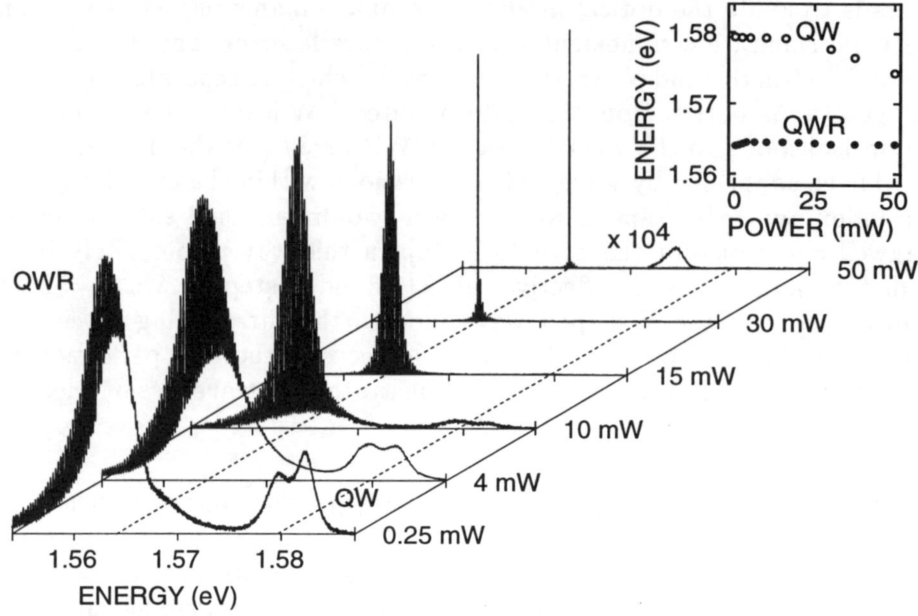

Figure 12. Photoluminescence (PL) spectra recorded below, at and above threshold for stimulated emission in the QWRs ($P_{th} \approx 10$ mW). Inset: Power dependency of the QW and QWR PL energies.

30 mW laser operation predominantly in a single longitudinal mode takes place. The laser output as a function of pump power is shown in Fig. 13 together with high-resolution spectra of the QWR emission. In contrast to the QW peak intensity which increases linearly or sublinearily with excitation power, superlinear behavior is observed.

A remarkable feature of the QWR laser is the striking insensitivity of the QWR emission wavelength on pump power over a range of almost three orders of magnitude (see inset of Fig. 12). In contrast, the QW luminescence line shows the typically observed redshift with increasing pump power of as much as 5 meV. This shift known as band-gap shrinkage in semiconductor lasers is consistent with exciton ionization and photoexcitation of a free electron-hole plasma which is subject to band-gap renormalization effects. It thus appears that in wires at intersecting QWs the 1D exciton gas phase is more stable against formation of an electron-hole plasma. This argument receives further support by the spectral constancy of the emission from QWR lasers which differ in their cavity lengths.

In order to further understand this behavior we estimate the carrier densities present in the QWRs and QWs. The highest excitation power of 50 mW used in the experiment corresponds to a power density of ≈ 3

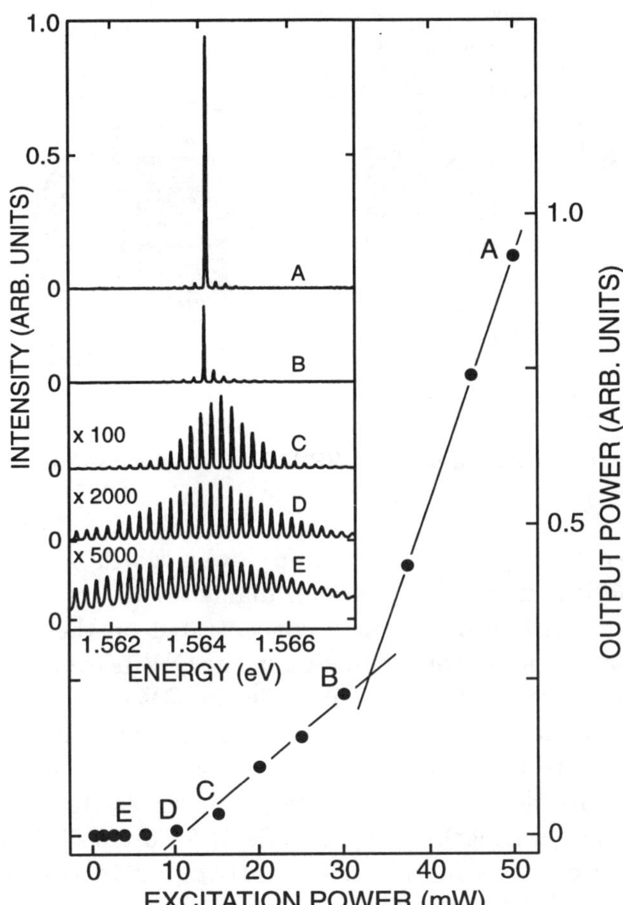

Figure 13. QWR laser output vs excitation power. High-resolution emission spectra recorded at different excitation levels are shown in the inset.

kW/cm^2. Assuming an absorption length of 0.5 μm in GaAs and a surface reflectivity R of 0.3, this corresponds to a rate of $\approx 10^{20}$ e-h pairs/cm^{2}s generated in the individual QWs. With a recombination time of 1 ns in the QWs [47] we expect a sheet carrier density of $\approx 10^{11}$ e-h pairs/cm^2. The narrowing of the free-particle band-gap due to band-gap renormalization for this carrier density is about 15 meV [4]. This value exceeds the QW exciton binding energy by about 5 meV, consistent with the notion that QW luminescence is due to exciton recombination with excitation less than 25 mW, and band-to-band above. Despite the much smaller volume of the QWRs their luminescence intensity equals or exceeds that of the QWs at all powers. This can be attributed to two effects: (i) Carriers, most of which are generated in the QWs drain into the QWR region. Assuming comparable lifetimes in the QWs and QWRs, carriers diffusing from 0.5–1 μm into the QWRs, as indicated by the SNOM measurements [39], would

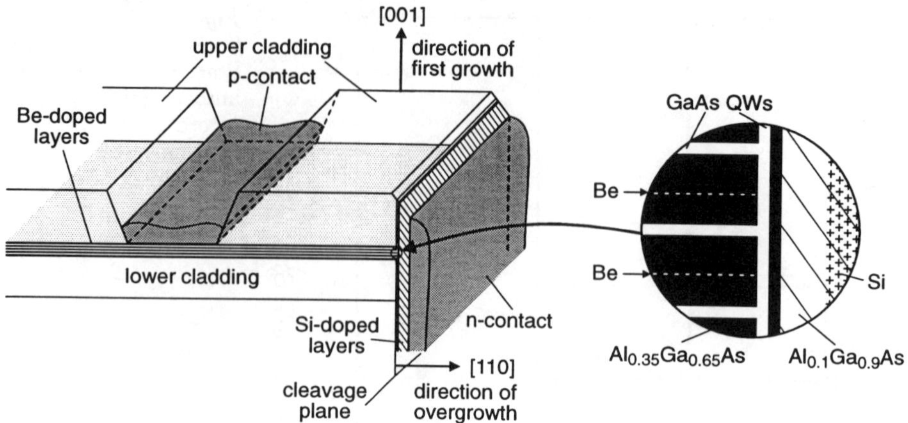

Figure 14. Schematic representation of the QWR laser diode (not to scale).

lead to $\approx 10^6$/cm carrier density at 10 mW pump power, which turns out to be approximately the inverse of a 1D exciton diameter. (ii) In 1D the oscillator strength is predicted to be strongly concentrated on the lowest exciton state [7]. Observation of excitonic emission in the QWRs even at the highest excitation levels, at which the 2D excitons in the QWs are ionized, clearly demonstrates the enhanced stability of the 1D exciton gas phase. This is in marked contrast to GaAs heterostructure lasers, which operate in the regime of a degenerate electron-hole plasma. However, excitonic gain has been identified as the gain mechanism in ZnSe QW lasers [48]. The enhanced stability of the excitons in this case is attributed to the large exciton binding energies in II/VI materials.

4.3. DIODE LASERS

A schematic view of the QWR laser structure suitable for current injection is shown in Fig. 14. In this case the MQW region (15 periods) with well and barrier thicknesses of 7 and 58 nm, respectively, is δ-doped using Be. Except for these Be doping spikes (2×10^{11} Be cm^{-2}) in each $Al_{0.35}Ga_{0.65}As$ barrier located 34 nm below or 24 nm above each QW (see magnified portion of Fig. 14), the whole layer sequence as well as the substrate is again undoped. Note we have slightly off-centered the δ-dopant layers in order to compensate for expected Be surface segregation along the growth direction [49]. The post-cleave growth sequence consists of a 7 nm GaAs QW (undoped) followed by a 7 nm $Al_{0.35}Ga_{0.65}As$ barrier (undoped), a 43 nm undoped $Al_{0.1}Ga_{0.9}As$ setback, a 124 nm doped $Al_{0.1}Ga_{0.9}As$ layer (2×10^{18} Si cm^{-3}), a 1 μm wide $Al_{0.5}Ga_{0.5}As$ cladding layer (2×10^{18} Si cm^{-3}) and

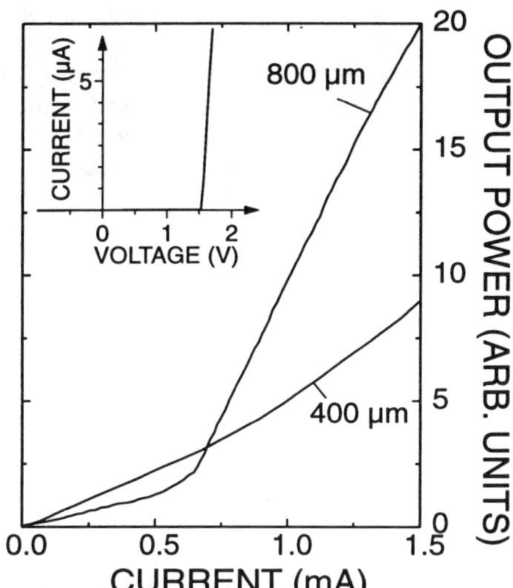

Figure 15. Light vs current characteristic of two QWR lasers with cavity lengths of 400 and 800 μm. Inset: Current vs voltage characteristic of the 400 μm long device.

a 10 nm GaAs cap(2×10^{18} Si cm^{-3}). This doping scheme leads to the formation of a linear *p-n* junction, i.e. a junction where holes and electrons meet along a line instead of a plane, in the vicinity of the QWRs.

Figure 15 compares the light versus current (L/I) characteristic of two QWR lasers with cavity lengths of 400 and 800 μm and uncoated mirrors measured at 4.2 K. The 800 μm long device shows clear superlinear behavior for currents in excess of about 0.5 mA, indicating a threshold current of less than 0.6 mA. The L/I curve of the shorter laser (400 μm) deviates much more weakly from a strictly linear relationship. This weak superlinear behavior suggests gain saturation in the QWRs whereby the higher gain required to overcome the internal and mirror losses of the shorter cavity is just exceeded. Extrapolation of the superlinear branch of the L/I curve to zero output power leads to a threshold current of about 0.4 mA for the 400 μm long device. The current versus voltage characteristic of the 400 μm long QWR laser sample is depicted in the inset of Fig. 15. It exhibits a typical diode curve with a turn on voltage of about 1.55 V, a leakage current below this voltage of less than 100 nA and a differential resistance in forward bias of a few hundred ohms. The breakdown voltage under reverse bias condition is larger than 10 V.

The spontaneous emission spectra of the 400 μm long QWR laser sample operated at a current of 0.2 mA is shown in Fig. 16. The PL response of the QWR laser structure before overgrowth, i.e. of the MQW layer only, has been also included in this figure as a dashed line for comparison. The QWR laser diode emission is characterized by a single peak which exhibits a shoulder on the high energy side. By comparison with the MQW reference

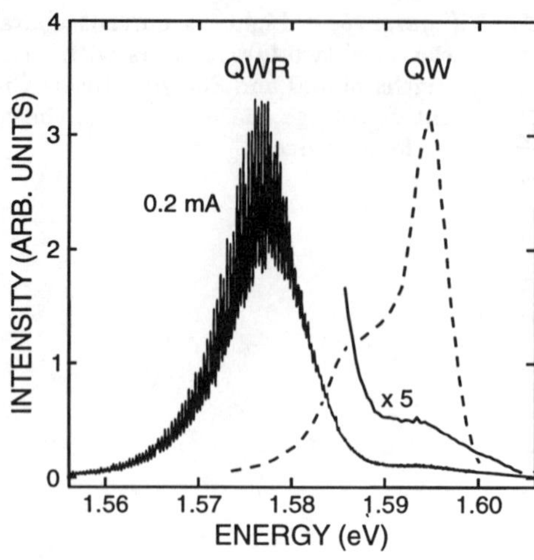

Figure 16. Spontaneous emission spectrum of the 400 μm long QWR laser structure recorded $\approx$ 50% below threshold (solid line) and photoluminescence response of the MQW structure before overgrowth (dashed line).

spectrum the latter can be unambiguously identified as the luminescence originating from the MQWs formed during the first growth step. The luminescence energy of the single QW formed during overgrowth is expected to be close to that of the MQWs due to nominally identical well widths. The strong peak in the QWR laser emission centered at around 1.577 eV can, thus, only be attributed to radiative recombination from 1D states in the QWRs. The fact that the emission from our laser at injection levels considerably below threshold is almost completely dominated by optical transitions in the QWRs emphasizes the highly efficient way charge carriers are injected using this approach. Similarly to the results on optical excited QWR laser structures, closely spaced FP oscillations corresponding to different longitudinal modes within the optical cavity are superimposed onto the QWR response. At an injection current of 0.4 mA, the contrast ratio in these oscillations already exceeds the theoretically expected ratio for an empty cavity, $(1 + R)^2/(1 - R)^2 = 3.45$, using a mirror reflectivity of $R = 0.3$. Thus the QWR laser already develops net gain at this injection current. Higher injection levels of 0.8 and 1.5 mA result in progressive narrowing of the laser emission envelopes as well as of the individual FP peaks. The corresponding threshold current to meet the FP contrast ratio criterion given above is 0.55 mA for the 800 μm long QWR laser cavity.

4.4. EFFECT OF A STRONG MAGNETIC FIELD ON THE LASING CHARACTERISTIC

The first attempts to study the operation of QWR and quantum box lasers were based on Landau quantization of carriers [1, 50]. As a result of Lorentz force confinement electrons and holes in heterostructure lasers can move freely only in the direction of the magnetic field and quasi–QWR structures

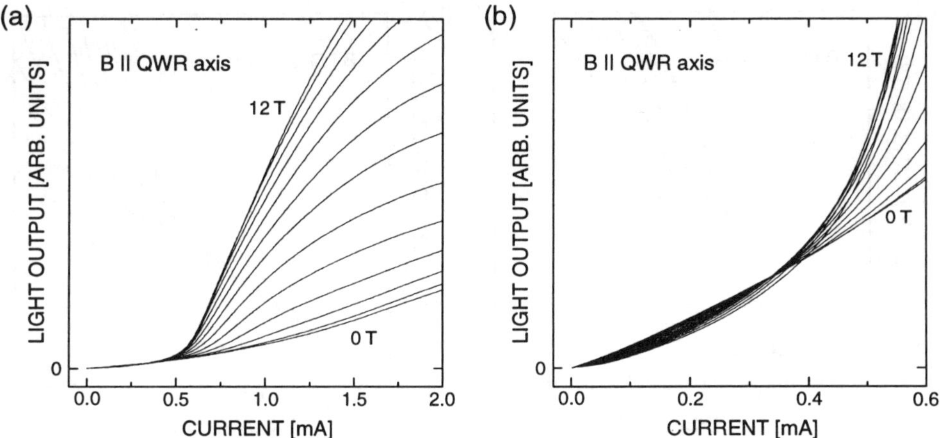

Figure 17. Light vs current characteristic of a 400 μm long QWR laser in a magnetic field oriented parallel to the QWR axis. The magnetic field was increased from $B = 0$ to 12 T in increments of 1T.

are obtained. Accordingly, application of a magnetic field in the growth direction of a QW laser leads to full quantization of the charge carriers and the properties of 0D lasers can be simulated. In agreement with theory these experiments indeed demonstrate a reduced temperature sensitivity of the threshold current [1, 51] as well as an enhanced modulation bandwidth [52] and a reduction of the spectral linewidth [53, 54]. However, the emitted intensity and the threshold current of the lasers was almost unaffected by the application of magnetic fields up to $B = 30$ T.

Here we study "true" 1D lasers incorporating "rigid" QWR potential wells under strong magnetic fields oriented parallel and perpendicular to the QWR axis. Figure 17(a) shows the evolution of the light output from the 400 μm long T-QWR diode laser presented previously with increasing magnetic field for a magnetic field direction parallel to the axis of the QWRs. For $B = 0$ T the light versus current (L/I) characteristic of this device deviates only weakly from a strictly linear relationship. With increasing magnetic field the emitted light intensity above this threshold current (I_{th}) shows a pronounced increase of about a factor of 8 for the highest magnetic field of $B = 12$ T at 1 mA injection current and clear superlinear behavior is observed. At the same time the electrical diode characteristic of the device remains unaffected by the application of the magnetic field. A more detailed measurement of the L/I characteristics in the vicinity of the threshold current of about 0.4 mA is shown in Fig. 17(b). Again, an increase in the emitted intensity above threshold can be seen. However, all L/I curves cross at about 0.4 mA, i.e. light emission is reduced for currents $I < I_{th}$, and is enhanced for currents $I > I_{th}$. In order to understand this

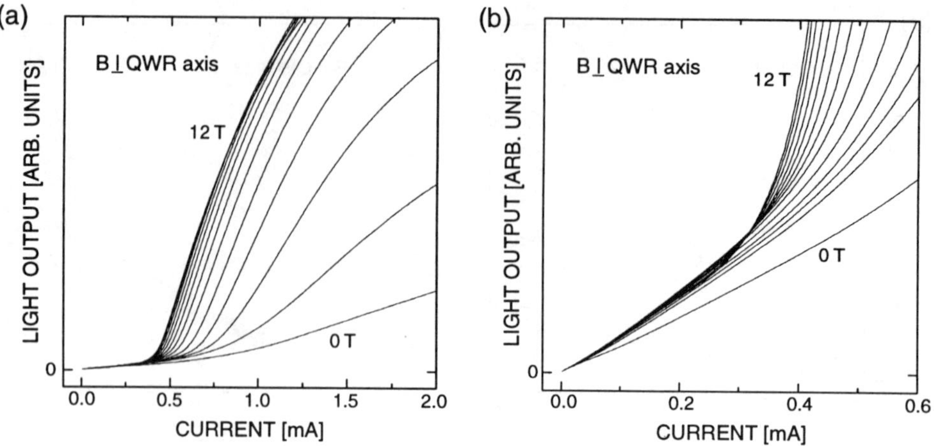

Figure 18. Light vs current characteristic of the same laser studied in the parallel magnetic field configuration (see Fig. 17) using a magnetic field direction along the [110] overgrowth direction, i.e. perpendicular to the QWR axis. The magnetic field was increased from $B = 0$ to 12 T in increments of 1T.

behavior we have simulated the L/I characteristic using a standard rate equation model, modified for a laser which operates from localized excitons, which are bound at minima in the QWR potential [55]. The results of the simulation very well reproduce the measured L/I curves shown in Fig. 17. This strongly suggests that the effect of a strong magnetic field applied parallel to the QWR axis is to modulate the gain. This is equivalent to an increase of the oscillator strength for the transitions between the 1D states. However, since confinement of excitons to 1D is already present at $B = 0$ as a result of the T-shaped potential in our samples, substantial effects due to magnetic confinement can be only expected if the cyclotron radius becomes comparable to the exciton Bohr radius of about 10 nm. The observed increase in the laser emission is maximal for a field strength of about 6 T, a value which indeed corresponds to a magnetic length of $(\hbar/eB)^{1/2} = 10$ nm. For higher fields the effect saturates as predicted by theory. This saturation of total gain is a peculiarity which arises from the excitonic nature of the emission in our laser. Without the blocking effect which is due to the fact that two excitons cannot be localized on the same site the light output is only weakly influenced by changes in gain. In particular, in an "ideal" laser, above threshold the differential slope of the L/I curve is independent of the gain since all the current is converted to light.

For a magnetic field orientation along the direction of overgrowth ([110]), i.e. perpendicular to the QWR axis, a completely different behavior is observed. The corresponding L/I curves recorded on the same sample are shown in Fig. 18. In this case the emitted intensity is already doubled for

$B = 1$ T, a field strength which has only little effect on the laser characteristic in the parallel magnetic field configuration (see Fig. 17). Magnetic fields exceeding 4 T, however, are not accompanied by an appreciable increase in the light output for this field direction. In addition, the behavior of the QWR laser below threshold is different as can be seen in more detail in Fig. 18(b). Up to $B = 6.5$ T the light output increases. At higher fields, crossing of the L/I curves is observed. Although the emission intensity starts to decrease for such strong magnetic fields, its absolute value always exceeds the $B = 0$ T value. Similar to the parallel magnetic field configuration the electrical diode characteristic of the device remains unaffected by the application of the magnetic field. Following our argumentation to interpret the results obtained with the magnetic field applied in the direction of the QWR axis, it is very unlikely that the pronounced increase in the emission intensity at low magnetic fields is due to magnetic confinement. In addition, according to our model the increase in light output observed below threshold contradicts an increase in gain. An alternative approach to describe these results must rely on a mechanism which leads to a more efficient pumping of localized 1D exciton states, which then can radiatively recombine and emit more light at all injection levels. One possibility would be the interaction of the electron and hole spins with the magnetic field. If we assume that the electrons and holes generated by current injection are initially not spin polarized, the proportion of singlet (1S) to triplet excitons ($2P^0$, $2P^-$, $2P^+$) in the QWRs should be roughly 1:4. Since only the singlet excitons are optically active, changes in the relative populations induced by the Zeeman effect would also influence the light output. This would lead to a rescaling of the L/I curves, because for any injection current independent of the laser threshold the emission intensity would either increase or decrease. Such a rescaling behavior, which is characterized by the absence of L/I curve crossings near threshold is observed in the low magnetic field data shown in Fig. 18(b). One problem with this interpretation is, that the splitting in energy between the singlet and triplet states is usually less than $k_B T$, so in thermal equilibrium the changes in population would be small. If the rate of equilibration of the populations was slow compared to the radiative lifetime, a population difference, large enough to explain the increase in the emission intensity, might result. The spin–flip times which are only known for 2D systems are factors of 2 to 4 shorter than the radiative lifetime [56, 57]. In the absence of a calculation of the electron Landé g factors for the two magnetic field orientations in the QWR structure and a reliable value for the spin–flip time it is, however, difficult to judge the validity of our interpretation. Interestingly, this effect should not be observable in conventional 2D lasers which operate in the regime of a degenerate electron–hole plasma, where the spin–flip rates are so fast that spin effects are unimportant. Another possibility to explain the increase in emission with applied magnetic field would be the suppression of carrier diffusion along the wire axis. While for a magnetic field direction parallel to the QWR axis the cyclotron orbits are located in a plane perpendicular to this axis, cyclotron motion for a magnetic field orientation along the direction of overgrowth inhibits free carrier transport along the QWR axis. Since the operation of our T-QWR laser critically depends on the presence of localized excitons this effect which supports the formation of such excitons should also influence the laser characteristic of the device. However, in order to elucidate the nature of the unusual behavior of the T-QWR lasers under strong magnetic fields, further investigations both experimentally and theoretically have to be performed.

5. Conclusions and prospects

Cleaved edge overgrowth has proven to be a powerful technique for the fabrication of quantum wires with applications in optics and, not subject of this article, transport. The T-geometry quantum wires represent a model system for other 1D structures with typical dimensions comparable to the exciton Bohr radius. For these structures excitonic effects become increasingly important and strongly influence the optical properties. As a result stimulated emission in T-QWR lasers is not due to recombination of electrons and holes in a highly dense plasma, but due to transitions between 1D exciton states. As an extension of this work, we propose the use of InGaAs for the QW material. In addition to the larger band offsets achievable in this way, elastic strain relaxation at the T-intersection of such QWs is expected to increase the 1D confinement energies. The growth of strained layers would also enable us to prepare self assembled InAs quantum dots during overgrowth. Since the cleavage face of a pseudomorphically strained multilayer system represents a substrate with a laterally modulated lattice constant these 0D structures can minimize their strain energy by preferential nucleation in areas which provide a larger lattice constant.

References

1. Arakawa, Y. and Sakaki, H. (1982) Multidimensional Quantum Well Lasers and Temperature Dependence of its Threshold Current, *Appl. Phys. Lett.*, **40**, 939–941.
2. Arakawa, Y. and Yariv, A. (1986) Quantum Well Lasers—Gain, Spectra, Dynamics, *IEEE J. Quantum Electron.*, **QE-22**, 1887–1899.
3. Asada, M., Miyamato, Y. and Suematsu, Y. (1986) Gain and the Threshold of Three-Dimensional Quantum-Box Lasers, *IEEE J. Quantum Electron.*, **QE-22**, 1915–1921.
4. Schmitt-Rink, S., Chemla, D. S. and Miller, D. A. B. (1989) Linear and nonlinear properties of semiconductor quantum wells, *Advan. in Phys.*, **38**, 89–188.
5. Degani, M. H., Hipólito, O. (1987) Exciton binding energy in quantum-well wires,

Phys. Rev. B, **35**, 9345–9348.
6. Bányai, L., Galbraith, I., Ell, C. and Haug, H. (1987) Excitons and biexcitons in semiconductor quantum wires, *Phys. Rev. B*, **36**, 6099–6104.
7. Ogawa, T. and Takahara, T. (1991) Optical absorption and Sommerfeld factors of one-dimensional semiconductors: An exact treatment of excitonic effects, *Phys. Rev. B*, **44**, 8138–8156.
8. Ivanov, A. L. and Haug, H. (1993) Existence of Exciton Crystals in Quantum Wires, *Phys. Rev. Lett.*, **71**, 3182–3185.
9. Weisbuch, C. and Vinter, B. (1991) Quantum Semiconductor Structures, Academic Press, San Diego, CA, pp. 191–193.
10. Kapon, E., Hwang, D. M. and Bhat, R. (1989) Stimulated Emission in Semiconductor Quantum Wire Heterostructures, *Phys. Rev. Lett.*, **63**, 430–433.
11. Tsuchiya, M., Gaines, J. M., Yan, R. H., Simes, R. J., Holtz, P. O., Coldren, L. A. and Petroff, P. M. (1989) Optical Anisotropy in a Quantum-Well-Wire Array with Two-Dimensional Quantum Confinement, *Phys. Rev. Lett.*, **62**, 466–469.
12. Nötzel, R., Ledentsov, N. N., Däweritz, L., Hohenstein, M. and Ploog, K. (1991) Direct Synthesis of Corrugated Superlattices on Non-(100)-Oriented Surfaces, *Phys. Rev. Lett.*, **67**, 3812–3815.
13. Tsukamoto, S., Nagamune, Y., Nishioka, M. and Arakawa, Y. (1993) Fabrication of GaAs quantum wires ($\approx$ 10 nm) by metalorganic chemical vapor selective deposition growth, *Appl. Phys. Lett.*, **63**, 355–357.
14. Brunner, K., Bockelmann, U., Abstreiter, G., Walther, M., Böhm, G., Tränkle, G. and Weimann, G. (1992) Photoluminescence from a Single GaAs/AlGaAs Quantum Dot, *Phys. Rev. Lett.*, **69**, 3216–3219.
15. Prins, F. E., Lehr, G., Burkard, M., Schweizer, H., Pillkuhn, M., H. and Smith, G. W. (1993) Photoluminescence excitation spectroscopy on intermixed GaAs/AlGaAs quantum wires, *Appl. Phys. Lett.*, **62**, 1365–1367.
16. Leonard, D., Krishnamurthy, M., Reaves, C. M., Denbaars, S. P. and Petroff P. M. (1993) Direct formation of quantum-sized dots from uniform coherent islands of InGaAs on GaAs surfaces, *Appl. Phys. Lett.*, **63**, 3203–3205.
17. Kirstaedter, N. et al. (1994) Low Threshold, large T0 Injection Laser Emission from (InGa)As Quantum Dots, *Electron. Lett.*, **30**, 1416–1417.
18. Störmer, H. L., Gossard, A. C. and Wiegmann, W., unpublished.
19. Pfeiffer, L. West, K. W., Stormer, H. L., Eisenstein, J. P., Baldwin, K. W., Gershoni, D. and Spector, J. (1990) Formation of a high quality two-dimensional electron gas on cleaved GaAs, *Appl. Phys. Lett.*, **56**, 1697–1699.
20. Nötzel, R., Däweritz, L., Ledentsov, N. N. and Ploog, K. (1992) Size quantization by faceting in (110)-oriented GaAs/AlAs heterostructures, *Appl. Phys. Lett.*, **60**, 1615–1617.
21. Schubert, E. F., Pfeiffer, L., West, K. W., Luftman, H. S. and Zydzik, G. J. (1994) Si δ-doping of <011>-oriented GaAs and AlGaAs grown by molecular-beam epitaxy, *Appl. Phys. Lett.*, **64**, 2238–2240.
22. Pfeiffer, L., West, K. W., Stormer, H. L. and Baldwin, K. W. (1989) Electron Mobilities exceeding 10^7 cm^2/Vs in modulation-doped GaAs, *Appl. Phys. Lett.*, **55**, 1888–1890.
23. Kean, A. H., Holland, M. C. and Stanley, C. R. (1993) Growth of (Al,Ga)As structures on (110)-GaAs by MBE, *J. Cryst. Growth*, **127**, 904–907.
24. Schedelbeck, G., Wegscheider, W., Bichler, M. and Neumann R., unpublished.
25. Pfeiffer, L., Störmer, H. L., Baldwin, K. W., West, K. W., Goñi, A. R., Pinczuk, A., Ashoori, R. C., Dignam, M. M. and Wegscheider, W. (1993) Cleaved Edge Overgrowth for quantum wire fabrication, *J. Cryst. Growth*, **127**, 849–857.
26. Sakaki, H. (1990) Quantum Boxes, Quantum Wires, and In-Plane Superlattices: Their Impact in Device Physics and Required Breakthroughs in Materials Science, in *Springer Series in Solid-State Sciences: Localization and Confinement of Electrons in Semiconductors*, **Vol. 97**, edited by Kuchar, F., Heinrich, H. and Bauer, G.,

Springer, Berlin, pp. 4–5.
27. Yacobi, A., Stormer, H. L., Baldwin, K. W., Pfeiffer, L. N. and West, K. W. (1996) Magneto-transport spectroscopy on a quantum wire, *Solid State Commun.*, **101**, 77–81.
28. Störmer, H. L., unpublished.
29. Chang, Y.-C., Chang, L. L. and Esaki, L. (1985) A new one-dimensional quantum well structure, *Appl. Phys. Lett.*, **47**, 1324–1326.
30. Bastard, G. and Marzin, J. Y. (1994) Hartree-like calculations of energy levels in quantum wires *Solid State Commun.*, **91**, 39–43.
31. Kiselev, A. A. and Rössler, U. (1996) Towards optimization of T-shaped quantum structures, *Semicond. Sci. Technol.*, **11**, 203–206.
32. Pfeiffer, L., Baranger, H., Gershoni, D., Smith, K. and Wegscheider, W. (1995) Binding of electrons and holes at quantum wires formed by T-intersecting quantum wells, in *Low Dimensional Structures Prepared by Epitaxial Growth or Regrowth on Patterned Substrates*, edited by Eberl, K., Petroff, P. M. and Demeester, P., Kluwer Academic, Dordrecht, pp. 93–100.
33. Langbein, W., Gislason, H. and Hvam, J. M. (1996) Optimization of the confinement energy of quantum wire states in T-shaped GaAs/AlGaAs structures, *Phys. Rev. B*, **54**, 14595–14603.
34. Andreani, L. C. and Pasquarello, A. (1990) Accurate theory of excitons in GaAs-$Ga_{1-x}Al_x$ quantum wells, *Phys. Rev. B*, **42**, 8928–8938.
35. Mathieu, H., Lefebre, P. and Christol, P. (1992) Simple analytical method for calculating exciton binding energies in semiconductor quantum wells, *Phys. Rev. B*, **46**, 4092–4101.
36. Gershoni, D., Weiner, J. S., Chu, S. N. G., Baraff, G. A., Vandenberg, J. M., Pfeiffer, L. N., West, K., Logan, R. A. and Tanbun-Ek, T. (1990) Optical Transitions in Quantum Wires with Strain-Induced Lateral Confinement *Phys. Rev. Lett.*, **65**, 1631–1634.
37. Goñi, A. R., Pfeiffer, L. N., West, K. W., Pinczuk, A., Baranger, H. U. and Stormer, H. L. (1992) Observation of quantum wire formation at intersecting quantum wells, *Appl. Phys. Lett.*, **61**, 1956–1958.
38. Wegscheider, W., Pfeiffer, L. N., Dignam, M. M., Pinczuk, A., West, K. W., McCall, S. L. and Hull, R. (1993) Lasing from Excitons in Quantum Wires, *Phys. Rev. Lett.*, **71**, 4071–4074.
39. Grober, R. D., Harris, T. D., Trautman, J. K., Betzig, E., Wegscheider, W., Pfeiffer, L. and West, W. (1994) Optical spectroscopy of a GaAs/AlGaAs quantum wire structure using near-field scanning optical microscopy, *Appl. Phys. Lett.*, **64**, 1421–1423.
40. Greene, R. L., Bajaj, K. K. and Phelps, D. E. (1984) Energy levels of Wannier excitons in GaAs-$Ga_{1-x}Al_xAs$ quantum well structures, *Phys. Rev. B*, **29**, 1807–1812.
41. Wegscheider, W., Pfeiffer, L., Dignam, M., Pinczuk, A., West, K. and Hull, R. (1994) Lasing in lower-dimensional structures formed by cleaved edge overgrowth, *Semicond. Sci. Technol.*, **9**, 1933–1938.
42. Someya, T, Akiyama, H. and Sakaki, H. (1995) Laterally Squeezed Excitonic Wave Function in Quantum Wires, *Phys. Rev. Lett.*, **74**, 3664–4667.
43. Akiyama, H., Someya, T. and Sakaki, H. (1996) Optical anisotropy in 5-nm-scale T-shaped quantum wires fabricated by the cleaved-edge overgrowth method, *Phys. Rev. B*, **53**, 4229–4232.
44. Someya, T, Akiyama, H. and Sakaki, H. (1996) Enhanced Binding Energy of One-Dimensional Excitons in Quantum Wires, *Phys. Rev. Lett.*, **76**, 2965–2968.
45. Gislason, H., Sørensen, C. and Hvam, J. M. (1996) Asymmetric GaAs/AlGaAs T-wires with large confinement energies *Appl. Phys. Lett.*, **69**, 3248–3250.
46. Wang, W.-K. unpublished.
47. Feldmann, J., Peter, G., Göbel, E. O., Dawson, P., Moore, K., Foxon, C. and Elliot,

R. J. (1987) Linewidth Dependence of Radiative Exciton Lifetimes in Quantum Wells, *Phys. Rev. Lett.*, **59**, 2337–2340.
48. Ding, J., Jeon, H., Ishihara, T., Hagerott, M. and Nurmikko, A. V. (1992) Excitonic gain and laser emission in ZnSe-based quantum wells, *Phys. Rev. Lett.*, **69**, 1707–1710.
49. Schubert, E. F., Kuo, J. M., Kopf, R. F., Jordan, A. S., Luftman, H. S. and Hopkins, L. C. (1990) Fermi-level-pinning-induced impurity redistribution in semiconductors during epitaxial growth, *Phys. Rev. B*, **42**, 1364–1368.
50. Arakawa, Y., Sakaki, H., Nishioka, M., Okamoto, H. and Miura, N. (1983) Spontaneous Emission Characteristics of Quantum Well Lasers in Strong Magnetic Fields—An Approach to Quantum-Well-Box Light Source—, *Jap. J. Appl. Phys.*, **22**, L804–L806.
51. Berendschot, T. T. J. M., Reinen, H. A. J. M., Bluyssen, H. J. A., Harder, C. and Meier, H. P. (1989) Wavelength and threshold current of a quantum well laser in a strong magnetic field, *Appl. Phys. Lett.*, **54**, 1827–1829.
52. Arakawa, Y., Vahala, K., Yariv, A. and Lau, K. (1985) Enhanced modulation bandwidth of GaAlAs double heterostructure lasers in high magnetic fields: Dynamic response with quantum wire effects, *Appl. Phys. Lett.*, **47**, 1142–1144.
53. Arakawa, Y., Vahala, K., Yariv, A. and Lau, K. (1986) Reduction of the spectral linewidth of semiconductor lasers with quantum wire effects—Spectral properties of GaAlAs double heterostructure lasers in high magnetic fields, *Appl. Phys. Lett.*, **48**, 384–386.
54. Vahala, K., Arakawa, Y. and Yariv, A. (1987) Reduction of the field spectrum linewidth of a multiple quantum well laser in a high magnetic field—spectral properties of quantum dot lasers, *Appl. Phys. Lett.*, **50**, 365–367.
55. Wegscheider, W., Pfeiffer, L. N., West, K. W., Littlewood, P., Narayan, O., Hagn, M., Dignam, M. M. and Leibenguth, R. E. (1996) Strong magnetic field dependence of laser emission from quantum wires formed by cleaved edge overgrowth, *Solid-State Electron.*, **40**, 1–6.
56. Maialle, M. Z., de Andrada e Silva, E. A. and Sham, L. J. (1993) Exciton spin dynamics in quantum wells, *Phys. Rev. B*, **47**, 15776–15788.
57. Vinattieri, A., Shah, J., Damen, T. C., Kim, D. S., Pfeiffer, L. N., Maialle, M. Z. and Sham, L. J. (1994) Exciton dynamics in GaAs quantum wells under resonant excitation, *Phys. Rev. B*, **50**, 10868–10879.

RAMAN SCATTERING IN SEMICONDUCTORS WITH WAVELENGTH SCALE DIELECTRIC MODULATION

Bernard Jusserand, Claus Dahl, Alex Fainstein and Florent Perez
France Télécom, CNET, Laboratoire de Bagneux,
196 Avenue Henri Ravera, 92220 Bagneux, France

In the presence of strong modulations, either parallel or normal to the surface, of the dielectric constant at the scale of the wavelength, light scattering properties of semiconductor structures are strongly modified. We illustrate this observation from mainly two examples: the deep modification of the apparent selection rules for plasmons in deep etched wires and the strong enhancement of the Raman scattering external efficiency in microcavities.

1. Introduction

Light scattering by elementary excitations in semiconducting crystals is characterized by the conservation of energy and wavevector during the interaction in the compound, by polarization selection rules resulting from the symmetry of the interaction and by an internal cross section reflecting the amplitude of the matrix elements describing the microscopic Raman process [1]. When studied in the backscattering configuration in a semi-infinite sample with a planar surface, modifications of these properties due to the refraction of light, both incident and scattered, at the sample surface are easily established. Most of the consequences are currently taken into account when analysing experimental results.

The wavevector transferred to the excitation can be divided into components q_z and q_{xy}, normal and parallel to the surface respectively,

$$q_z = \frac{4\pi n}{\lambda}\left(1 - \frac{\sin^2\theta}{2n^2}\right)$$

$$q_{xy} = \frac{4\pi}{\lambda}\sin\theta$$

G. Abstreiter et al. (eds.), Optical Spectroscopy of Low Dimensional Semiconductors, 157–178.

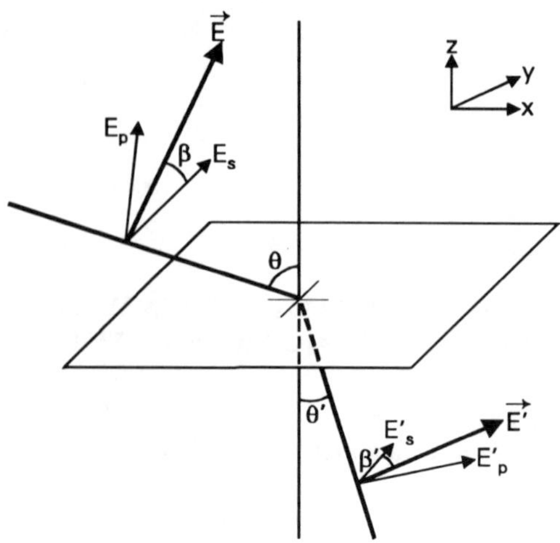

Figure 1. Scattering geometry on a planar surface

where we have neglected the small difference between the incident and scattered wavelength. n is the refractive index at this wavelength and θ the ingoing and outgoing angle, measured outside the sample. q_z is only weakly dependent on θ and much larger than q_{xy}. For bulk systems, the transferred wavevector is thus assumed to be normal to the surface with a constant magnitude $\frac{4\pi n}{\lambda}$. For low-dimensionnal systems, parallel to the surface, only $q_{//}$ is conserved and its variation as a function of θ is used to determine the dispersion of the excitations around zone-center.

To determine the apparent selection rules from the internal ones, one has to analyse the modification of the electric field orientation when traversing the sample surface. Assuming a general orientation, defined by two angles θ and β (see Fig.1), the electric field has to be divided into two components E_p and E_s parallel and perpendicular to the plane of incidence respectively. Being the coefficient of transmission for p and s components (T_p and T_s) generally different, β is modified when crossing the surface. The scalar product of internal fields corresponding to electric fields perpendicular outside the sample, defined by angles θ,β and θ,β+π/2, reads:

$$\sin\beta\cos\beta(T_p^2 - T_s^2)$$

This quantity vanishes when either β=0,π/2 or when θ=0. In the latter case, T_p=T_s. In the general case, however, one has to expect an apparent relaxation of the polarization selection rule. Though the most usual experimental geometries, involving s or p waves only, do not suffer this modification, it is clearly observed for intermediate polarizations. This is illustrated in Fig.2, where a two-dimensionnal (2D) plasmon, allowed in parallel polarisation is absent in the sp configuration but is recorded when the two polarizations are simultaneously turned by 45°. These modifications of the polarization across the surface, though having marginal consequences for planar surfaces, are essential to analyse Raman spectra when the surface is deeply corrugated. This will be illustrated in Part 2 for light scattering by 1D plasmons in deep etched wires.

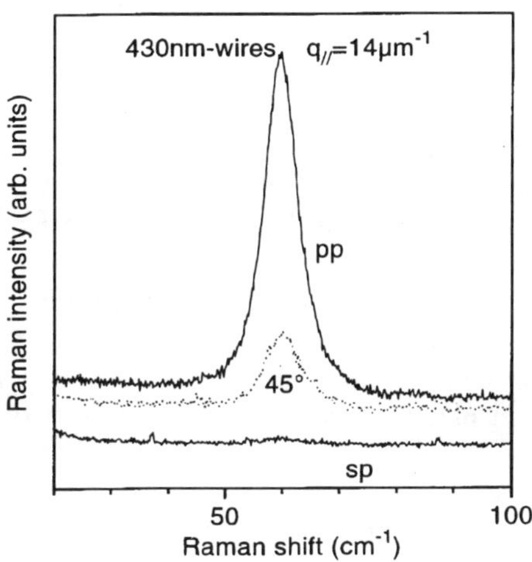

Figure 2. Selection rules for 2D plasmons at intermediate angles of incidence

When crossing the surface, the light intensity is also modified, due to reflectivity. Simple corrections are usually introduced for the ingoing and the outgoing light, depending on the incident angle and polarization. For samples with moderate reflectivity, this correction plays a minor role. Brewster angle and p polarization is often used for the incident light to maximize the Raman signal. When R becomes very large, such as in metals, this can however prevent the observation of Raman signals. On the other hand, light scattering from thin transparent layers deposited on a substrate with a different refractive index can be modified due to reflection on this interface. Constructive or destructive interferences take place depending on the thickness of the layer and the sign of the index jump at the back interface [2]. These interferences can give rise to dramatic modifications of the Raman external efficiency when the reflectivities become close to unity. We will present in Part 3 an illustration of these effects for light scattering by phonons in planar microcavities.

2. Light scattering by quasi-one-dimensionnal plasmons in deep etched wires

Collective excitations of electron gases have been studied since a very long time in doped semiconductors [3]. Based on the modulation doping technique, high mobility gases with a quasi-2D character have been realized, which display very narrow plasmon lines. Contrary to the bulk case, the 2D-plasmon energy vanishes at zone-center and displays a significant dispersion in the wavevector range accessible in a light scattering experiment. As shown in the introduction, q_{xy} can be varied as a function of the angle of incidence θ between 0 and a typical value of $1.5x10^6$ cm^{-1}. Furthermore, recent progress in the nanostructuration of semiconductors has allowed a lateral confinement of the electron gases to be defined in the layer plane [4]. In the presence of lateral confinement, 2D plasmon splits into several branches with one-dimensionnal dispersion. Among the different techniques of lateral nanostructuration, deep etching has proven to be a very powerful and versatile one [5]. Lateral sizes below 100nm have been reported. In such samples, the surface is corrugated at a scale comparable to the wavelength both parallel and normal to the surface and the related modulation of the refractive index is large. This part will be mostly devoted to the observation and modelling of the modifications of light scattering due to this corrugation. We will first recall some features of the plasmon excitations, as deduced within a classical model [6]. This model applies when the confinement dimensions are sufficiently large as compared to the Fermi wavelength of electrons, which holds for most of the wires fabricated by deep etching.

2.1. PLASMON DISPERSION: THEORY AND EXPERIMENTS

In the classical model, the electron gas, described by a time dependent density distribution n(x,y,z,t) and the associated electrostatic potential ϕ(x,y,z,t), must satisfy the two following equations:

$$\Delta\phi = \frac{e}{\varepsilon}n \qquad \text{(Poisson equation)}$$

and

$$\frac{\partial^2 n}{\partial t^2} = -\frac{e}{m}\vec{\nabla}\cdot(n\nabla\phi) \qquad \text{(conservation equation)}$$

Assuming a small density fluctation n_1 at frequency ω around the equilibrium density n_0, one can linearize the two differential equations and eliminate n_1, using the integral form of the Poisson equation. This results in the following integro-differential equation on the fluctuation of the electrostatic potential ϕ_1:

$$\omega^2\phi_1(\vec{r}) = -\frac{e^2}{4\pi\varepsilon m}\iiint\frac{d^3\vec{r}\,'}{|\vec{r}-\vec{r}\,'|}\vec{\nabla}\cdot(n_0(\vec{r}\,')\vec{\nabla}\phi_1(\vec{r}\,'))$$

The dielectric constant ε and the effective electron mass m are assumed to be constant over the whole system. The roots ω of this equation are the plasmon energies of the electron system. For infinite 3D systems, ϕ_1 is a bulk plane wave with a wavevector q and the integral is proportionnal to the Fourier transform of the Coulomb potential which amounts to $4\pi/q^2$. The plasmon energy reads:

$$\omega^2 = \frac{e^2 n_0}{\varepsilon m}$$

In infinite 2D systems, located in the x,y plane, the same treatment applies to $\phi 1(r//,z=0)$ but the 2D Fourier transform $(2\pi/q)$ appears instead of the 3D one. The plasmon energy

$$\omega^2 = \frac{e^2 n_0}{2\varepsilon m} q$$

becomes dispersive.

When lateral confinement is introduced along the x direction, numerical solutions of this eigenvalue problem must be considered [7], which present specific difficulties due to the strong non-local character of the Coulomb interaction. We show in Fig.3 the dispersion along the wire axis y of confined plasmons in wires with a constant equilibrium density across the whole wire. The dispersion looks qualitatively similar to the one obtained from simply quantizing q_x in the 2D dispersion [8] but the absolute energies are significantly reduced as compared to this approximate model because of the long range character of the interactions. We also show the variation of the potential across the wire for a few modes at $q_y=0$. These modes display alternating parities and a mixed character of confined and edge states, which results from the lateral boundary conditions.

This theory quantitatively reproduces the dispersion of the plasmon modes observed by Raman scattering [8-10]. We show in Fig.3 an illustration of this agreement for a wire of lateral width a=180nm. Details on the fabrication technique, the sample parameters and the experimental conditions are given in Ref.10. For all the wires investigated, the fitted values of the free parameters of the model n_0 and a are very close to the nominal ones. This demonstrates the excellent control of the fabrication process, including the passivation of etch defects at lateral surfaces. Nevertheless, this results are surprising, considering the well established pinning of the Fermi level at the surface of GaAs. In fact, the typical depletion length in 2D electron sytems with comparable electron densities amounts to 500nm [11]. This should prevent the existence of any free carrier in most of the wires which we have studied. The origin of this discrepancy is still under investigation.

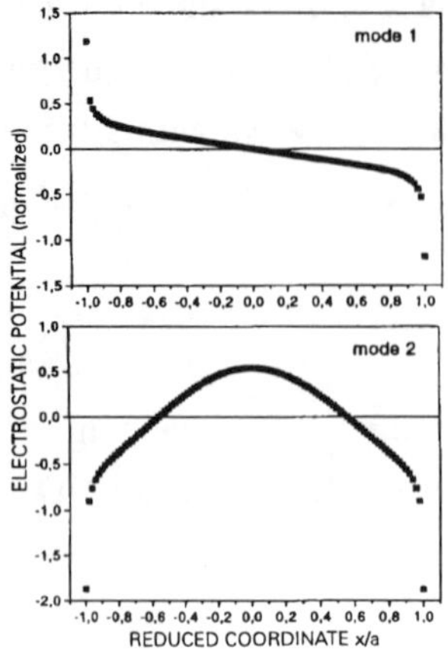

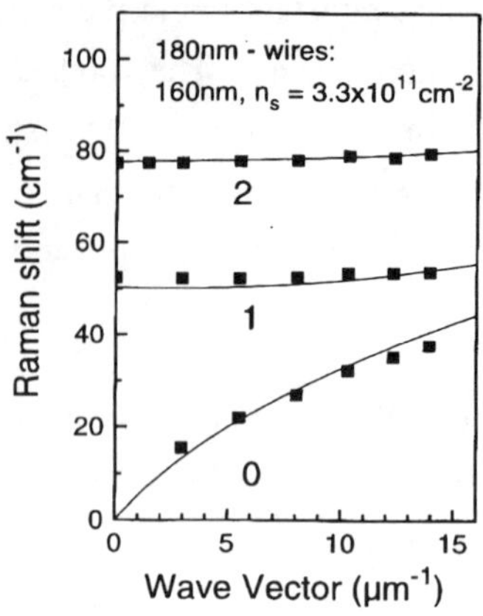

Figure 3. Plasmon dispersion calculated within the classical model and compared to the measured frequencies (right), distribution of the electrostatic potential across the wire for two different modes at zone center (left).

2.2. SELECTION RULES: OBSERVATIONS AND QUALITATIVE MODEL

We have also observed in these wires unusual Raman selection rules [12]. They are illustrated in Fig.4, where the Raman spectra on the same wire array are shown for two different geometries: the wavevectors of the incident and scattered photons are either parallel or perpendicular to the wire axis. In the former case, plasmons with finite wavevector are involved, while in the $q_\perp$ configuration zone center 1D plasmons couple to the light. The energies are thus slightly shifted down when going to the $q_\perp$ configuration for $n \geq 1$, while mode 0 is no longer observable. The striking feature in these experimental results lies in the selections rules, which are shown in Fig.4 for both geometries in parallel and crossed polarizations. For $q_\perp$, all the plasmons modes are allowed in parallel polarization and forbidden in crossed polarization. For $q_{//}$, even and odd plasmons alternatively appear in parallel and crossed polarization respectively. These selection rules are in total contradiction with the well established plasmon selection rules. Firstly, they are only allowed in parallel polarization because only spin flip transitions, unaffected by the direct part of the Coulomb interaction, are Raman active in crossed polarization. This rule has been always verified in 2D systems. Moreover, they display parity conservation in some conditions. This is the counterpart of the wavevector conservation when a symmetric confinement potential with a typical width a replaces the translationnal invariance. When the wavevector q along the confinement direction is such that $qa \ll 1$, one can neglect spatial dispersion in the Raman cross section and only even excitations exhibit a non-vanishing integrated activity. This second rule is not usually considered in the analysis of light scattering in modulation doped quantum wells, because the confining potentials are not usually symmetric along the growth direction z and because $q_z a>1$.

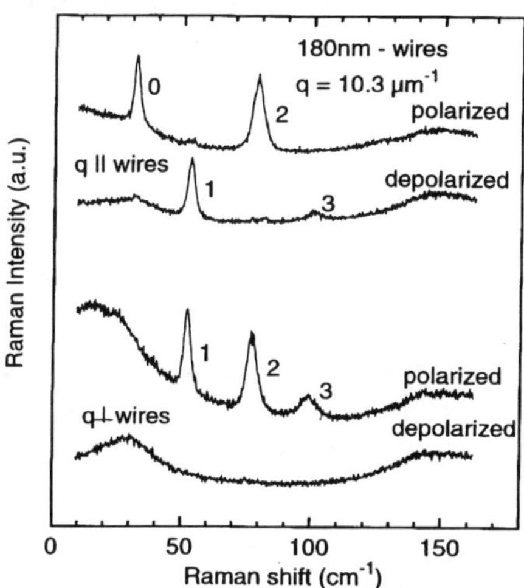

Figure 4. Selection rules for plasmons determined on the same wire array as in Fig.3 in four different configurations described in Fig.5.

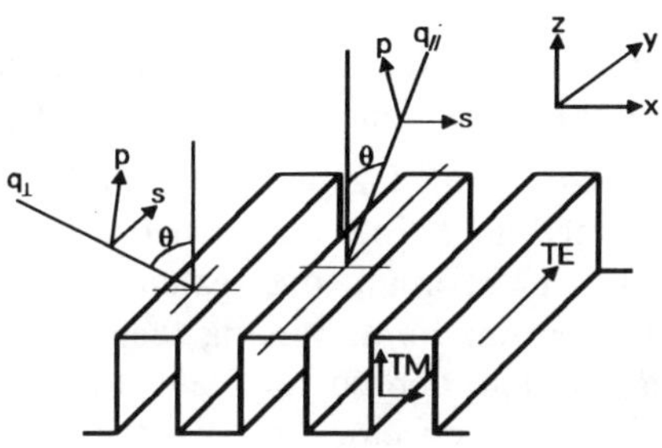

Figure 5. Scattering geometry on a 1D corrugated surface.

In deep etched wires with lateral sizes comparable to the wavelength of the light, plane waves are no longer the correct eigenmodes and must be replaced by transverse electric (TE) modes with the electric field along the wire axis y and transverse magnetic (TM) modes with the electric field within the (x,z) plane. z is the direction normal to the surface and x the direction of the modulation. The scattering geometries are illustrated in Fig.5. Based on a qualitative analysis of the field continuity at the boundary between air and the etched layer, we are able to understand some of the new selection rules. For $q_\perp$, the orientation of the electric field for s and p waves outside coincides with the one of TE and TM waves inside the grating, respectively. Hence, a sp configuration gives rise to a TETM one with perpendicular electric fields within the wire. Based on the microscopic Raman selection rules at the location of the electron gas, plasmon scattering remains forbidden in this configuration in agreement with the observations. On the contrary, for $q_{//}$, s and p waves do not generate pure TE and TM ones but linear combinations of both. Thus, for external crossed polarizations, the local fields are no longer crossed by symmetry inside the wires and plasmons are likely to become active in this configuration, which indeed is observed.

Moreover, the lateral confinement potential is symmetric by construction. One has to expect parity conservation when $q_x a \ll 1$. This is the case for $q_{//}$ because q_x exactly vanishes and only even plasmons are observed in parallel polarization. On the contrary for $q_\perp$, $q_x a$ is non vanishing and all the plasmons become Raman active in parallel configuration. The last observation which remains to be explained (odd plasmons in crossed polarization for $q_{//}$) also reflects that parity conservation is verified by plasmon scattering in etched wires. This will be illustrated next from the calculated field distribution within the wire.

2.3. LOCAL FIELD CALCULATION

Such a calculation of the distribution of the electromagnetic field inside a dielectric grating has been first introduced in the field of semiconductor wires by Bockelmann for normal incidence [13]. On the basis of the results, he was able to qualititavely explain some features of the excitonic absorption and luminescence in quantum wires. Here we extend this calculation for oblique incidence either parallel ($q_{//}$) or perpendicular($q_\perp$) to the wire axis. In this calculation, one has to consider a three-layer sample, a thin slice of the periodically stratified medium being sandwiched between air and the substrate. In a first step, using the Bloch theorem, one calculates the eigenmodes for light propagation within the stratified medium. Among them, one selects the modes which are likely to be excited by the chosen incident field, taking into account the conservation of the in-plane wavevector and of the electric and magnetic field orientations. These modes consist in both propagating and evanescent waves along z. A subset is then selected to act as a finite basis for an approximate developpement of the general solution of the problem and the corresponding field distributions are Fourier transformed along x to write down continuity equations of the electric and magnetic field magnitudes at the boundaries with air and substrate respectively. After elimination of the transmitted and reflected fields, the inhomogeneous linear problem is finally solved by numerical inversion. We illustrate the results in Fig.6 where we show the field distibution across the period of the wire grating at the depth corresponding to the location of the electron gas in the studied samples. The geometrical parameters used in the calculations are typical of the samples studied in this work and s an p excitations have been considered for $q_{//}$. In agreement with our qualitative discussion, all the components are non-vanishing and display well-defined parities. Moreover, each of them display opposite parities for s and p excitations respectively. This explains the observed activation of odd plasmons in the sp configuration.

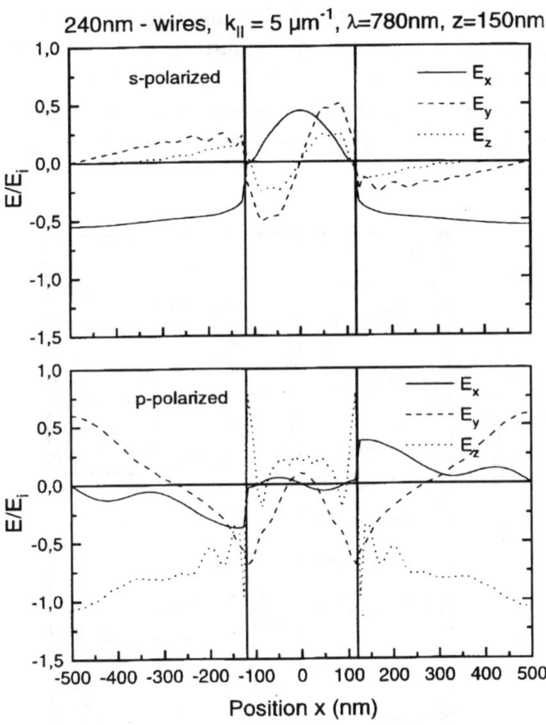

Figure 6. Field distribution accross the wire array at the location of the electron gas for q// and both s and p incident polarizations.

The same calculation may be used to discuss the intensity of plasmon lines for $q_{\perp}$ when the parity conservation is relaxed. We indeed observed very strong, complex and specific variations of the intensities of the different plasmon lines as a function of the magnitude of $q_{\perp}$. These observations, shown in Fig.7, can not be explained by simply considering an increasing activation of the lines forbidden at $q_{\perp}=0$ and a corresponding decrease of the allowed ones. This only applies when $q_{\perp}a \ll 1$. In our case, where $q_{\perp}a$ is of the order or 1, strong modifications of the field distribution in the wires are predicted from the calculation. Using the expression of the Raman activity in terms of the density-density response function, modified to apply to a system without translationnal invariance, one predicts strong, complex and specific variations of the Raman activity of the different confined plasmons, as observed experimentally. However, quantitative agreement is still completely lacking. Owing to the strong oscillating character of the quantities involved in the calculation, this could be due to an excessive simplification of the unit cell with respect to the real sample: we have assumed a rectangular profile and we have neglected the oxyde layer with an intermediate index of refraction in the field calculation. Moreover, we have used a constant equilibrium density with abrupt boundary conditions in the charge density model, which increases the contribution of the electric field at the boundaries to the integrated Raman activity. Further calculations and a critical reexamination of the various approximations in our model are clearly needed to improve the quantitative description of the experiments.

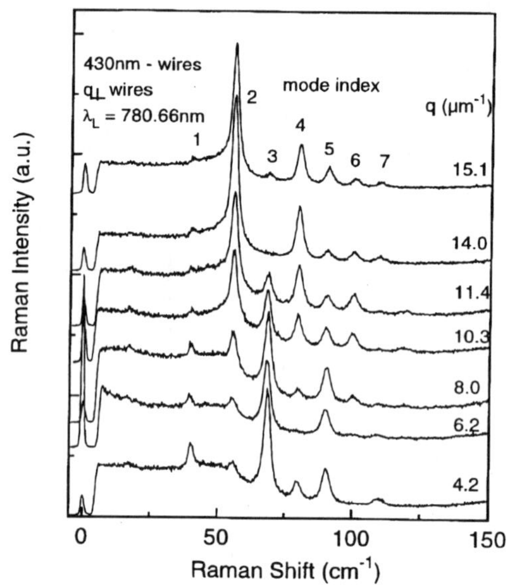

Figure 7. Variation as a function of the wavevector magnitude in the $q_\perp$ configuration of the plasmon line intensities.

To summarize, we have shown in this part that the electromagnetic field distribution in deep etched wires is strongly modified both in intensity and orientation with respect to the external field. We have explained on the basis of this modification most of the apparently striking features which we have observed in the light scattering spectra by confined plasmons in wires. Such modifications have to be carefully taken into account before drawing any conclusions on modifications of the microscopic cross section due to the reduction of the dimensionnality in nanostructured samples.

3. Light scattering enhancement in planar microcavities

Studied first in the domain of atomic physics [14,15] and, more recently, in the field of semiconductors [15,16], optical microcavities have demonstrated their ability to strongly modify the excitation and emission characteristics of atoms embedded in them leading, for instance, to increased luminescence and reduced thresholds for laser gain. It has been clearly settled that the enhancement of the optical field at the core of the cavity, between two mirrors with reflectivities very close to unity, affects *both* first-order in light-mater processes: emission *and* absorption. By the same token, second-order processes in which one or more photons are absorbed and/or emitted should be equally affected. One of such processes is Raman scattering, where one photon is absorbed and another emitted, both differing in the energy and, eventually, wavevector of an excitation left behind. Enhancement of Raman scattering due to photon field modifications has indeed been already reported in micrometer-size droplets acting as quasi-spherical optical cavities, including the observation of up to fourtenth-order Stokes peaks and of reduced thresholds for stimulated gain [17]. Also, field confinement effects at the Stokes wavelength have been reported for liquide benzene in a piezoelectrically controled microcavity [18]. We will discuss here the modifications of the Raman scattering cross-section in monolithic semiconductor planar microcavities. Before reviewing experimental results and presenting a quantitative model of the scattering efficiency, we will recall some properties of the Perot-Fabry resonators and of the Bragg reflectors which are used to realize high reflectivity mirrors.

3.1. THE IDEAL PEROT-FABRY

It has been known since a very long time that a Perot-Fabry isolates some resonant photon modes from all other non-resonant ones and enhances the magnitude of the electric field in those modes (see Fig.8). The wavelength selectivity of the resonator is defined by the finesse which expresses the ratio between the width of the resonance peak and the separation between successive resonance peaks [19]. Assuming perfect mirrors with the same reflectivity R on both sides of the cavity, it reads:

$$F = \frac{\pi\sqrt{R}}{1-R}$$

On the other hand the internal field intensity at the resonance peak is enhanced by a factor: $\frac{1+R}{1-R}$, while the intensity at the mean position between two peaks is reduced by the same factor. Assuming very large reflectivities (1-R<<1), both quantities coincide up to a constant factor close to unity and the measured finesse is generally used to characterize the field enhancement in a resonator. The wavelength of the resonance peaks is given by:

$$\lambda_m = \frac{2nd\cos\theta}{m} \qquad (1)$$

where n is the index of refraction and d the thickness of the resonator. θ is the angle of incidence inside the resonator. This expression exhibits discrete modes along z, labelled by m, each of them corresponding to a continuum of wavelengths characterized by θ.

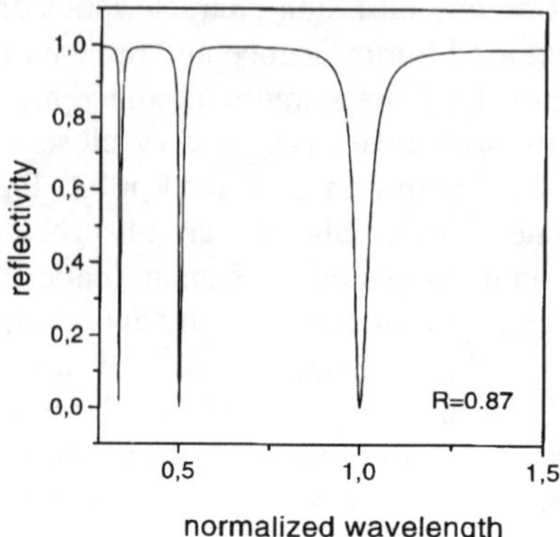

Figure 8. Wavelength dependence of the reflectivity for a Perot-Fabry resonator.

If one inserts an emitting dipole at the center of the resonator, the radiation pattern of the emitter will be modified. To determine the new radiation pattern, one has to solve the Maxwell equations for the stratified medium with a source term due to the dipole located in the cavity [20,21]. It is then convenient to transform the dipole emission from spherical to cylindrical coordinates (ρ,θ,z), using the Sommerfeld identity. Assuming for instance an horizontal electric dipole, the electromagnetic field due to the emission of the dipole can be deduced from the two following independent components:

$$E_z = E_0 \cos\theta \int_0^\infty dk_\rho k_\rho^2 J_1(k_\rho \rho) \mathrm{sign}(z) e^{-ik_z|z|}$$

$$H_z = H_0 \sin\theta \int_0^\infty dk_\rho \frac{k_\rho^2}{k_z} J_1(k_\rho \rho) e^{-ik_z|z|}$$

where E_0 and H_0 are related to the dipole oscillator stength and k_r and k_z are the wavevector components in cylindrical coordinates. When added to the general plane wave solutions, developed on the same cylindrical waves, these lead, through application of the interface boundary conditions and far field requirements, to an inhomogeneous linear system in these coordinates, the inversion of which provides the field distribution in the resonator.

The emission of the dipole becomes highly directive along the resonating modes of the cavity, when they exist at the emission wavelength. Assuming a dipole located at the center of a Perot-Fabry resonator, emitting at the resonance wavelength for normal emission, the emission intensity is enhanced along this direction by the same factor as the incident field is in the cavity [22]. The directivity of the emission is also related to the finesse by the following expression

$$\frac{\delta\lambda}{\lambda} \approx \sin\theta \frac{\delta\theta}{\theta} \qquad (2)$$

3.2. MICROCAVITIES WITH BRAGG REFLECTORS

Almost total reflectivity mirrors are needed to build up cavities with high finesse. In semiconductors, and for epitaxial growth requirements, Bragg reflectors are generally used to obtain such a strong confinement. They consist of a periodic stacking of double layers of thicknesses $a_{1,2}$ and refractive indexes $n_{1,2}$. Using the transfer matrix method, an analytical expression has been determined [23] for the reflectivity of a Bragg reflector with N unit cells, which reads:

$$R = \frac{1}{1 + \frac{1-r}{r}\left(\frac{\sinh K}{\sinh NK}\right)^2}$$

where r is the reflectivity of the unit cell and K is given by:

$$2\cos(K) = \cos\frac{2\pi n_1 a_1}{\lambda}\cos\frac{2\pi n_2 a_2}{\lambda} + \frac{1}{2}\left(\frac{n_1}{n_2} + \frac{n_2}{n_1}\right)\sin\frac{2\pi n_1 a_1}{\lambda}\sin\frac{2\pi n_2 a_2}{\lambda}$$

for normal incidence. One can easily deduce that, for large N, R approches 1 exponentially with N, with a prefactor K which is maximum when the two layers have quarter-wave thicknesses and the largest refractive index contrast. The same applies for oblique incidence, provided that the layer thicknesses are divided by the cosine of the internal angle of incidence. Using typically 15 periods of GaAs/AlAs unit cells, reflectivities as high as .998 can be achieved. The feasibility of even higher reflectivities is prevented by the difficulty to increase the index difference with the usual III-V compounds and the limitations in the growth duration.

Using mirrors with such reflectivities separated by an half-wavelength layer, microcavities with a finesse above 1000 have been realized. The corresponding field enhancement is of the order of 30 [15]. The methods which we have discussed previously to calculate the internal field and the emission pattern within a Perot-Fabry remain valid for such microcavities, but including a reflectivity varying both with the wavelength and the angle of incidence. Alternatively, the Bragg mirrors can be included from the beginning in the calculation of the field distribution and the radiation pattern. This inclusion does not introduce much complications in the calculation, based on the transfer matrix method. It allows to determine the distribution inside the mirror layers of the quantities of interest. The results are illustrated in Fig.9.

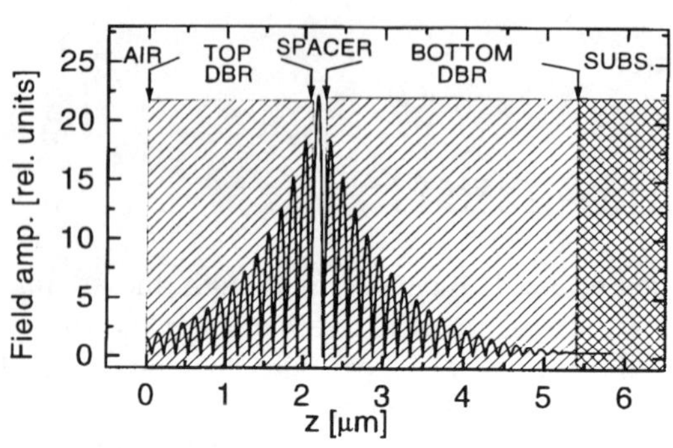

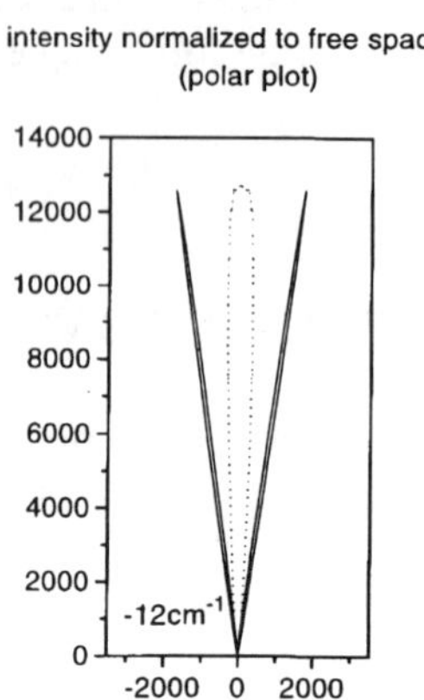

Figure 9. a) Distribution of the electric field at the resonance in the microcavity studied below (sample A). b) radiation pattern (polar plot) for a dipole field at resonance (dashed line) and shifted up by 12 cm^{-1} (full line) respectively.

3.3. DOUBLE OPTICAL RESONANCE FOR LIGHT SCATTERING IN MICROCAVITIES

We have mentioned in the introduction that the external Raman efficiency has to be corrected by the reflectivity of both incident and Stokes fields at the sample surface. Multiple reflections leading to photon confinement in a microcavity also modify the scattering efficiency. The Raman cross section contains two *squared* matrix elements of light-matter interaction, due to the incoming and outgoing photon vertexes. These matrix elements are proportional to the electric field amplitude of the laser and the Stokes photon, respectively. Consequently, if these amplitudes are enhanced in a cavity, the Raman scattering efficiency should be *fourth order* in the enhancement factor.

Contrary to luminescence experiments in microcavities which are usually performed by excitation above the structure stop-band [15], the excitation energy is not a free parameter in a Raman experiment: *two* photons with energies differing by a *fixed* amount (given by the excitation under study) must get into, or go out from, the structure. A double resonance condition can still be met by exploiting the continuum of in-plane optical modes. In fact, if a photon propagating along the cavity axis resonates at an energy ω_S, one incident at an angle θ_0 (measured in air) will resonate at a *larger* energy given approximately by Eq.1, provided that an effective index n_{eff} is used to transform external angles into internal ones. Because of the large penetration of the resonant field inside the Bragg mirrors, this index does not equal the cavity one but takes some intermediate value. When $(\omega_i - \omega_s)$ is tuned to an excitation of the system, the conditions for the observation of optical double resonant Raman scattering are obtained.

Such a large enhancement has been reported recently [24] in a structure consisting of an AlAs half-wave spacer enclosed by GaAs/AlAs DBR's (20 pairs below, 13.5 above) with two 12-nm-wide $In_{0.14}Ga_{0.86}As$ QW's located at its center (sample A). A finesse of ~300 was estimated for this structure from reflectivity measurements [25]. The sample was grown by molecular beem epitaxy on a GaAs wafer and the growth conditions were such that the thickness of all layers varies with position enabling the tuning of the cavity mode by displacing the spot under consideration.

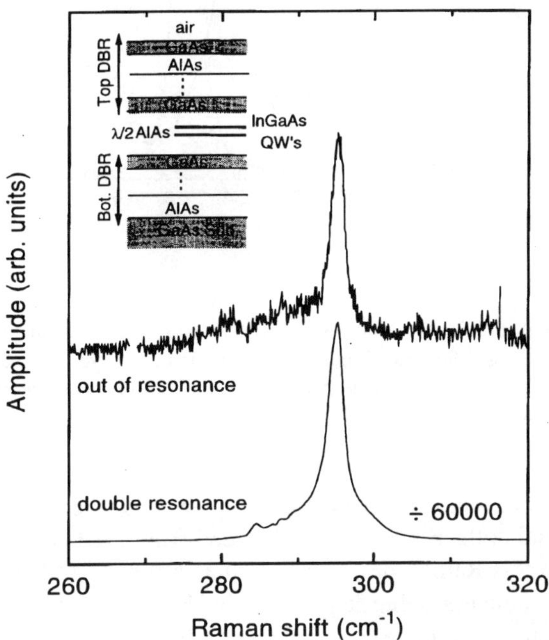

Figure 10. Raman spectra taken in crossed configuration on the full cavity sample with a spot position corresponding to double optical resonance and out of resonance conditions respectively.

Raman spectra taken on this sample are presented in Fig.10, for θ_0=54° with crossed (allowed) polarizations and at an excitation energy of 1.35 eV, *below* the QW's gap to avoid the QW's luminescence. For the bottom spectrum, the angle of incidence and the spot under observation in the cavity sample correspond to a double-resonant condition tuned to the energy of the bulk-GaAs longitudinal optical (LO) phonon, seen as the largest peak in Fig.10 (~295 cm^{-1}). A spectrum taken from the same sample but completely detuned by displacing the spot under observation, is also shown. Almost five orders of magnitude difference in the Raman intensity is observed.

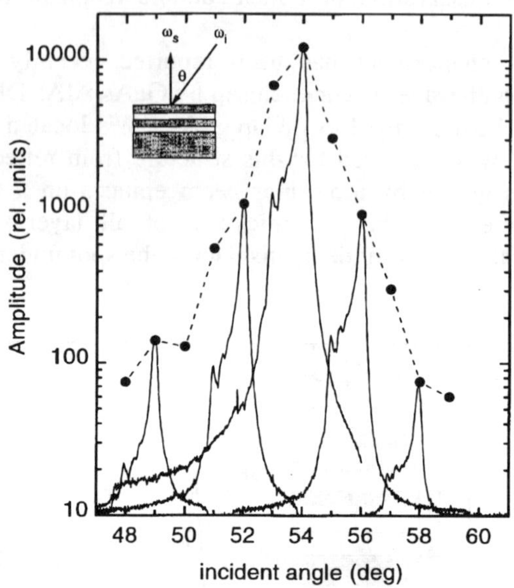

Figure 11. Dependence of the Raman scattering amplitude (full dots) as a function of θ_0, for the cavity tuned along z to the LO-peak. The dashed line is only a guide to the eye. Raman spectra, compressed in the horizontal scale and shifted so that the larger peak coincides with the respective incidence angle, are also shown for illustration.

The Raman scattering enhancement in a planar microcavity is critically sensitive on both incidence and collection angle, as deduced from Eq.1. In Fig.11, we present the amplitude of the largest Raman peak in Fig.10 as a function of the angle of incidence θ_0, all the other parameters being kept constant. More than two orders of magnitude variations of the Raman amplitude are observed by rotating the incidence angle less than 4°, thus going from double to single resonance (the Stokes photon is always kept in resonance). The angular dependence HWHM of the order of 1° is consistent, according to Eq.2, with the reflectivity peak HWHM measured for the same structure (~0.3 nm) [27]. Similar results have been also reported for the collection angle. In fact, such small ingoing and outgoing acceptances are likely to limit the observed enhancement with respect to the maximum calculated one, because of the finite apertures used in the Raman set-up.

3.4. QUANTITATIVE MODEL FOR THE RAMAN EFFICIENCY

To give a quantitative estimate of the Raman cross section in microcavities, a model of the eigenvectors of the phonons involved in the Raman process would be needed. Moreover, as the photon field decays slowly in the Bragg mirrors, this modelling should include the vibrations in the whole sample. A detailed analysis of these vibrations has been reported recently [26] (see futher below). However, to emphasize the significant information on the cavity effect, it is more convenient to assume a vibrationnal amplitude constant over the cavity and vanishing outside. The cross-section is then approximated as the product of the internal field intensity (at the laser frequency and for a chosen incidence angle and the radiation intensity (at the fixed Stokes frequency and emission angle), both integrated throughout the cavity length. Using this approch, one recovers qualitatively the fourth power dependence on the field amplitude which we introduced previously. Moreover, both internal intensity and radiation distribution are easily obtained along the lines of the above presented theory.

To improve the validity of this approach, better adapted samples have been designed in which the cavity exhibits specific phonons [26,27]. It is made of a GaAs/AlAs superlattice, filling up the whole cavity, enclosed with GaAlAs/AlAs Bragg reflectors. Both GaAs-like LO phonons and folded acoustic phonons in the superlattice appear at frequencies outside of the vibrationnal range of the reflectors. In order to collect a large set of experimental data and to compare the relative intensities to the model, two different samples have been studied with and without the top Bragg reflector. The second sample (half-cavity) exhibits a weaker, but still significant, enhancement of the Raman cross section as well as a larger angular acceptance. Moreover, for both samples, double resonance experiments, obtained in quasi-backscattering, have been compared to single resonance ones. In the latter case, the Stokes photons are collected at the edge of the layer, thanks to the guiding property of the cavity structure in these new samples, while the excitation is done along the cavity axis. Hence, only the incident photon is enhanced by cavity confinement.

The comparison between measured and calculated Raman intensities is shown in Fig.12 for the GaAs-like LO phonon in the superlattice and in the four different experimental situations [27,28]. The only free parameter is the absolute intensity at the maximum resonance for one of the four curves (we have chosen the single resonance on the half-cavity). The overall agreement is quite good, in particular for the half-cavity. For the full cavity, the measured enhancement is smaller than the expected one, even when the finite collection aperture is taken into account, as done here. This residual discrepencancy has been attributed to some fluctuations of the layer thicknesses in the real sample, which lower the real finesse with respect to the nominal one, and to distribution of angles due to the slight focussing of the incident beam. The full-cavity structure is much more sensitive to the latter correction because of its smaller angular acceptance. Moreover, the fact that this acceptance decreases with increasing average incidence angle could explain the better agreement obtained in the single resonance experiment, performed at normal incidence. Finally, the finite size of the laser spot on a sample with an intentionnal gradient of thicknesses could give an alternative explanation to these observations.

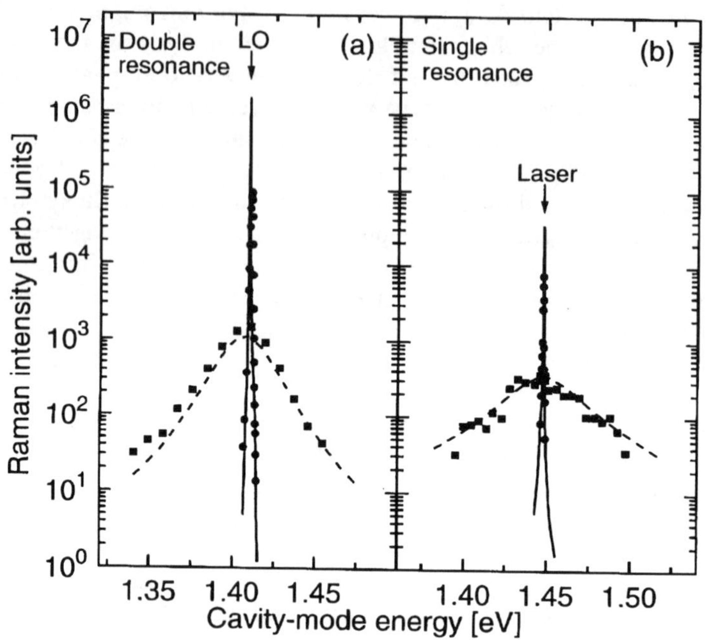

Figure 12. Comparison between measured and calculated variation around the resonance of the scattering intensity for both single and double resonance conditions on both full (circles) and half (squares) cavity samples.

3.5. SELECTIVE SPECTROSCOPY IN MICROCAVITIES

Thanks to the high finesse of the optical cavities, one is able to selectively enhance some spectral features in the Raman spectra with respect to the others. This is illustrated in Fig.13 for sample A. Changing the relative energies of the cavity mode and the laser, either by tuning the laser or by moving the sample, one can bring to exact resonance with the cavity mode at normal incidence different outgoing energies corresponding to different vibrations in the sample, while the ingoing photon remains within the cavity resonance at fixed angle. Even stronger selectivity could be obtained by adjusting this angle to achieve double resonance for each Raman shift, at the price of an increased experimental complexity. The rich spectrum of phonons is associated to interface optical vibrations in the structure and a detailed analysis can be found in Ref.26. Here, we want to highlight the possibility to extract some weak lines from complex Raman spectra.

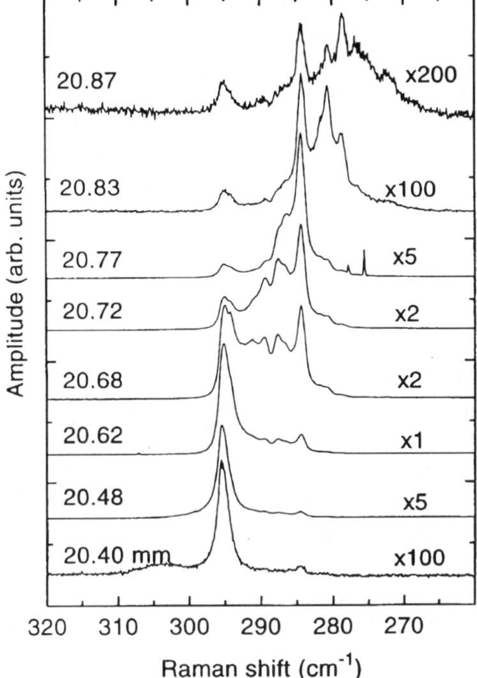

Figure 13. Raman spectra on sample A for different positions around the one where double resonance condition at the LO energy is realized.

To summarize this part, we have shown that huge enhancements of the Raman cross sections have been obtained in microcavities, which can be reasonably reproduced quantitatively by a simple combination of the incident field enhancement and the radiation redistribution. We would like to point out the potentialities of the results presented here. On one hand, we have demonstrated the cavity geometry as a promissing tool to study the excitations in weakly scattering objects, such as could be single interfaces or small quantities of quantum dots. Moreover, the possibility of selectively amplifying chosen regions of otherwise intense signals enables, for instance, the observation of subtle features in the spectra. Furthermore, Raman scattering by well characterized and localized excitations, which can be done easily with micrometer resolution, may serve as a localized probe of the field distribution in photonic microstructures. On the other hand, it would be very interesting to search for stimulated Raman processes in these structures, which should no doubt present strongly reduced thresholds opening the way to their use in optical devices.

4. Conclusion

We have shown in this lecture that light scattering properties can be significantly modified when the dielectric properties of the sample are modulated at the wavelength scale. These extrinsic modifications, in terms of selection rules and cross section, have to be taken into account when analysing Raman active excitations in nanostructures and carefully distinguished from intrinsic modifications due to electron confinement. On the other hand, they can be turned to great advantage when manipulating the local field in the structures to selectively enhance or inhibit specific spectral or spatial distributions and, hence, to increase the sensitivity and selectivity of Raman scattering experiments.

5. References:

1. For a review, see Cardona, M. (1982) Resonance Phenomena, in M. Cardona and G. Güntherodt (eds.), *Light Scattering in Solids, Vol. 2*, Springer, Berlin, pp. 19-178.
2. Connell, G.A.N., Nemanich, R.J. and Tsai, C.C. (1980) Interference enhanced Raman scattering from very thin absorbing films, *Appl. Phys. Lett.* **36**, 31-33. We also mention the broad field of Surface Enhanced Raman Scattering (SERS). One of the explanations for this effect is associated to local field effects at rough surfaces. For a review, see for instance Otto, A. (1984) Surface-enhanced Raman scattering: "Classical" and "Chemical" origins, in M. Cardona and G. Güntherodt (eds.), *Light Scattering in Solids, Vol. 4*, Springer, Berlin, pp. 289-418.
3. For a review, see Pinczuk, A. and Abstreiter, G. (1989) Spectroscopy of free carrier excitations in semiconductor quantum wells, in M. Cardona and G. Güntherodt (eds.), *Light Scattering in Solids, Vol. 5*, Springer, Berlin, 153-211.
4. For a review, see for instance Sakaki, H. and Noge, H. (eds) (1994) *Nanostructures and quantum effects*, Springer, Berlin.
5. Marzin, J.Y., Izrael, A. and Birotheau, L. (1994) Optical properties of etched GaAs/GaAlAs quantum wires and dots, *Solid State Electron.* **37**, 1091-1096.
6. Fetter, A.L. (1986) Magnetoplasmons in a two-dimensionnal electron fluid: disk geometry, *Phys.Rev.B* **33**, 5221-5227.
7. Eliasson, G., Wu, J.W., Hawrylak, P. and Quinn, J.J. (1986) Magnetoplasma modes of a spatially periodic two-dimensional electron gas, *Solid State Commun.* **60**, 41-44.
8. Egeler, T., Abstreiter, G., Weimann, G., Demel, T., Heitmann, D., Grambow, P. and Schlapp, W. (1990) Anisotropic plasmon dispersion in lateral quantum-wire superlattice, *Phys.Rev.Lett.* **65**, 1804-1807.
9. Strenz, R., Rosskopf, V., Hirler, F., Abstreiter, G., Böhm, G., Tränkle, G. and Weimann, G. (1994) Confined plasmons in shallow etched quantum wires, *Semicond. Sci. Technol.* **9**, 399-403.
10. Dahl, C., Jusserand, B. and Etienne, B. (1996) Raman scattering by plasmons in deep etched quantum wires, *Solid State Electron.* **40**, 261-264.

11. Dahl, C., Manus, S., Kotthaus, J.P., Nickel, H. and Schlapp, W. (1995) Edge magnetoplasmons in single two-dimensionnal electron disks at microwave frequancies: determination of the lateral depletion length, *Appl.Phys.Lett.* **66**, 2271-2273.
12. Dahl., C., Jusserand, B. and Etienne, B. (1995) Selection rules in Raman scattering by plasmons in quantum wires, Phys.Rev.B **51**, 17211-17214.
13. Bockelmann, U. (1991) Polarization dependent optical wave fields in grating structures, *Europhys.Lett.* **16**, 601-606.
14. See, for instance, Thompson, R.J., Rempe, G. and Kimble, H.J. (1992) Observation of normal mode splitting for an atom in an optical cavity, *Phys.Rev.Lett.* **68**, 1132-1135.
15. For a recent, comprehensive review, see Burstein, E. and Weisbuch, C. (eds.) (1995) *Confined electrons and photons,* Plenum, New York.
16. Weisbuch, C., Nishioka, M., Ishikawa, A. and Arakawa, Y. (1992) Observation of the coupled exciton photon mode in a semiconductor quantum microcavity, *Phys.Rev.Lett.* **69**, 3314-3317.
17. Lin, H.B., Eversole, J.D. and Campillo, A.J. (1992) Continuous wave stimulated Raman scattering in microdroplets, *Optics Lett.* **17**, 828-830.
18. Cairo, F., De Martini, F.and Murra, D. (1993) QED-vacuum confinement of inelastic quantum scattering at optical frequencies: a new perspective in Raman spectroscopy, *Phys.Rev.Lett.* **70**, 1413-1416.
19. Born, M and Wolf, E (1970) *Principle of Optics, Fourth Edition,* Pergamon Press, Oxford, pp. 323-329.
20. Kong, Jin Au (1986) *Electromagnetic wave theory*, Wiley, New York, pp. 309-319.
21. Abram, I. and Oudar, J.L. (1995) Spontaneous emission in planar semiconductor microcavities displaying Rabi splitting, *Phys.Rev.A* **51**, 4116-4122.
22. Kastler, A. (1962) Atomes à l'intérieur d'un interféromètre Perot-Fabry, *Applied Optics* **1**, 17-24.
23. Yeh, P., Yariv, A. and Hong, C.-S. (1977) Electromagnetic propagation in periodic stratified media. I. general theory, *J.Opt.Soc.Am.* **67**, 423-438.
24. Fainstein, A., Jusserand, B. and Thierry-Mieg, V. (1995) Raman scattering enhancement by optical confinement in a semiconductor planar microcavity, *Phys.Rev.Lett.* **75**, 3764-3767.
25. Abram, I., Iung, S., Kuszelewicz, R., Le Roux, G., Licoppe, C., Oudar, J.L., Bloch, J.I., Planel, R. and Thierry-Mieg, V. (1994) Nonguiding half-wave semiconductor microcavities displaying the exciton-photon mode splitting, *Appl. Phys.Lett.* **65**, 2516-2518.
26. Fainstein, A. and Jusserand, B. (1996) Phonons in semiconductor planar microcavities: a Raman scattering study, to appear in *Phys.Rev.B.*
27. Fainstein, A., Jusserand, B. and Thierry-Mieg, V. (1996) Raman efficiency in planar microcavities, *Phys.Rev.B.* **53**, R13287-13290.

28. Fainstein, A. and Jusserand, B. Single and double optical resonant Raman scattering in semiconductor planar microcavities, *Proceedings of 23th International conference on the physics of semiconductors*, Berlin, 1996 (to be published).

BAND-GAP RENORMALIZATION IN QUASI-ONE-DIMENSIONAL SYSTEMS

B. TANATAR
Department of Physics,
Bilkent University,
06533, Ankara, Turkey

1. Introduction

A dense electron-hole plasma forming in a semiconductor under intense laser excitation comprises an interesting many-body system. Because of the exchange effects and the screening of the Coulomb interaction, the single-particle properties are renormalized. A notable phenomenon is the band-gap renormalization as a function of the plasma density which is important to determine the emission wavelength of coherent emitters as being used in semiconductors.[1] As a substantial amount of carrier population may be induced by optical excitation, the renormalized band gap can affect the excitation process in turn and lead to optical nonlinearities. In this paper we investigate the density dependence of the band-gap renormalization (BGR) in quasi-one-dimensional (Q1D) photoexcited semiconductors. Under high optical excitation the band gap for 2D and bulk systems is found to decrease with increasing plasma density due to exchange-correlation effects. The observed band gaps are typically renormalized by $\sim$ 20 meV within the range of plasma densities of interest which arise chiefly from the conduction band electrons and valence band holes. In the Q1D structures based on the confinement of electrons and holes, the electron-hole plasma is quantized in two transverse directions, thus the charge carriers essentially move only in the longitudinal direction. Recent progress in the fabrication techniques such as molecular-beam epitaxy (MBE) and lithographic deposition have made possible the realization of such quasi-one-dimensional systems.[2] Band-gap renormalization as well as various optical properties of the Q1D electron-hole systems have been studied[3, 4, 5, 6, 7] similar to

G. Abstreiter et al. (eds.), Optical Spectroscopy of Low Dimensional Semiconductors, 179–190.

the bulk (3D) and quantum-well (2D) semiconductors[8, 9, 10, 11] where generally good agreement with the corresponding measurements[12] exist.

Our main motivation comes from the recent experiments of Cingolani *et al.*[13, 14] in which they investigated the carrier density dependence of a quasi-one-dimensional electron-hole plasma confined in GaAs quantum wires using luminescence spectra. Comparing the band-gap data with the available calculations, Cingolani *et al.*[13] pointed out the need for a more realistic calculation. Density dependence of the BGR in Q1D systems was first considered by Benner and Haug[3] within the quasi-static approximation as previously employed for 2D and 3D systems.[8, 9, 10, 11] In a detailed study that appeared recently Hu and Das Sarma[4] also calculated the BGR, neglecting the hole population and considering an electron plasma confined in the lowest conduction subband only. The results of Hu and Das Sarma[4] are rather close to the experimental data.[13]

In this study our aim is to calculate the BGR using a statically screened approximation which is based on the RPA. We employ the temperature dependent, static, RPA dielectric function and address the question of validity of using the plasmon-pole approximation to it. We investigate the temperature dependence of the BGR at various electron-hole plasma densities and quantum well widths. We also discuss the effects of electron-phonon interactions.

The rest of this paper is organized as follows. In the next section we give a brief outline of the static screening approximation (quasi-static approximation). In Section III we present our results for the BGR in Q1D electron-hole plasmas and compare them with the experiments. Finally, we conclude with a brief summary of our main results.

2. Theory

For the Q1D system we consider a square-well of width a with infinite barriers. It may be built from a Q2D quantum-well (grown in the z-direction) by introducing an additional lateral confinement. We assume that effective mass approximation holds and for GaAs take $m_e = 0.067m$, and $m_h = 0.2m$, where m is the bare electron mass. Note that we have chosen the hole effective mass to reproduce on average the dispersion of the four topmost 1D subbands to conform with the experimental analysis of Ref. 13. The effective Coulomb interaction between the charge carriers is given by[4]

$$V(q) = \frac{2e^2}{\epsilon_0} \int_0^1 dx\, K_0(qax) \left[(1-x)[2+\cos(2\pi x)] + \frac{3}{2\pi}\sin(2\pi x)\right] , \quad (1)$$

in which $K_0(x)$ is the zeroth-order modified Bessel function of the second kind, and ϵ_0 is the lattice dielectric constant. Due to the presence of an electron-hole plasma, assumed to be in equilibrium, the bare Coulomb interaction is screened. The equilibrium assumption is justified since the laser pulse durations are typically much longer than the relaxation times of the semiconductor structures under study. Defining the statically screened Coulomb interaction as $V_s(q) = V(q)/\varepsilon(q)$, we consider the dielectric function in the random-phase-approximation (RPA)

$$\varepsilon(q) = 1 - 2V(q) \sum_{i,k} \frac{f_i(k) - f_i(k+q)}{\epsilon_i(k) - \epsilon_i(k+q) + i\eta}, \tag{2}$$

where the index $i = e, h$, and $\epsilon_i(k) = \hbar^2 k^2 / 2m_i$ are the bare single-particle energies. In most previous studies, the dielectric function $\varepsilon(q)$ was further simplified by the plasmon-pole approximation. Here we use the full static RPA at finite temperature without resorting to any approximations and discuss in the following section the validity of the plasmon-pole approximation.

Assuming a homogeneously distributed electron-hole plasma in thermal equilibrium the electron and hole distribution functions are written as

$$f_i(k) = \frac{1}{e^{\beta(\epsilon_i(k) - \mu_i^0)} + 1}, \tag{3}$$

where $\beta = 1/k_B T$ and μ_i^0 are the inverse carrier temperature and (unrenormalized) chemical potential of the different species, respectively. The plasma density N determines μ_i^0 through the normalization condition $N = 2\sum_k f_i(k)$.

Adopting the quasi-static approximation[8, 9] which amounts to neglecting the recoil effects relative to the plasma frequency in the full frequency dependent expressions, we decompose[8, 9] the electron and hole self-energies into screened exchange (sx) and Coulomb hole (Ch) terms: $\Sigma_i(k) = \Sigma_i^{\mathrm{sx}}(k) + \Sigma_i^{\mathrm{Ch}}$, where

$$\Sigma_i^{\mathrm{sx}}(k) = -\sum_{k'} V_s(k-k') f_i(k'), \quad \text{and} \quad \Sigma_i^{\mathrm{Ch}} = \frac{1}{2} \sum_{k'} [V_s(k') - V(k')]. \tag{4}$$

The above set of equations may be derived[8] from the dynamical self-energy expressions by neglecting all recoil energies with respect to the plasma frequency. As in the case of 2D and 3D calculations[8, 9, 10, 11] we assume that the BGR results from rigid bandshifts; i.e., the self-energies depend only weakly on wave vector k. The band-gap renormalization is then given by

$$\Delta E_g = E_g' - E_g = \Sigma_e(0) + \Sigma_h(0), \tag{5}$$

namely the electron and hole self-energies calculated at the respective band edges. Within the same spirit, we calculate the renormalized total chemical potential of the electron-hole plasma using

$$\mu_T = \sum_i [\mu_i^0 + \Sigma_i(k_F)] \,, \tag{6}$$

in which $k_F = \pi N/2$ is the Fermi wave vector. The self-energy part in the above expression is also called the exchange-correlation contribution $\mu_{\rm xc}$ to the chemical potential.

3. Results and discussions

In Fig. 1 we show the results of our calculation for the BGR (indicated by the solid curve) as a function of the electron-hole plasma density N. In order to make a ready comparison with the experimental results of Cingolani *et al.*[13] (shown by full circles), we have evaluated ΔE_g for a quantum wire of width $a = 500$ Å, at $T = 100$ K. The investigated[13] quantum wires were fabricated by plasma etching from quantum-well structures with lateral widths of 600 ± 50 Å. The experimental data have been collected over a whole set of spectra at various carrier temperatures. As we shall demonstrate below the BGR is not very sensitive to the temperature and we obtain rather good agreement with the experimental results. The zero-temperature calculation of Hu and Das Sarma[4] also represents well the Cingolani *et al.*[13] data, rendering the insensitivity of ΔE_g to temperature in the range of densities reported. The agreement between our calculated results and the experiment appears to be rather good. However, we caution that the experimental data points[13] were extracted from the observed luminescence spectra by assuming a free-carrier model. Our calculations indicate the importance of Coulomb effects. Thus a more refined line-shape analysis would be required to render the comparison more meaningful. To assess the relative importance of the various contributions to ΔE_g we display in Fig. 2 the screened-exchange (dashed line) and Coulomb hole (dotted line) parts of the self-energy. It is observed that the Ch term dominates for low densities ($N \leq 3 \times 10^5\,\text{cm}^{-1}$) but the total ΔE_g is mainly determined by the sx term at high densities. This situation is somewhat different than the case in 2D systems, where the Ch term approximates the band-gap renormalization satisfactorily for the relevant density regime.[15] On the other hand, a recent work[6] on InGaAs/InP quantum wires made use of the unscreened exchange energy (Hartree-Fock) only to account for the subband renormalizations.

The dotted curve in Fig. 1 gives the BGR calculated within the plasmon-pole approximation to the dielectric function using the same parameters. In

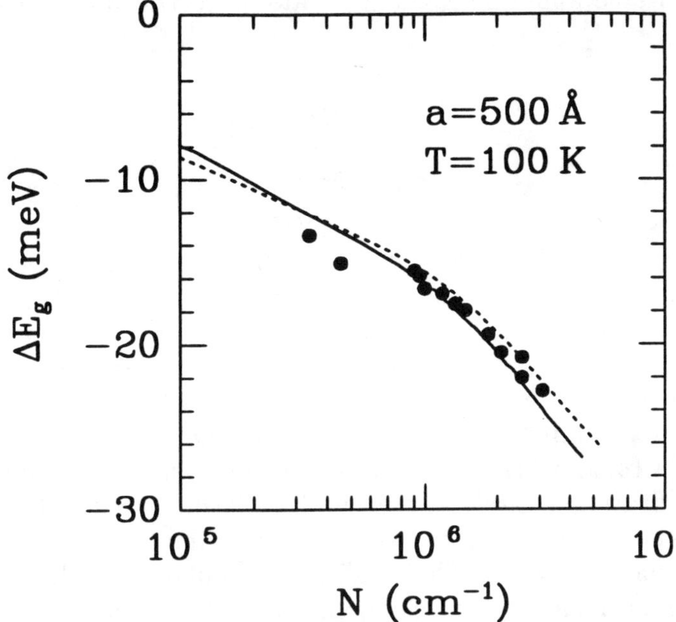

Figure 1. The calculated band-gap renormalization ΔE_g as a function of the electron-hole pair density. Full circles are the experimental results from Ref. 13. Solid and dotted lines are calculated with the full RPA dielectric function and the plasmon-pole approximation to it, respectively.

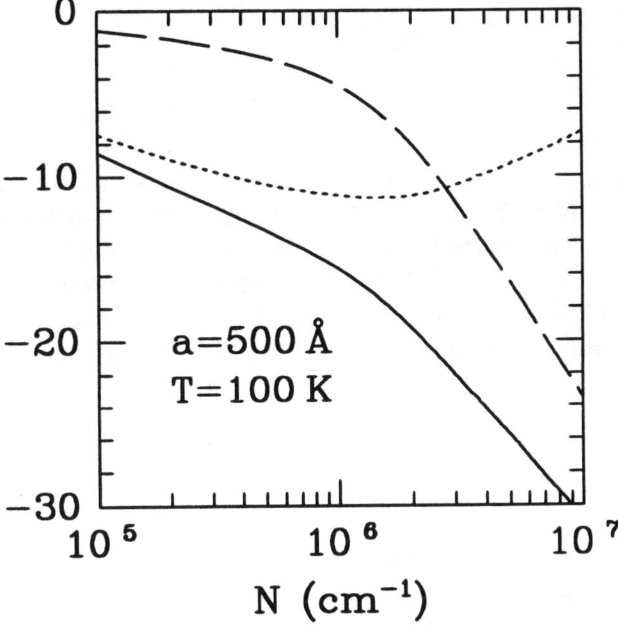

Figure 2. The sx (dashed line) and Ch (dotted line) contributions to the total band-gap renormalization (solid line) for $a = 500$ Å wide quantum wire at $T = 100$ K.

the plasmon-pole approximation the static dielectric function is expressed as[3]

$$\varepsilon(q) = 1 + \frac{\omega_p^2}{\frac{Nq^2}{\mu\kappa} + \left(\frac{q^2}{2\mu}\right)}, \tag{7}$$

where the plasmon frequency for the Q1D system is $\omega_p^2 = (N/\mu)V(q)$, and the screening parameter is $\kappa = \sum_i \partial N/\partial\mu_i^0$. Here $\mu^{-1} = m_e^{-1} + m_h^{-1}$ is the reduced mass. This is essentially the approach taken by Benner and Haug[3], where they use a parabolic confinement potential. The plasmon-pole approximation consists of ignoring the weight of single-particle excitations and assuming that all the weight of the dynamic susceptibility $\chi_0(q,\omega)$ is at an effective plasmon energy ω_p. It correctly describes the static and long wavelength limits of the full RPA expression. Most BGR calculations [c.f. Refs. 8,9,10, and 11] are performed in the plasmon-pole approximation and its justification is rarely addressed. Das Sarma *et al.*[16] have found significant deviations of the plasmon-pole approximation from the full RPA results in quantum wells. The qualitative similarity of dotted and solid curves in Fig. 1 demonstrates the applicability of plasmon-pole approximation in Q1D systems in contrast to Q2D systems as found by Das Sarma et al.[16] We have calculated the $\varepsilon(q)$ within the temperature dependent RPA and the plasmon-pole approximation and found that they are quite similar. The RPA calculation is performed using Eq. (2) at a finite temperature, since the thermal electron and hole distribution functions $f_i(k)$ are used. The temperature dependence of $\varepsilon(q)$ in the plasmon-pole approximation comes from the screening parameter κ. Our calculations indicate that the plasmon-pole approximation becomes better for large T.

In Fig. 3 we show the temperature dependence of the band-gap renormalization ΔE_g in the Q1D electron system. The solid lines indicate BGR for a system at $N = 10^5\,\mathrm{cm}^{-1}$ with $a = 100\,\text{Å}$ (lower curve) and $a = 500\,\text{Å}$ (upper curve). The dotted lines are for $N = 10^6\,\mathrm{cm}^{-1}$ with $a = 100\,\text{Å}$ (lower curve) and $a = 500\,\text{Å}$ (upper curve). The results shown in Fig. 3 were calculated using the full RPA dielectric function at finite temperature, but we found that plasmon-pole approximation also works quite well. Hu and Das Sarma[4] have also investigated the temperature dependence of the BGR within the leading-order dynamical screening approximation (*GW* approximation). Our statically screened approximation yields qualitatively similar results suggesting dynamical screening is not significant in the range of plasma densities of experimental interest.

We have evaluated the renormalized chemical potential of the electron-hole plasma including the exchange-correlation contribution as set out in the previous section. The total chemical potential μ_T of the Q1D electron-hole system with well width $a = 600\,\text{Å}$, for $k_BT = 8$ and $16\,\mathrm{meV}$ as

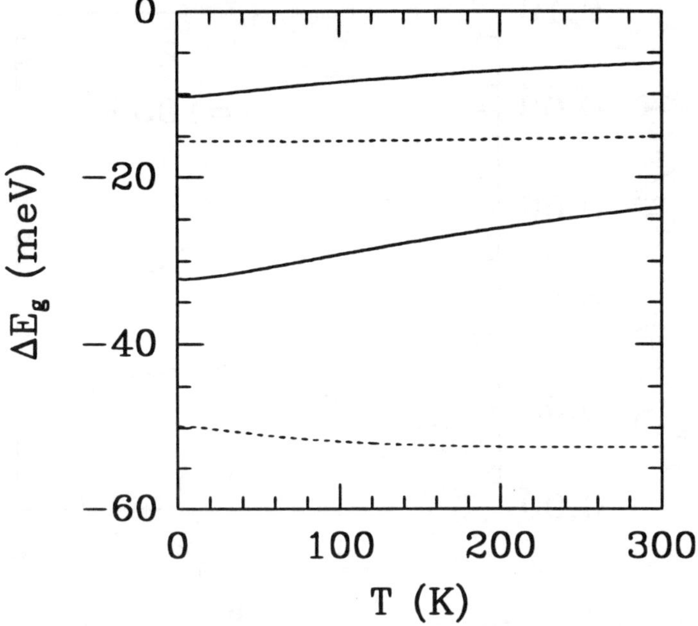

Figure 3. The temperature dependence of the band-gap renormalization for $N = 10^5$ (solid lines) and $N = 10^6\ \mathrm{cm}^{-1}$ (dotted lines). The upper and lower curves are for $a = 500$ and $a = 100$ Å wide quantum-well wires.

a function of the plasma density is in qualitative agreement[5] with the experimental results of Cingolani *et al.*[13] obtained for similar parameters [c.f. Fig. 4 of Ref. 13]. We found that there is a quantitative disagreement with the experiment especially for large densities, which may be attributed to the subband effects. Our calculations provide some indication about the rigid bandshift assumption. Using Eq. (6), we have calculated the exchange-correlation part of the chemical potential with both $\Sigma(k = 0)$ and $\Sigma(k = k_F)$ and found no notable difference which suggests that the bandshifts occur rigidly. Figure 4 exhibits the band-gap renormalization at the band-edge ($k = 0$) and at k_F, for quantum-wire widths $a = 500$ Å (solid line) and $a = 1000$ Å (dashed line).

There seems to be a discrepancy in the band-gap renormalization between theory and experiment at high densities for Q2D structures. The origin of this general disagreement is not well understood. Several attempts to improve the theory, particularly the multisubband population case, did not change the qualitative behavior of the BGR. To explore the existence of similar behavior it would be interesting to perform experiments in Q1D structures at higher densities.

We now discuss the effects beyond the RPA, the local-field corrections to the BGR in quantum wires. Writing the dielectric function as $\varepsilon(q) =$

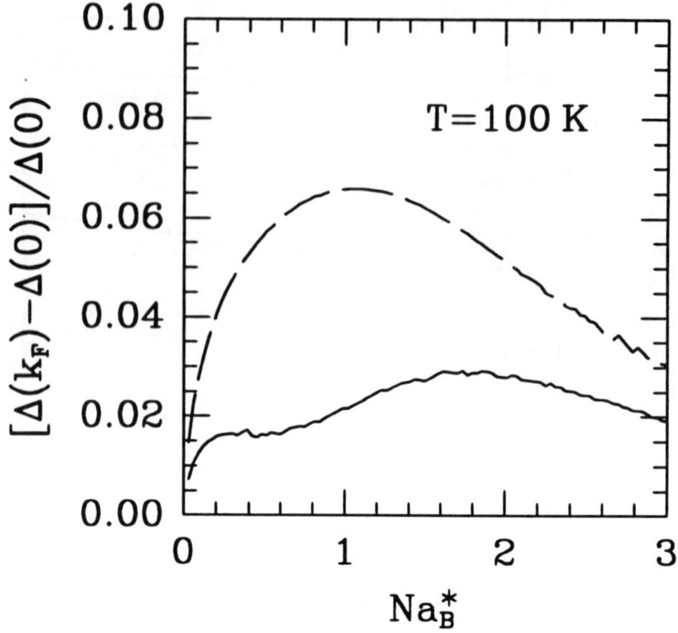

Figure 4. The total self-energy calculated at $k = k_F$ and at the band-edge ($k = 0$) as a function of the plasma density (we scale the density using the effective Bohr radius $a_B^* = \epsilon_0 \hbar^2/\mu e^2$ in terms of the reduced mass μ and dielectric constant ϵ_0). The solid and dotted curves indicate $a = 500$ Å and $a = 1000$ Å wide quantum-wires.

$1 - V(q)\Pi(q)[1 - G(q)]$, where $G(q)$ is the static local-field factor and $\Pi(q)$ is the static polarizability, we may account for the vertex corrections to $\Pi(q)$ in the mean-field sense. Recently, Schuster, Ell and Haug[17] considered finite-temperature vertex corrections in the form of second-order exchange contribution to the self-energy in 2D and 3D electron-hole plasmas. We use the equivalent of Hubbard approximation for $G(q)$ in one-dimension to obtain[18]

$$G(q) = \frac{1}{2}\frac{V(\sqrt{q^2 + k_F^2})}{V(q)} . \tag{8}$$

The physical nature of the Hubbard approximation is such that it takes exchange into account and corresponds to using the Pauli hole in the calculation of the local field correction between the particles of the same kind. Coulomb correlations are omitted. In this simple form, the static local-field factor $G(q)$ is temperature independent. Fig. 5 shows the BGR for quantum wires of various lateral widths at $T = 100\,\mathrm{K}$ with and without the local field corrections. The solid curves are calculated with the local field correction whereas the dotted curves give the RPA ($G(q) = 0$). We observe that within the simple Hubbard approximation to $G(q)$, the BGR

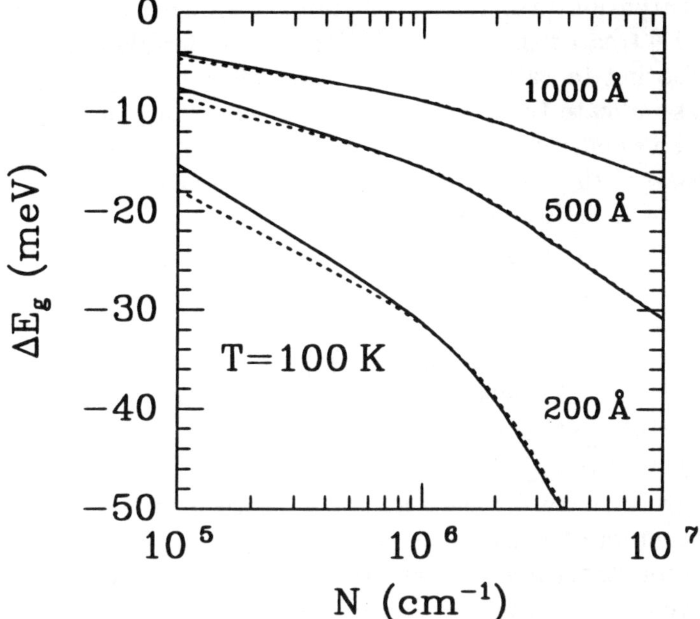

Figure 5. Effects of local-field corrections on the band-gap renormalization as a function of the plasma density at $T = 100$ K. Dotted and solid lines are calculated with and without local-field corrections, respectively.

deviates from the RPA result as the quantum-well width decreases. The difference in BGR with and without $G(q)$ is more appreciable for lower densities. We argue that in general RPA is a good approximation for high densities, but requires modifications for low densities. In order to assess a reliable measure of corrections beyond RPA, better approximations to the local field factor $G(q)$ are needed. Returning to the issue of discrepancy in the calculated ΔE_g and measurements for high densities in Q2D systems, it seems unlikely that improvements of RPA could yield satisfactory agreement. Elaborate calculations of Schuster et al.[17] give an indication in this direction. Clearly, more experimental measurements of the type reported by Cingolani et al.[13, 14] covering a wide range of plasma densities are necessary to resolve these questions.

As pointed out earlier, the band-gap renormalization accounts for the optical nonlinearities in the photoexcited semiconductor structures. The theoretical description of these phenomena in Q2D and 3D systems has been reviewed by Haug and Schmitt-Rink.[19] It would be interesting to examine the nonlinear optical properties of Q1D electron-hole systems using a similar approach. It is also possible to investigate the effects of electron-phonon interaction on the band-gap renormalization. Electron-phonon contribution to the BGR in Q1D systems were calculated by Güven and Tanatar[20] within perturbation theory and variational approaches. Recently, Dan and Bechstedt[21] treated the carrier-carrier and carrier-phonon interactions on an equal footing to calculate the phonon effects in Q1D systems. These calculations indicate the importance of the contribution of phonons to the BGR and their density dependence. The BGR due to carrier-carrier interactions increases with plasma density, whereas the carrier-phonon interactions tend to decrease ΔE_g at high densities.[20, 21]

For the Q1D electron system we have used the model developed by Hu and Das Sarma[4] which introduces an additional confinement to an infinite square-well. There are various other models of the quantum-well wire structures using parabolic confining potentials, geometrical reduction of dimensionality. The general trends obtained here for the plasma density and temperature dependence should be valid irrespective of the details of the model chosen.

Although we have carried out our numerical calculations for the material parameters of GaAs, the same formalism may be applied to other semiconductor structures such as InAs, GaSb, AlAs, etc. It would be desirable to have experimental results of the BGR for different Q1D semiconductor materials to compare with theoretical calculations. In 2D and 3D, a somewhat universal dependence of ΔE_g on plasma density is established largely independent of the band structure details. Whether a similar general behavior exists in Q1D electron-hole systems would be settled as more photoluminescence experiments become available.

4. Summary

We have found that the static plasmon-pole approximation to the dielectric function yields very close results to the full RPA expression. The temperature dependence of the BGR is weak for densities $N \sim 10^6\,\mathrm{cm}^{-1}$. Local-field corrections employed within the Hubbard approximation decreases the BGR at low densities especially when the lateral width of the quantum wire is small. The calculated renormalized chemical potential of the electron-hole plasma qualitatively differs from the measurement which may be attributed to the subband effects.

Extension of our calculations to cases where more than one subband is populated would be interesting for comparison with future experiments. It was recently found by Ryan and Reinecke[22] that in Q2D systems the intersubband interactions make significant contribution to the band-gap renormalization. More experimental results are needed in quantum wires to discuss fully the various aspects of BGR. Given the importance of the Coulomb interaction, it would be useful to analyze the experimental results with more refined line-shape models.

We gratefully acknowledge the partial support of this work by the Scientific and Technical Research Council of Turkey (TUBITAK) under Grant No. TBAG-AY/77, and fruitful discussions with Professors R. Cingolani, E. Kapon, and C. Sotomayor-Torres.

References

1. Schmitt-Rink S., Chemla D. S., and Miller D. A. B., (1989) Linear and nonlinear optical properties of semiconductor quantum wells, Adv. Phys. **38**, 89; Cingolani R. and Ploog K., (1991) Frequency and density dependent radiative recombination processes in III-V semiconductor quantum wells and superlattices, Adv. Phys. **40**, 535.
2. Plaut A. S., Lage H., Grambow P., Heitmann D., von Klitzing K., and Ploog K., (1991) Direct magneto-optical observation of a quantum confined one-dimensional electron gas, Phys. Rev. Lett. **67**, 1642; Goni A. R., Wiener J. S., Calleja J. M., Dennis B. S., Pfeiffer L. N., and West K. W., (1991) One-dimensional plasmon dispersion and dispersionless intersubband excitations in GaAs quantum wires, Phys. Rev. Lett. **67**, 3298.
3. Benner S. and Haug H., (1991) Plasma-density dependence of the optical spectra for quasi-one-dimensional quantum well wires, Europhys. Lett. **16**, 579; (1993) Influence of external electric and magnetic fields on the excitonic absorption spectra of quantum-well wires, Phys. Rev. B **47**, 15 570.
4. Hu B. Y.-K. and Das Sarma S., (1992) Many-body properties of a quasi-one-dimensional semiconductor quantum wire, Phys. Rev. Lett. **68**, 1750; (1993) Many-body exchange-correlation effects in the lowest subband of semiconductor quantum wires, Phys. Rev. B **48**, 5469.
5. Tanatar B., (1996) Band-gap renormalization in quasi-one-dimensional electron-hole systems, J. Phys. Condens. Matter **8**, 5997.
6. Wang K. H., Bayer M., Forchel A., Ils P., Benner S., Haug H., Pagnod-Rossiaux Ph., and Goldstein L., (1996) Subband renormalization in dense electron-hole plasmas in $In_{0.53}Ga_{0.47}As$/InP quantum wires, Phys. Rev. B **53**,10 505.
7. Grundman M., Christen J., Joschko M., Bimberg D., and Kapon E., (1995) Bandgap renormalization in quantum wires, in D. J. Lockwood (ed.), *The Physics of Semiconductors*, World Scientific, Singapore, p. 1675.
8. Haug H. and Schmitt-Rink S., (1984) Electron theory of the optical properties of laser-excited semiconductors, Prog. Quantum Electron. **9**, 3.
9. Schmitt-Rink S., Ell C., Koch S. W., Schmidt H. E., and Haug H., (1984) Subband-level renormalization and absorptive optical bistability in semiconductor multiple quantum well structures, Solid State Commun. **52**, 123; Haug H. and Koch S. W., (1989) Semiconductor laser theory with many-body effects, Phys. Rev. A **39**, 1887.
10. Ell C., Blank R., Benner S., and Haug H., (1989) Simplified calculations of the optical spectra of two- and three-dimensional laser-excited semiconductors, J. Opt. Soc. Am. B **6**, 2006.
11. Ell C., Haug H., and Koch S. W., (1989) Many-body effects in gain and refractive-index spectra of bulk and quantum-well semiconductor lasers, Opt. Lett. **14**, 356.
12. Tränkle G., Lach E., Forchel A., Scholz F., Ell C., Haug H., Weimann G., Griffiths G., Kroemer H., and Subbanna S., (1987) General relation between band-gap renormalization and carrier density in two-dimensional electron-hole plasmas, Phys. Rev. B **36**, 6712; Tränkle G., Leier H., Forchel A., Haug H., Ell C., and Weimann G., (1987) Dimensionality dependence of the band-gap renormalization in two- and three-dimensional electron-hole plasmas in GaAs, Phys. Rev. Lett. **58**, 419; Bongiovanni G. and Staehli J. L., (1989) Properties of the electron-hole plasma in GaAs-(Ga,Al)As quantum wells: the influence of the finite well width, Phys. Rev.

B **39**, 8359; Weber C., Klingshirn C., Chemla D. S., Miller D. A. B., Cunningham J. E., and Ell C., (1988) Gain measurements and band-gap renormalization in GaAs/$Al_xGa_{1-x}As$ multiple-quantum-well structures, Phys. Rev. B **38**, 12 748; Kulakovskii V. D., Lach E., Forchel A., and Grützmacher D., (1989) Band-gap renormalization and band-filling effects in a homogeneous electron-hole plasma in $In_{0.53}Ga_{0.47}As$/InP single quantum wells, Phys. Rev. B **40**, 8087.

13. Cingolani R., Rinaldi R., Ferrara M., La Rocca G. C., Lage H., Heitmann D., Ploog K., and Kalt H., (1983) Band-gap renormalization in quantum wires, Phys. Rev. B **48**, 14 331.
14. Cingolani R., Lage H., Tapfer L., Kalt H., Heitmann D., and Ploog K., (1991) Quantum confined one-dimensional electron-hole plasma in semiconductor quantum wires, Phys. Rev. Lett. **67**, 891.
15. Güven K. and Tanatar B., (1996) Simplified calculations of band-gap renormalization in quantum-wells, Superlatt. Microstruct. **20**, 81; and references therein.
16. Das Sarma S., Jalabert R., and Eric Yang S.-R., (1989) Band-gap renormalization in quasi-two-dimensional systems induced by many-body electron-electron and electron-phonon interactions, Phys. Rev. B **39**, 5516; (1990) Band-gap renormalization in semiconductor quantum wells, *ibid.* **41**, 8288.
17. Schuster S., Ell C., and Haug H., (1992) Vertex correction to the single-particle energy renormalization in three- and two-dimensional electron-hole plasmas, Phys. Rev. B **46**, 16 167.
18. Gold A. and Ghazali A., (1990) Analytical results for semiconductor quantum-well wire: plasmons, shallow impurity states, and mobility, Phys. Rev. B **41**, 7626.
19. Haug H. and Schmitt-Rink S., (1985) Basic mechanisms of the optical nonlinearities of semiconductors near the band edge, J. Opt. Soc. Am. B **2**, 1135.
20. Güven K. and Tanatar B., (1995) Phonon renormalization effects in photoexcited quantum wires, Phys. Rev. B **51**, 1784; (1996) Variational approach for phonon renormalization effects in photoexcited quantum wires and quantum wells, phys. status solidi b **197**, 369.
21. Dan N. T. and Bechstedt F., (1996) Optical phonon effects in quasi-one-dimensional semiconductor quantum wires: band-gap renormalization, Physica B **219&220**, 47.
22. Ryan J. C. and Reinecke T. L., (1993) Band gap renormalization in semiconductor quantum wells, Superlatt. Microstruct. **13**, 177; (1993) Band-gap renormalization of optically excited semiconductor quantum wells, Phys. Rev. B **47**, 9615.

OPTICAL PROPERTIES OF 1D QUANTUM STRUCTURES.

[a),b)]R.RINALDI and [b)]R.CINGOLANI
a) Istituto Nuovi Materiali per l'Elettronica del CNR,
Dipartimento di Scienza dei Materiali, Universita' degli Studi di Lecce,
via Arnesano , 73100 Lecce (Italy)
b) Istituto Nazionale di Fisica della Materia- Unita' di Lecce,
Dipartimento di Scienza dei Materiali, Universita' degli Studi di Lecce,
via Arnesano , 73100 Lecce (Italy)

1. Introduction

In this lecture we will present a comprehensive discussion of the optical properties of semiconductor quantum wires of different geometrical shape and potential profile. In the last ten years there has been an impressive development of nanotechnologies, aimed to the achievement of ultranarrow semiconductor wires (well below 100 nm) in large area planar arrays, exhibiting strong quantization of the electron states . The basic goal of this worldwide effort is to exploit the modification of the density of states, electron wavefunctions and non-linear polarizability induced by the reduction of dimensionality, for the realization of a new generation of nanodevices with low-threshold, high gain, and strong non-linear response.

In this contribution we will briefly describe the current methods for the fabrication of quantum wire (QWR) structures, then we will discuss the linear and non-linear optical processes involving one-dimensional (1D) electronic states. Finally, a short overview about the present status of quantum wire devices will be given.

G. Abstreiter et al. (eds.), Optical Spectroscopy of Low Dimensional Semiconductors, 191–211.

2.Methods for the fabrication of quantum wire heterostructures.

Several methods are currently used for the synthesis of quantum wire heterostructures [1]. The simplest rectangular-shape wires are realized by holographic or e-beam litography of a quantum well structure followed by wet or reactive ion etching. The resulting quantum wire heterostructures are schematized in Fig.1a (left hand side). The wires have rectangular section and due to the confining potentials along the growth direction (z-axis) and the lateral direction (y-axis) two rectangular wells originate (right hand side of Fig.1a). The main drawback of this method is the formation of large defected areas along the wire sidewalls, resulting in a dramatic deterioration of the optical properties of the structure (due to surface recombination and localization) [2]. Furthermore, it is almost impossible to integrate such free-standing structures in a device.

This problem is partly circumvented by means of the epitaxial regrowth of the patterned region, which originates a sort of buried quantum wire structure. The main advantage of this method is the lack of free surfaces limiting the optical performances of the structure. More recently, the epitaxial regrowth of substrates patterned along high-index crystallographic directions has been shown to be a succesfull method for the fabrication of high quality quantum wire structures of small lateral size and excellent optical quality. This is the case of the V-shaped wires and T-shaped wires, obtained by regrowing a quantum well at the bottom of a groove of suitable shape and orientation (V-shaped QWR), or along the in-situ cleaved (110) facet of a quantum well (T-shaped QWR), respectively. The V-shaped wires can be grown by MBE or MOCVD on patterned substrates of GaAs or InP [3] . The patterned substrates are usually obtained by means of holographic lithography and anisotropic wet chemical etching. The groove sidewalls lie on the (311) or (111) crystallographic planes. The typical groove periodicity is of the order of 250nm

RECTANGULAR QUANTUM WIRES

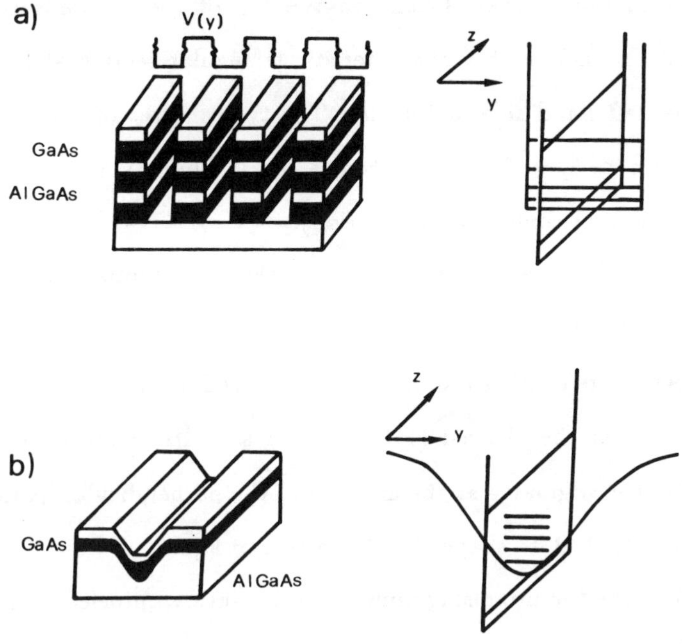

V-SHAPED QUANTUM WIRES

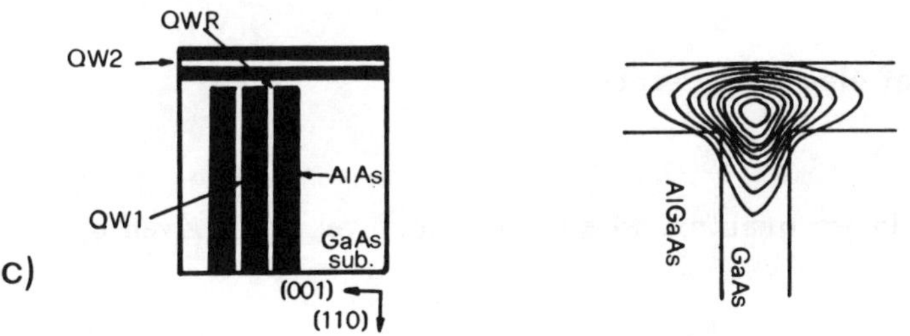

T-SHAPED QUANTUM WIRES

Fig. 1 - Schematic diagram of the rectangular (a), V-shaped (b), and T- shaped (c) quantum wire heterostructures (left hand side) and of the respective lateral potential shape (right hand side).

and the depth of the groove ranges between 130 and 160nm. The quantum well deposited into the grooves shrinks on the top of the sidewalls resulting in one-dimensional channels with a characteristic triangular section at the bottom of the grooves (see left hand side in Fig.1b). The typical sizes of these QWR are of the order of $14nm \leq L_y \leq 22nm$ and $6nm \leq L_z \leq 10nm$ (where z is the growth axis and y is the lateral confinement direction).

The shape of the vertical and lateral confining potential are shematized in the right hand side of the Fig.1b.

T-shaped wires are synthesized directly in the MBE chamber by growing a quantum well structure on the cleaved (110) edge of a multi quantum well specimen [4]. The 1D channel originates at the crossing of the perpendicular quantum well layers assuming the typical T- shape (left hand side of fig.1c).

Though these are the promising approaches for future optoelectronic applications, a number of other methods have been developed, namely, ion-implantation followed by thermal annealing, strain-induced formation of quantum wires, direct MBE growth on tilted substrates, etc. For a detailed description of the different technologies the reader is demainded to Ref. [1].

3. Theory of electronic states

In a semiconductor quantum wire the ground level gap is given by

$$E_n = E_g + E_{e,n_z,n_y} + E_{h,n_z,n_y} - E_b \tag{1}$$

where E_g is the gap of the bulk cristal, $E_{e(h),n_z,n_y}$ is the confinement energy of electrons (holes) of quantum numbers n_z and n_y (along the z and y-axis), and E_b is the binding energy of the exciton. In the simplest case, the confinement energies are evaluated by separating the confinement potential. This is possible when both the confining potentials along y and z (see Fig.1) are rectangular, or when the confinement along the y-direction is substantially smaller than that along z

(perturbative approach) [1]. In this case the envelope wavefunction for the conduction band states is:

$$\Psi_e = \phi_e(y) \cdot L_x^{1/2} exp(ik_x x) \chi_e(z) \tag{2}$$

the eigenenergies and the eigenfunctions are the solution of the following equation

$$\left\{ \frac{\hbar^2}{2m}(-\partial_y^2 + k_x^2) + V_y(y) - E_{xy} \right\} \phi_e(y) = 0 \tag{3}$$

$$\left\{ -\frac{\hbar^2}{2m}\partial_z^2 + V_z(z) - E_z \right\} \chi_e(z) = 0 \tag{4}$$

The function $V_y(y)$ represents the lateral confining potential,which depends on the geometrical shape of the wire, and has to be known precisely in order to get a reliable evaluation of the confinement energy. Rectangular QWRs generate eigenstates given by $E_{e(h),n_y} = \frac{\hbar^2}{2m_{e(h)}}(\frac{n_y \pi}{L_y})^2$. V-shaped wires, whose lateral potential is actually parabolic-like, generate a set of eigenstates of constant energy splitting, like in a harmonic oscillator [5]. The lateral potential of V-shaped wires is well reproduced by the function :$V_e(y) = -\frac{\Delta E_e}{cosh^2(y/2W)}$ with typical depth ΔE_e of the order of 150-180meV for electrons (W is the wire width). For a more refined modeling of the electronic states the lateral confinement cannot be considered as a small perturbation. In this case the potential $V_{y,z}$ is no longer separable along the two directions and the full two dimensional Schroedinger equation including the exact shape of the confining potential has to be solved [6]. This kind of calculation strongly relies on the exact determination of the geometric profile of the wires. This is usually obtained by cross-sectional transmission electron microscopy, from which the actual shape of the potential can be determined. The typical contour plot of the charge density of the first two electron subbands in a $GaAs/[(AlAs)_4/(GaAs)_8]$ V-shaped QWR of width $W \simeq 20$ nm is depicted in Fig.2.

For the case of the InGaAs wires grown on non-planar (V-shaped) GaAs or AlGaAs substrates one has to take into account the additional effect of strain [7, 8]. Strain on vicinal surfaces has a two-fold effect: first, it modifies the energy gap of the crystal according to well known Pikus and Bir Hamiltonian, and, second, it generates a piezoelectric field in the structure. This phenomenon occurs because of the non-vanishing off-diagonal components of the strain tensor along vicinal directions, which generate local dipoles responsible for the anisotropic piezoelectric field. The piezoelectric potential at the QWR interfaces varies from zero at the bottom of the groove (corresponding to the 100 direction), to some $10^4 V/cm$ on the top of the sidewalls. The main consequences of this field are the position dependent (y-dependent) potential offset for the electrons and holes, which modifies the quantization energies, and the large internal electric field which modifies the electro-optic behavior of the structure [7].

All the considerations reported above are valid for electrons and, at first approximation, for holes. However, one has to remember the complex structure of the valence band with degeneracy and mixing of the heavy and light hole bands. An accurate calcuation of the the valence band dispersion in 1D requires at least a 4x4 Hamiltonian (Kohn-Luttinger) [9]. More advanced calculations based on the LCAO and tight binding have been recently reported. However, this issue is beyond the scope of this contribution and the reader is demainded to other key-references [10, 11].

4. Linear Optical Properties of QWR

The optical spectroscopy of QWRs has evidenced two different confinement regimes depending on the actual size of the quantum wire. In the weak confinement regime (i.e. when $L_y > 3a_0$, where a_o is the exciton Bohr radius) the optical properties of the QWR exhibit a mixed 2D and 1D character, and the optical response of the QWR is dominated by the quantization of the center of mass motion of the exciton. In this case the wide lateral potential of the large wire does not influence directly the electron and hole envelope functions. The main effect of the lateral boundaries is to confine the envelope function of the exciton as a whole. The motion of the center of mass (c.m.) of the exciton is thus limited by the lateral sidewalls where

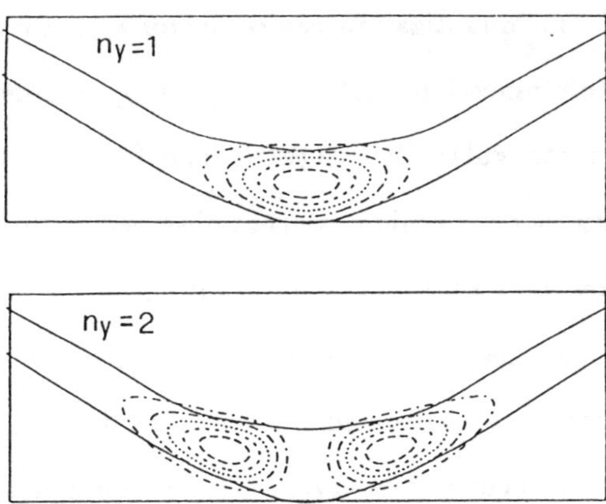

Fig. 2 - Contour plots of the $n_y = 1$ and $n_y = 2$ electron charge densities for a V-shaped wire.

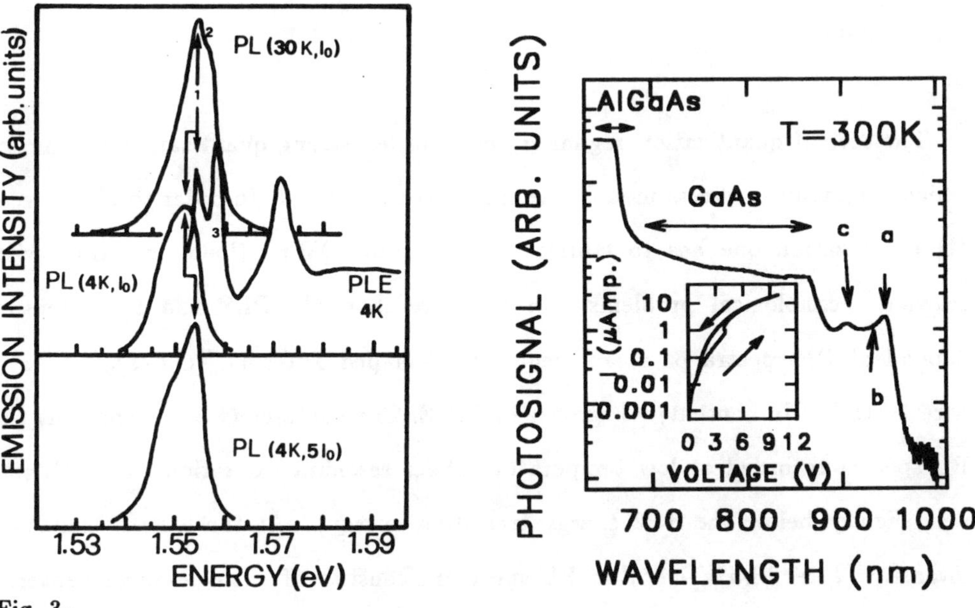

Fig. 3 -

Temperature- and intensity dependent photoluminescence (PL) and photoluminescence excitation (PLE) spectra of a rectangular GaAs wire ($L_y = 60nm$). The middle curves have been recorded at 4K . $I_0 = 0.1Wcm^{-2}$. The upper PL spectrum was recorded at 30K and at the lowest excitation intensity , whereas the lowest PL spectrum was recorded at 4K and under higher excitation intensity . The numbers label the transitions among quantized states of the excitonic center-of-mass. The arrows indicate the Stokes shift between the PL and PLE spectra.

standing waves and c.m. quantization can occur for $k_{exc} = n\pi/L_y$ [12, 13]. In Fig.3 we report the photoluminescence (PL) and photoluminescence excitation (PLE) spectra of a 60nm rectangular GaAs QWR. The PLE spectrum shows a twofold splitting of the low energy excitonic peak (not observed in the quatum well reference sample) and a Stokes-shift of about 3 meV with respect to the low intensity cw-luminescence. This Stokes shift is found to disappear with a little increase of temperature or excitation intensity and is ascribed to the presence of excitonic states localized at the free-surfaces of the QWR. The double peak in the PLE is due to the first and third c.m. states of the exciton. The quantized center of mass states have a discrete energy spectrum given by $E_{n_{c.m.}} = 1554.5 + Cn^2_{c.m.}(meV)$, where C is a constant.

The second quantization regime is the so-called strong quantization regime, in which confinement effects modify the single particle states. In order to obtain real 1D quantization one has to fabricate very narrow QWRs (below 50 nm), with relevant technological problems. In Fig.4 we show the PLE and the intensity dependent PL spectra of a representative V-shaped $InGaAs/[(GaAs)_8/(AlAs)_4]$ wire (with 0.1 In content) [7], grown by MBE. The continuous lines represent the PL spectra recorded at low temperature under resonant excitation ($\lambda = 842nm$, i.e. slightly below the n=2 energy transition energy). At the lowest excitation intensity ($I \simeq 13mW/cm^2$)the PL spectrum consists of a single band centered around 1.4516eV with a linewidth of 10 meV. With increasing excitation intensity the emission intensity increases linearly without any shift of the emission line. Off-resonant excitation (dashed curves in Fig.4) clearly results in a very efficient band filling, already visible at low power excitation density. At $I = 10Wcm^{-2}$ three bands are visible in the spectra due to recombination processes involving the first three quantized subbands (with quantum numbers n_y). The band separation shows almost constant energy splitting, as expected from the quasi-parabolic shape of the lateral confining potential. The PLE exhibits several structures. The ground-level state appears as a bump in the low energy tail of the PLE line shape, whereas at higher energies the $n_y = 2$ and $n_y = 3$ bands show a larger joint density of states, as it is expected from the crescent shape of the wire. The spectral separation of the PLE structures is consistent with that measured in the PL spectra and the Stokes shift amounts to about 5.5meV, which is consistent with the half width at half maximum of the PL band. Recently unprecedentedly well resolved polarization dependent PLE spectra, exhibiting sharp exciton resonances, have been reported for V-shaped wires grown by MOCVD [14] and T-shaped wires grown by MBE [15].

The linear-absorption spectra of GaAs wires have been calculated taking into account electron-hole correlation effects (Hartree-Fock model) in the solution of the semiconductor Bloch equations [6]. The inclusion of a level broadening comparable to the spectral broadening results in a theoretical linear-absorption spectrum with broad features, that perfectly reproduces the experimentally observed PLE spectra, giving exciton binding energies usually enhanced by about 25 % with respect to the 2D case.

A powerful experimental method to obtain information on the exciton binding energy is the analysis of magnetoluminescence spectra of QWRs. In Fig.5 we display the temperature dependent diamagnetic shift of the PL measured in GaAs and InGaAs V-shaped wires (symbols). This is compared to the theoretically calculated shift of the magnetoexcitons and of the free-carriers in the wire (lines) [16]. The ground-state-exciton binding energy of the investigated GaAs wires was independently determined to be about 12meV by two-photon absorption spectroscopy [17] (see next section). Therefore, for temperatures of the order of 150K excitons are expected to be ionized and the emission process must involve free carriers. This is clearly seen in Fig. 5b). At 4 K the diamagnetic shift is rather small, and it is

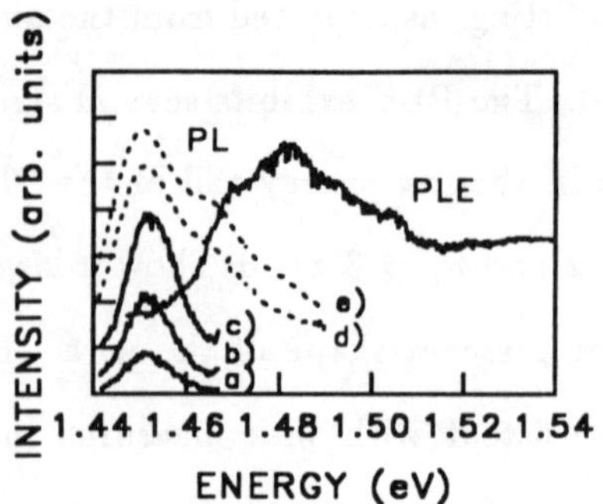

Fig. 4 - PLE and PL spectra of an $In_{0.12}Ga_{0.88}As/[(AlAs)_4/(gaAs)_8]$ V-shaped quantum wire under cw excitation at 10K. The solid PL curves were recorded under quasi resonant excitation (842nm) at power densities of (a) $I = 750mWcm^{-2}$; (b) $I = 2.1Wcm^{-2}$; (c) $I = 6.2Wcm^{-2}$. The dashed lines were recorded under off-resonant excitation (514nm) at power densities of (d) $I = 7.5Wcm^{-2}$ and (e) $I = 15Wcm^{-2}$. The detection wavelength for the PLE curve was set to 857.5nm and the measured excitation power at 842nm was $6.2Wcm^{-2}$.

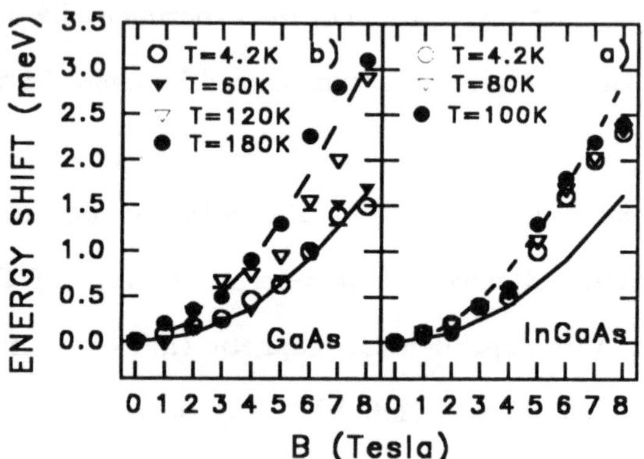

Fig. 5 - (b) Diamagnetic shift of the $n_y = 1$ luminescence band of a V-shaped GaAs wire at different temperatures (symbols) . Lines represent the theoretically evaluated shifts for excitons (solid lines) and for free carriers (dashed lines). - (a) Same as in (b) for a InGaAs V-shaped wire.

well reproduced by the excitonic diamagnetic shift evaluated theoretically (solid lines in Fig. 5). The shift remains excitonic at low temperatures (40 K). Around 120 K we observe some change in the diamagnetic shift, suggesting that excitons begin to dissociate and an exciton-free-carrier gas forms in the structure. Above 120 K the shift of the magnetoluminescence increases considerably with increasing field, and follows the expected free-carrier shift (dashed lines in Fig. 5). In the case of InGaAs wires (Fig. 5a) the situation is somewhat more complicated. The internal piezolectric field reduces the exciton binding energy to a few meV, and the transition from excitonic to free-carrier recombination should occur at lower temperature than in GaAs. However, in this case the extension of the exciton wavefunction (a_o) is larger than the magnetic length $l = \sqrt{\frac{\hbar}{eB}}$, so that the magnetic interaction involves directly the individual electron and hole wavefunctions, resulting in a free-carrier-like shift. An excitonic shift should be observed in the opposite case (i.e. for large binding energies). From Fig.5 one can see that below 4 Tesla the calculated excitonic and free-carrier shifts are almost indistinguishable, and the comparison with the experimental data does not clarify the recombination mechanism. Above this field the experimental data and the theory indicate the occurrence of free-carrier recombination. This puts the upper limit of 4 Tesla to the occurrence of diamagnetic shift of excitonic origin in these InGaAs wires. Such a value roughly corresponds to an exciton binding energy of the order of 5 meV. This is in turn consistent with a reduction of the exciton binding energy due to the internal piezoelectric field [7].

Finally, we would like to make some considerations on the typical lifetime of excitons confined in 1D semiconductors . Most of the GaAs and InGaAs quantum wires have an intersubband separation of the order of 15-20meV for electrons. This is smaller than the characteristic LO phonon energy. Therefore one expects that photogenerated carriers do not efficiently relax at the ground level transition , within their lifetime, and preferentially recombine from the higher index states. This is consistent with the observation of band filling excitonic luminescence even under very low power excitation. This is shown in Fig.6 , where we plot the temporal evolution of the $n_y = 1$, $n_y = 2$, and $n_y = 3$ transitions for a 20 GaAs V-shaped quantum wire array. The $n_y = 1$ transition exhibits a rather long decay time, of the order of 360ps , with a long plateau region due to free- carrier formation coexisting with the exciton gas . The higher index transitions exhibit decreasing decay times (190ps and 120 ps, respectively) , indicating that relaxation occurs both through radiative recombination and non-radiative decay via acoustic phonon-interaction or intercarrier scattering [18] . At low power excitation , in the excitonic regime, the three decay times are almost equal and there is no plateau region in the decay curve of the $n_y = 1$ luminescence.

5. Non-Linear Optical Properties

The non-linear optical properties of quantum wires have been investigated by two photon absorption-induced photoluminescence excitation (TPA-PLE) spectroscopy [17, 19] and high excitation spectroscopy under stationary and transient conditions [5, 7, 20].

Two photon spectroscopy in quantum wires shows strongly anisotropic selection rules, depending on the relative orientation of the laser polarization vector ϵ with respect to the quantization direction y [17, 19, 21]. In the $\epsilon \parallel y$ geometry , $1s$ excitons associated with $\Delta n_y = \pm 1, \pm 3, ...$ transitions are allowed as final states of the non-linear absorption process. On the contrary , $2p$ excitons associated with $\Delta n_y = 0$ transitions are expected in the $\epsilon \perp y$ configuration. The complementarity of these selection rules with the one-photon process permits the measurement of the $2p - 1s$ splitting of 1D excitons. This splitting is important for the evaluation of the exciton binding energy [17, 19] . In Fig.7 the TPA-PLE spectra in the $\epsilon \parallel y$ and

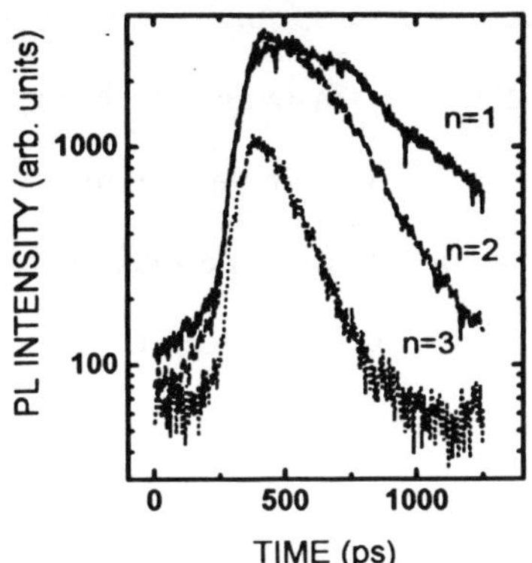

Fig. 6 - Temporal evolution of the $n_y = 1$, $n_y = 2$, and $n_y = 3$ luminescence for a 20 GaAs V-shaped quantum wire sample.

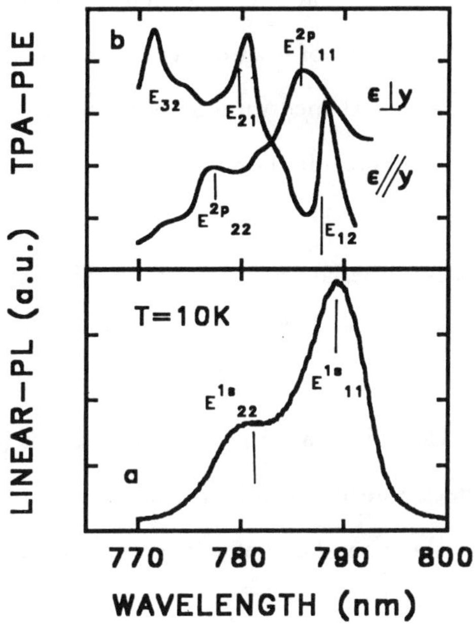

Fig. 7 - (a) Linear PL spectrum of a GaAs V-shaped wire measured at 10K under cw excitation (Ar^+ laser). - (b) TPA PLE of the same wire measured in the $\epsilon \perp y$ and $\epsilon \parallel y$ configurations. $1s$ and $2p$ label the final exciton state assocoted with electron (e) and hole (h) subbands of quantum number n_y in the wire. The vertical lines indicate the theoretically evaluated transition energies.

$\epsilon \perp y$ geometries of a GaAs V-shaped quantum wire is shown and compared with the linear PL spectrum. The spectra evidence sharp resonances corresponding to the transitions calculated according to the above mentioned selection rules (vertical lines). Assuming an hydrogenic exciton, the exciton binding energy obtained from the 1s-2p splitting is found to be 12 meV (i.e. about 25 % larger than in the corresponding (100) quantum well)).

Under high excitation intensity excitons are screened and the optical spectra are dominated by free carrier recombination. The large photogenerated density of electrons and holes causes strong many body effects like band gap renormalization and hot carrier generation, which have to be taken into account for the analysis of the optical spectra and for the design of laser devices based on quantum wires. Typical band filling spectra of GaAs V-shaped wires are reported in Fig.8. The spectra span over a large energy range and include the emission from the QWR, (bottom of the V-groove), from the lateral part of the quantum well along the sidewalls, and from the wire barrier (a GaAs/AlAs superlattice in this case). The important feature of these spectra is the strong band filling of the quantized states , which overwhelmes three interband recombination lines even at the lowest excitation density. With increasing the excitation intensity the emission saturates and broadens (due to hot carrier and lifetime broadening effects) until a featureless emission line-shape is observed. A quantitative analysis of the many-body effects in rectangular QWR can be found in Ref. [20].

6. Quantum Wire Devices

Great attention has been given to new devices based on low dimensional heterostructures Quantum wire laser based on three dimensional quantum wire arrays have been demonstrated [3, 22, 4].

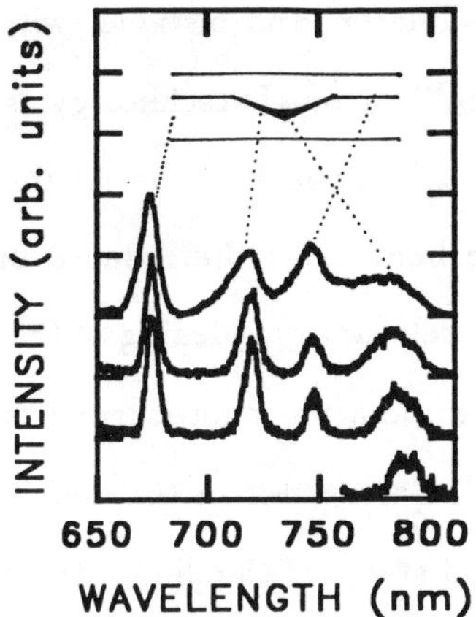

Fig. 8 - High excitation intensity PL spectra of a GaAs V-shaped wire heterostructure measured at (from bottom to top) $0.001I_0$, $0.05I_0$, $0.1I_0$ and $0.25I_0$, with $I_0 = 50kWcm^{-2}$. The lines indicate the sample regions originating the different luminescence bands.

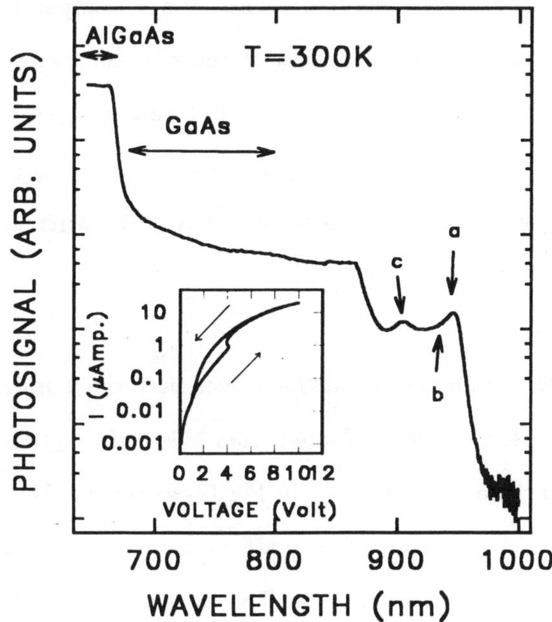

Fig. 9 - Room temperature photocurrent spectrum of an InGaAs p-i-n diode with a planar array of InGaAs V-shaped wires in the depletion region . The a, b, and c arrows indicate the $n_y = 1, n_z = 1$, $n_y = 2, n_z = 1$ and $n_y = 1, n_z = 2$ transition energies , respectively. Inset: Current-voltage curves at 80K recorded by sweeping upwards and downwards the voltage.

Non-linear quantum wire modulators and bistable switches have also been recently reported [23]. The main problem of 1D-technology is the reduction of disorder, inhomogeneity and size fluctuations which smear out the otherwise sharp 1D density of states, thus canceling the benefits of the reduced dimensionality. This can be immediatly inferred from the relevant broadening of the optical spectra of QWR. In Fig. 9 we report the room temperature photocurrent spectra of a p-i-n diode with a planar array of V-shaped InGaAs wires in the depletion region. The phocurrent reproduces the joint density of states of the diode structure: around 665nm there is the edge of the doped AlGaAs cladding layer. The second step in the photocurrent (around 872nm) is due to the GaAs barrier of the InGaAs wire. The wire resonances are located in the 900-1000nm spectral region, with two features ascribed to the n=1 and n=2 excitons [23]. Measurements of the temperature dependent I-V show strong hysteresis. The I-V curves of the quantum wire diode recorded by sweeping the current upwards and downwards follow a different path, originating a negative resistance region (S-shaped) when the current is increased (Fig. 9). This phenomenon has been explained assuming that the wires act like traps for the vertically transported carriers [23].

Electroluminescence measurements have been performed on these InGaAs diodes. The electroluminescence was collected from the surface of the sample, the bias being parallel to the growth axis [24]. In Fig.10 we report the electroluminescence spectra of an InGaAs quantum wire diode as a function of the pulsed injection current flowing in the structure at 80 K. The weakest detectable signal is observed for an injection current of 640μA at a repetition cycle of 67Hz and a duty cycle of 50%. With increasing the electrical pumping a broadening of the spectra and a red shift is observed. The broadening on the high energy side is due to band filling whereas the red-shift indicates the occurrence of band gap renormalization [3,4] and local sample heating at the highest applied currents. The dotted lines on the spectra represent the energy positions of the interband transitions obtained by a Gaussian deconvolution of the spectra. Though the strong broadening of the lines prevents the observation of distinct interband transitions, the transition energies obtained by the deconvolution are in excellent agreement with those calculated as explained in section 2.

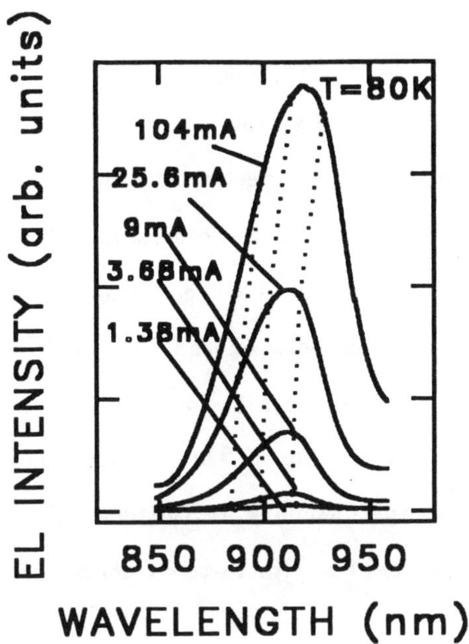

Fig. 10 - Electroluminescence (EL) spectra of a p-i-n diode with a planar array of InGaAs V-shaped wires in the depletion region recorded at 80K. The dashed lines indicate the energy positions of the interband transitions obtained after the deconvolution of the spectra. The interband splitting coincides with the theoretical value.

Finally, an anomalous quantum confined Stark (QCSE) effect has been measured in InGaAs V-shaped wires. Both the red shift and the quench of the exciton oscillator strength have been observed either in reverse or in forward bias, though with different switching voltages. This new effect is interpreted as a consequence of the vectorial composition of the internal piezoelectric field and the external field, leading to a net QCSE in both polarizations [23]. The rapid improvement of technology makes the future developement of this effect quite interesting for in-plane non-linear light modulators based on QWRs.

7. Conclusions

In summary we have reviewed recent optical and transport experiments performed on quantum wire structures of different shape and size. The technology of quantum wire is improving quite fast, and it is now possible to investigate heterostructures with clear 1D features in the optical spectra. Though some basic device has been recently proposed, much work has still to be done in order to make quantum wire devices really competitive with standard 2D technologies. This work is the outcome of a long–standing collaboration between the Optoelectronic Group at Material Science Department (Lecce University), the Micro-optoelectronic Institut of EPFL-Lausanne (Prof.F.K.Reinhart and Dr. U. Marti), and the former group of Max-Planck Institut fuer festkoerperforschung at Stuttgart (Prof. D.Heitmann and Dr. H.Lage). The authors are indebted with these collaborators.

8. References

[1] R.Cingolani and R.Rinaldi, " *Electronic States and Optical Transitions in Low-Dimensional Semiconductors* ", Rivista del Nuovo Cimento **16** (9),1 (1993)

[2] R.Cingolani, H.Lage, L.Tapfer, H.Kalt, D.Heitmann, and K. Ploog, "*Quantum*

Confined One-Dimensional Electron-Hole Plasma in Semiconductor Quantum Wires", Phys. Rev. Lett. **67**, 891 (1991)

[3] E.Kapon,*"Lateral Patterning of Quantum Well Heterostructures by Growth on Nonplanar Substrates"*, *Semiconductors and Semimentals* vol. 40, 259, Academic Press 1994

[4] W.Wegscheider, L.N.Pfeiffer, K.W. West, P.Littlewood, O.Narayan, M.Hagn, M.M.Dignam,and R.E.Leibenguth *Strong Magnetic Field Dependence of Laser Emission from Quantum Wires Formed by Cleaved Edge Overgrowth"* Solid State El. **40**, 1 (1996)

[5] R.Rinaldi, R.Cingolani, F.Rossi, L.Rota, M.Ferrara, P.Lugli, E.Molinari, U.Marti, D.Martin, F.Morier-Gemoud, F.K.Reinhart, *" Investigation of Quantum States in V-Shaped GaAs Quantum Wires"*, Inst. Phys. Conf. Ser. **136**, 233 (1993)

[6] F.Rossi, E.Molinari, R.Rinaldi, R.Cingolani, *" V-Grooved Quantum Wires as Prototypes of 1D-Systems : Single Particle Properties and Correlation Effects"*, Solid State Electron. **40**, 249 (1996)

[7] R.Rinaldi, R.Cingolani, L.DeCaro, M.Lomascolo, M.DiDio, L. Tapfer, U.Marti, and F.K.Reihart,*"Optical Spectroscopy of InGaAs/GaAs V-Shaped Quantum Wires"*, J. Opt. Soc. Am. **B13**, 1031 (1996)

[8] M.Grundmann, O.Stier, and D.Bimberg, *Symmetry Breaking in Pseudomorphic V-groove Quantum Wires"*, Phys. Rev. **B50**, 14187 (1994)

[9] U.Bockelmann and G.Bastard, *" Interband Optical Transitions in Semiconductor Quantum Wires: Selection Rules and Absorption Spectra"*, Europhys. Lett. **15**, 214 (1991)
U.Bockelmann and G.Bastard, *" Interband Absorption in Quantum Wires. I. Zero-Magnetic-Field Case"*, Phys. Rev. **B45**, 1688 (1992)

[10] D.S.Citrin and Yia-Chung Chang , *"Valence-Subband of GaAs/AlGaAs Quantum Wires: The Effect of Splitt-Off Bands"*, Phys. Rev. **B40**, 5507 (1989)

[11] G.Goldoni, F.Rossi, E.Molinari, A.Fasolino, R.Rinaldi, R. Cingolani, *" Valence Band Spectroscopy in V-grooved Quantum Wires"*, Appl. Phys. Lett. , in press (1996)

[12] H.Lage, D.Heitmann, R.Cingolani, P.Grambow, and K.Ploog, *Center-of-Mass Quantization of Excitons in GaAs Quantum-Well Wires"*, Phys. Rev. **B44**, 6550 (1991)

[13] R.Rinaldi, R.Cingolani, M.Ferrara, H.Lage, D.Heitmann, and K.Ploog, *Emission Properties of Quantum Well Wires under Stationary Conditions"*, Phys. Rev. **B47**, 7275 (1992)

[14] E.Kapon, G.Biasiol, D.M.Hwang, M.Walther, E.Colas, *"Self-Ordering Mechanism of Quantum Wires Grown on Nonplanar Substrates"* , Solid-State El. **40**, 815 (1996)

[15] H.Akiyama, T.Someya, and H.Sakaki, *" Dimensional Crossover and Confinement-Induced Optical Anisotropy in GaAs T-shaped Quantum Wires"*,Phys. Rev. **B53**, R10520 (1996)

[16] R.Rinaldi, P.V.Giugno, R.Cingolani, F.Rossi, E.Molinari, U.Marti, and F.K.Reinhart,*"Thermal Ionization of Excitons in V-Shaped Quantum Wires"*, Phys. Rev. **B53**, 13710 (1996)

[17] R.Rinaldi, R.Cingolani, M.Lepore, M.Ferrara, I.M.Catalano, F. Rossi, L.Rota, E.Molinari, P.Lugli, U.Marti, D.Martin, F.Morier-Gemoud, P. Ruterana, and F.K.Reinhart, *" Exciton Binding Energy in GaAs V-Shaped Quantum Wires"*,Phys. Rev. Lett. **73**, 2899 (1994)

[18] A.C.Maciel, J.F.Ryan, R.Rinaldi, R.Cingolani, M.Ferrara, U .Marti, D.Martin, F.Morier-Gemoud, and F.K.Reinhart, *"Hot Carrier Photoluminescence from GaAs V-Groove Quantum Wires"*, Semicond. Sci. Technol . **9**, 893 (1994)

[19] R.Cingolani, M.Lepore, R.Tommasi, I.M.Catalano, H.Lage, D .Heitmann, K.Ploog, A.Shimizu, H.Sakaki, and T.Ogawa, *" Two-Photon Absorption in GaAs Quantum Wires"* , Phys. Rev. Lett **69**, 1276 (1992)

[20] R.Cingolani, R.Rinaldi, M.Ferrara, G.C.LaRocca, H.Lage, D. Heitmann, K.Ploog, and K.Kalt,*" Band-gap Renormalization in Quantum Wires"*, Phys. Rev. **48**, 14331 (1993)

[21] A.Shimizu, T.Ogawa, and H.Sakaki, *" Two-Photon Absorption Spectra of Quasi-low-dimensional Exciton Systems "*, Phys. Rev. **B45**, 11338 (1992)

[22] S.Tiwari, G.D.Petitt, K. R. Milkove, F.Legoues, R.J.Davis, J.M.Woodall, *High Efficiency and Low Threshold Current Strained V- groove Quantum Wire Laser "*, Appl. Phys. Lett. **64**, 3536 (1994)

[23] R.Cingolani, R.Rinaldi, M.DeVittorio, L.Vasanelli, A.Cola,U. Marti, and F.K.Reinhart, *"Charge Trapping and Current-Voltage Bistability in InGaAs Quantum Wires"*, J. Appl. Phys. **80**, 936 (1996)

[24] R.Rinaldi, M.DeVittorio, R.Cingolani, U.Marti, amd F.K. Reinhart, Superlatt. and Microstruct., *" Electro-Optic Processes in InGaAs/GaAs Quantum Wires Grown by MBE on Patterned Substrates"*, in press (1996)

CONDUCTANCE IN NANOWIRES

H. MEHREZ AND S. CIRACI

Department of Physics,
Bilkent University,
06533, Ankara, Turkey

Abstract. This paper presents a detailed analysis of conductance and atomic structure in metal nanowires under tensile stress. We calculate the variation of conductance with the crossection of the constriction between two reservoirs, that is represented by three-dimensional circularly symmetric potentials. The absence of several observed features in the calculated conductance variations, in particular sudden jumps, suggests that the discontinuous rearrangements of atoms under stretch dominate the electron transport. To analyze the variations of atomic structure, we performed simulations based on the state of the art molecular dynamics simulations and revealed novel structural transformations. It is found that yielding and fracture mechanisms depend on the geometry, size, atomic arrangement and temperature. The elongation under uniaxial stress is realized by consecutive quasi elastic and yielding stages; the neck develops mainly by the implementation of a layer with a smaller crossection at certain stages of elongation. This causes to an abrupt decrease of the tensile force. Owing to the excessive strain at the neck, the original structure and atomic registry are modified; atoms show tendency to rearrange in closed-packed structures. In certain circumstances, a bundle of atomic chains or single atomic chain forms as a result of transition from the hallow site to the top site registry shortly before the break. The origin of the observed "giant" yield strength is explained by using results of present simulations and *ab initio* calculations of total energy and Young's modulus for an infinite atomic chain.

G. Abstreiter et al. (eds.), Optical Spectroscopy of Low Dimensional Semiconductors, 213–234.

1. Introduction

The optical, electronic and mechanical properties of condensed systems with nanometer dimensions can be rather different from those of the bulk. The effects of reduced size and dimensionality, in particular properties leading to novel devices have been subject matter of several research works in the last decade[1]. By pulling the tip of a scanning tunneling microscope (STM) after the nanoindentation[2-6] or by bending a mechanically controllable break junction[7], long metal wires with diameters in the range of few λ_F have been produced. As the crossection of the wire is reduced by stretching it continuously, the two terminal conductance G is measured. The measured conductance showed irregular and sudden changes with stretch s; the detailed behavior of which was not only sample specific, but also was varying in different scans of the same experiment. The abrupt falls of $G(s)$ became generally sharper and closer to the integer multiples of quantum conductance ($G_o = 2e^2/h$ per two spin) with decreasing crossection prior to the break of the neck. This has been taken as a manifestation of the quantization of conductance in atomic size wires (or connective necks). In contrast to various quantization arguments[5,6], some authors related abrupt variation of $G(s)$ to the irregular and discontinuous changes of the crossection of the wire $A(s)$ in the course of stretch[8-11]. Presently, these predictions were strengthened by the experimental results[12,13] that have been obtained by using a combination of STM and AFM (atomic force microscope). These experiments have shown that any sudden change in conductance is accompanied by an abrupt modification of atomic structure and hence the crossection of the neck. The yield strength σ_Y measured shortly before the break has been found much higher than that of the bulk.

The narrowest diameter of the nanowire prior to the break is only a few angstrom; it has the length scale λ_F, where discontinuous (discrete) nature of the metal dominates over its continuum description. For example, in this length scale, the level spacings of electrons ($0.1 - 1eV$) become easily resolved even at room temperature, and any change of atomic structure may lead to detectable changes in the related properties. For example the ballistic electron transport through nanowire is expected to be closely related with the atomic structure at the narrowest part of the connective neck. It becomes now clear that the mechanical properties, in particular the yielding mechanisms of the nanowires are quite different from those of bulk material.

We believe that the subject of dispute about the quantization of conductance emerges from the diverse criteria; this will be clarified in the present work by the analysis of ballistic transport in a nanoindentation and in an atomic size connective neck that is pulled between two electrodes. In this paper, we investigate the ballistic electron transport on the idealized wires which have circularly symmetric potential but short necks as in the experiment. Since the calculated variation of G with the radius does not exhibit sharp step structure, we seek the origin of the abrupt jumps in the structural changes of the wire during stretch. To this end, we carry out state of the art Molecular Dynamics (MD) studies which indicate a clear connection between the strain induced atomic rearrangement and conductance variations. We found novel atomic processes and atomic rearrangements induced by the applied stress. By using these results we provide a consistent interpretation for various structures of G observed in the course of stretch. While the step falls (or jumps) of $G(s)$ are associated with the sudden changes in the crossection, the spikes on the plateaus may be due to transient motion (migration) of atoms in the narrowest region of the neck or resonant tunneling through the constriction. The positive slope observed on the last plateau of Pt and Al wires[7] is due to increasing local density of states $\rho(r, E_F)$ at the neck which undergoes a structural transformation under extreme strain.

2. Conductance in the Nanowire

How the conductance of a mesoscopic object depends on its size and dimensionality has always been attractive, as well as intriguing subject for researchers. As early as 1957, Landauer[14] proposed that the conduction in a solid is a scattering event, and that transport is the consequence of the incident current flux. Based on the counting arguments of the transmission T, and reflection R, he derived his famous formula for the conductance $G = (2e^2/h)(T/R)$. Almost three decades ago Sharvin[15] pointed out the resistance of a ballistic channel (or point contact) and developed a formalism in the semiclassical regime. Nowadays, almost defect-free electronic devices have been fabricated, which have dimensions, in one or more directions, on the quantum scale. The two terminal conductance of such a quasi $1D$ device (or constriction) fabricated from the high mobility GaAs-$Al_{1-x}Ga_xAs$ heterojunction[16,17] (which have width w in the range of Fermi wavelength λ_F, and length d smaller than the electron mean free path) were found to change with w approximately in steps of $2e^2/h$. This observation has been interpreted as the quantization of the ballistic conductance. It became clear from subsequent studies[9,18,19] that the motion of electrons in the constriction is transversally quantized if $d \sim \lambda_F$, and then its level spacings become in the range of a few Kelvin, if w is in the range of λ_F. A n-fold degenerate current-carrying state becomes conducting and hence G increases by $2e^2n/h$ whenever its energy coincides with E_F; this situation is denoted as the opening of a n-fold channel. The variation of G

with w displays sharp steps of integer multiples of $2e^2/h$, and flat plateaus form between two consecutive steps provided that the potential in the constriction is uniform, temperature and bias voltage are low, and tunneling contribution is negligible. Otherwise, conduction channels mix significantly, the opening of channels are delayed, sharp step structure is smeared out and step heights are lowered[9].

The atomic size point contact was produced first by the STM: By displacing the tip towards the sample surface, the conductance were measured in a wide range covering tunneling and ballistic regimes. The experimental $logG$ versus tip displacement either exhibits a linear variation and then saturation at the first plateau before a jump, or it ends the linear variation directly by a jump[2].

Theoretical studies[8,20,21] showed that the two terminal conductance at the single atom contact shall be in the range of $2e^2/h$. Furthermore, it has been shown that the observed jumps of conductance originate from the discontinuous and abrupt change of crossection of the contact[8-10].

Better controls of dimensions[4-6] has led to a renewed interest in the transport through the connective necks produced by STM. The principle issue is to what extent the observed structures in the conductance versus stretch curve $G(s)$, comply with the quantization of ballistic conductance. It is expected that the atomic structure of the neck is rather irregular and the potential is nonuniform. The narrowest part of the neck, which determines conductance, is rather short. In those circumstances, the sharp quantization with marginal channel mixing should not occur. Of course the transversal motion of electrons in the neck is quantized with significant level spacing; the effect of this quantization is reflected to the transport. We strengthen our arguments by performing calculations of conductance for uniform constrictions, where the channel mixing shall be insignificant. We will show that the features of $G(s)$ or $G(A)$ for these idealized wires are rather different from the observed behavior.

3. Ballistic Free Electron Model

The nanowire which is connected to the left (L) and right (R) reservoirs (or metal electrodes) by a neck is viewed as a constriction represented by a $3D$ confining potential $V(\vec{r})$. A free electron of the L-reservoir with wave function Ψ_i, enters in the constriction with a finite momentum in the $z-$direction and evolves into current-carrying states[9]. These current-carrying states, either propagate to the R-reservoir or are occasionally scattered, and eventually reach the R-reservoir. The current-carrying states inside the constriction are determined by using multiple boundary matching method, and the conductance is calculated by evaluating the current oper-

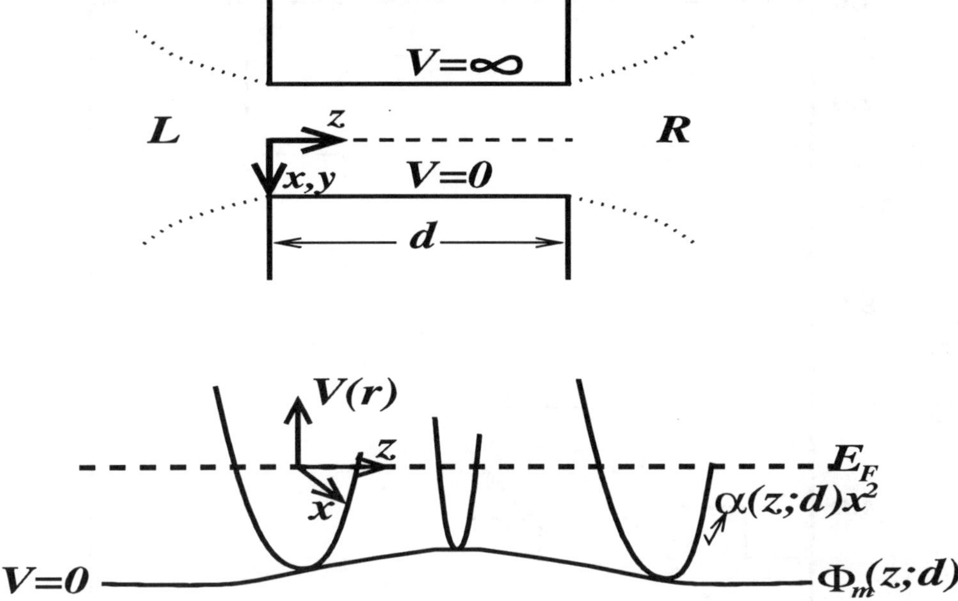

Figure 1. Schematic description of the constriction connected to the left (L) and right (R) reservoirs, which is used to model the electron transport through a nanowire. (a) The infinite wall cylindrical potential with abrupt (continuous line) and trumpet-like (dotted line) connections to the reservoirs. (b) Schematic description of the parabolic potential in the (xz)-plane

ator. As a simplest approximation, the infinite wall cylindrical potential is used for $V(\vec{r})$:

$$\begin{aligned} V(\vec{r}) &= 0 \text{ for } (x^2+y^2)^{1/2} < R(z) \\ V(\vec{r}) &= \infty \text{ otherwise} \end{aligned} \tag{1}$$

Here $R(z)$ is the radius of the constriction which depends on the z coordinate.

Our self-consistent field (SCF) pseudopotential calculations for infinite Al and Na atomic wires with optimized nearest neighbor distance, suggest that the parabolic potential form in the transversal direction is better suited for $V(\vec{r})$. The self-consistent pseudopotential is parameterized as follows:

$$V(\vec{r}) = \Phi_m(z;d) + \alpha(z;d)(x^2+y^2) \tag{2}$$

Here d is the length of the constriction (or neck), and Φ_m is the saddle point potential, which occurs for finite constriction. The two different approximations for the constriction potential are described in Fig.1. We note that these potentials with circular symmetry are very idealized description of the realistic potential inside the constriction. Because of circular

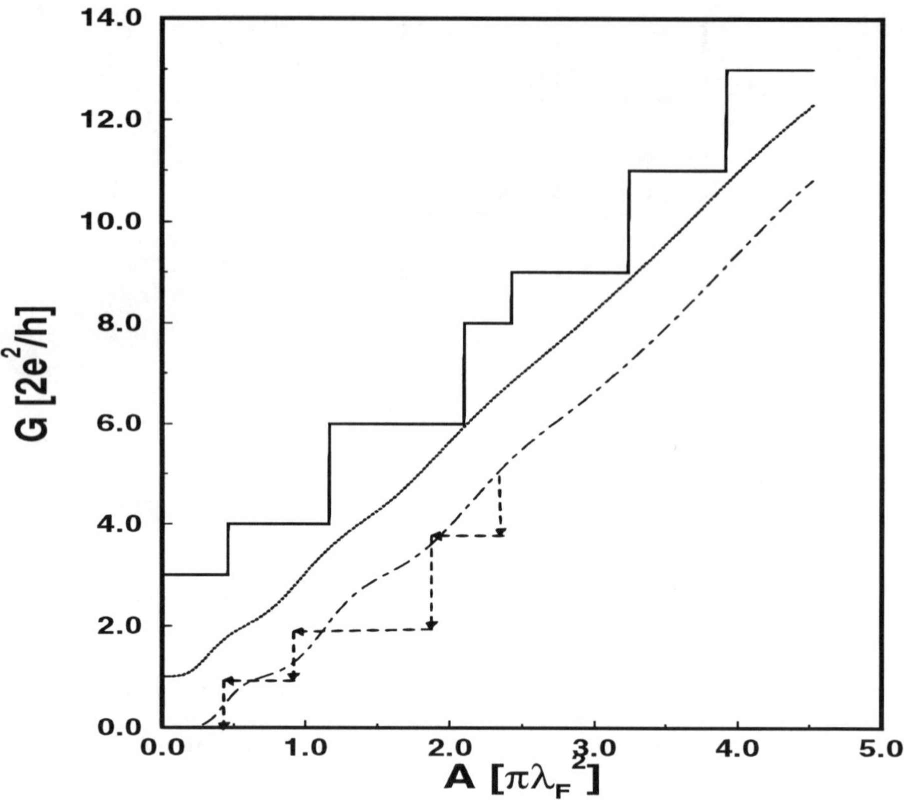

Figure 2. Conductance G versus crossection A for a constriction with an infinite wall cylindrical potential and abrupt connections to reservoirs. Continuous line is for the infinite and uniform neck; dotted line is the quantum Sharvin regime with $d \sim 0$. Uniform and finite neck having $d \sim \lambda_F$ corresponds to dash-dotted line. The discontinuous change of A during stretch give rise to sharp steps as schematically described by dashed lines. The curves are up-shifted for clarity.

symmetry, one can obtain analytical solutions for the wave functions in the transversal $(xy)-$plane. For example, in the parabolic confinement given by Eq.2, the wave functions can be separable in the transversal $(xy)-$plane and in the longitudinal $z-$direction, and the transversal ones become $2D$ harmonic oscillator wave functions.

By using the above forms of the potentials with the abrupt or trumpet-like connections to the reservoirs, we calculate the variation of the conductance as a function of the crossection. By comparing the calculated curves $G(A)$ with the conductance versus stretch curves obtained experimentally, we aim to draw conclusions on the quantization of ballistic conductance in nanowires or connective necks. Figure 2 illustrates our results for G versus crossection A for the infinite wall cylindrical potential with uniform radius R that is connected abruptly to the reservoirs. The dotted curve corresponds to Sharvin's conductance of a contact with $d \sim 0$. Owing to $d \ll \lambda_F$, the quantization is not completed and step structure is smeared out by tunneling. On the other hand, as $d \to \infty$, the current carrying states have the same quantization at every point along the constriction. Tunneling and mixing of channels do not take place and hence, the variations of G with A displays sharp step structure. Here, any quantized state that coincides with the Fermi level opens a conduction channel and contributes to the ballistic conductance by $2ne^2/h$; n being the degeneracy of the state. In an intermediate situation, i.e $d \sim \lambda_F$, the effect of quantization of transversal state in the neck becomes apparent in the $G(A)$ curve; but neither the steps are sharp nor the the plateaus are flat. Calculations by using uniform parabolic potential with abrupt connections to L- and R-reservoirs and with $d = \lambda_F/2$ and $d = \lambda_F$, lead to essentially the same conclusions.

Next we consider a constriction potential which has abrupt connection from one end and trumpet-like connection from the other end. Such a potential may correspond to the nanoindentation of an STM tip into a flat metal surface. The crossection and the length of the contact can be calculated as a function of the displacement of the tip towards the surface by using continuum contact mechanics or simply by assuming incompressible material in the nanoindentation. Figure 3 shows variation of G with the crossection A and with the push s for different cone angles 2α. We note that the contact made by a tip having relatively larger cone angle, has shorter length d and hence relatively weaker step structure.
Two important approximations are made for the above calculations:
i) Within the continuum approximation, we assumed that the crossection of the neck varies continuously. This may be a reasonable approximation for a metal wire which has large diameter ($2R \gg \lambda_F$). However, as we show in the next section, the crossection of an atomic size wire changes discontinuously.
ii) The potentials we used in calculating G are too idealized and have circular symmetry. The simulation of stretch in the next section show that the crossection of the wire is not circular, and its surface is rather rough.

The constrictions we studied above have weak features in the $G(A)$ curves that are different from the experimental curves. If the structure of experimental $G(A)$ or $G(s)$ is attributed to the quantized conductance the conditions of the above constrictions would be even more suitable for perfect quantization. Apparently, the earlier interpretation of the experimental results have to be revised. In what follows, we show that the discontinuous rearrangement of atoms and structural transformations are essential for a better understanding of the observed $G(s)$ curves.

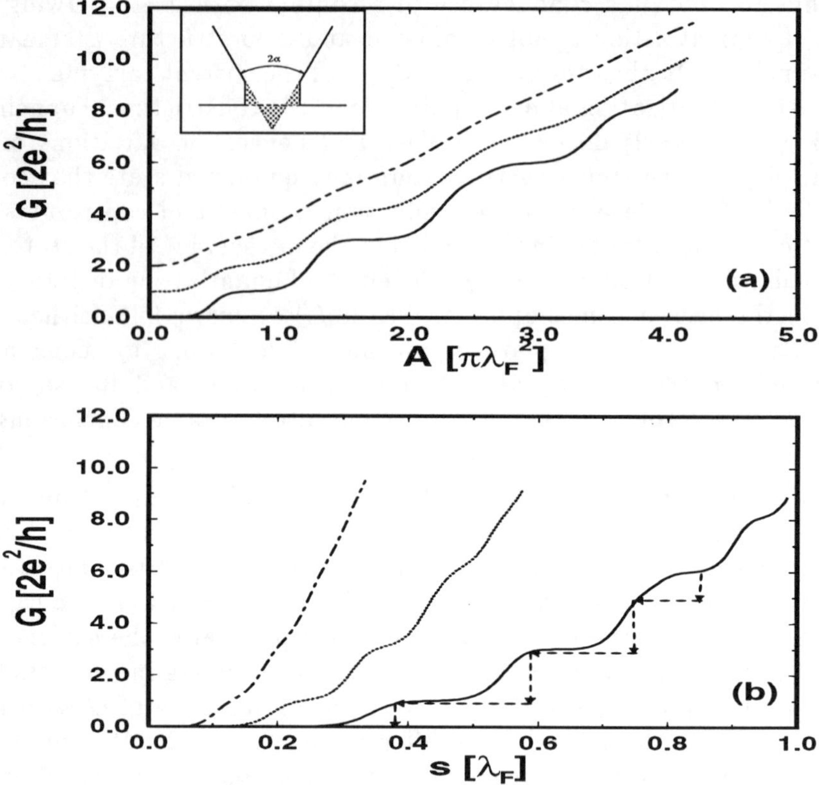

Figure 3. (a) G versus A for a contact formed by an STM tip indenting the substrate surface. (b) same curves plotted as a function of the displacement s of the tip towards sample. Dash-dotted, dotted and continuous curves correspond to the cone angles $2\alpha = 120^o, 90^o$ and 60°, respectively.

4. MD-Simulations

4.1. DESCRIPTION OF MD-SIMULATIONS

In this section, we study yielding and fracture mechanisms of nanowires that are pulled by an external agent. To understand the origin of these mechanisms and abrupt force variations[12,13], we perform an extensive analysis of atomic structure in the course of pulling. In particular, we follow the motion of the neck atoms and examine their coordination numbers and the structure of atomic layers during the abrupt force variations. We use embedded atom (EA) potential[22-24], and carry out simulations based on the MD method for the Cu nanowires made from (001) and (111) atomic planes[25]. Recently, the MD simulation is proven to be powerful method to

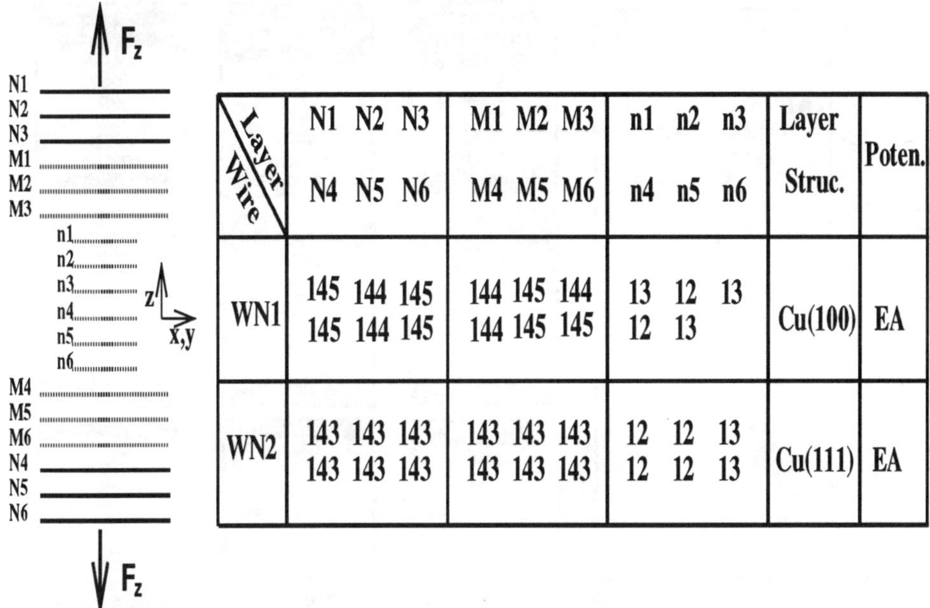

Layer / Wire	N1 N2 N3 / N4 N5 N6	M1 M2 M3 / M4 M5 M6	n1 n2 n3 / n4 n5 n6	Layer Struc.	Poten.
WN1	145 144 145 / 145 144 145	144 145 144 / 144 145 145	13 12 13 / 12 13	Cu(100)	EA
WN2	143 143 143 / 143 143 143	143 143 143 / 143 143 143	12 12 13 / 12 12 13	Cu(111)	EA

Figure 4. Schematic description of the Cu nanowires. The layers labeled by N, M and n are described in the text. The number of atoms in each layer and their $2D$ structure before pulling are indicated. The $x-$ and $y-$axis lie in the atomic [(001) or (111)] planes. The z axis coincides with the axis of the wire; the tensile force $\mathbf{F}_z$ is applied along the z-axis. EA stands for the embedded atom potential.

reveal the atomic rearrangement under tensile stress[3,5,25-27]. The wires we studied have two ends that are connected by a neck, and have quasi-circular crossections. Their structures are summarized in Fig.4. Last three layers at both ends (N_1, N_2, N_3 and N_4, N_5, N_6) are robust: All the atoms in these layers are translated solidly along the stretch direction only by the increment Δl; otherwise they are kept fixed during the MD-steps. These fixed three layers are assumed to be connected to the external agent that applies the tensile stress. Atoms in the following three layers adjacent to the fixed ones (M_1, M_2, M_3 and M_4, M_5, M_6) and those of the neck ($n1, n2, n3$...) are fully relaxed during the MD-steps. The nanowire indicated by $WN1$ is formed from Cu(001) atomic layers. The wire itself is represented by a periodically repeating system in the $(xy)-$plane. The z-axis is taken to be parallel to the axis of the nanowire. $WN2$ is made from Cu(111) atomic planes, and it is investigated at $T = 300K$ and $T = 150K$. The pulling (stretch) is realized by displacing the fixed layers (N's) solidly along the $z-$direction by the increment of length Δl ($0.1\AA$). After each increment, the atoms of the wires (M's and n's) are relaxed to find their new positions. Simulations are performed with time steps of $0.1ps$ Between two

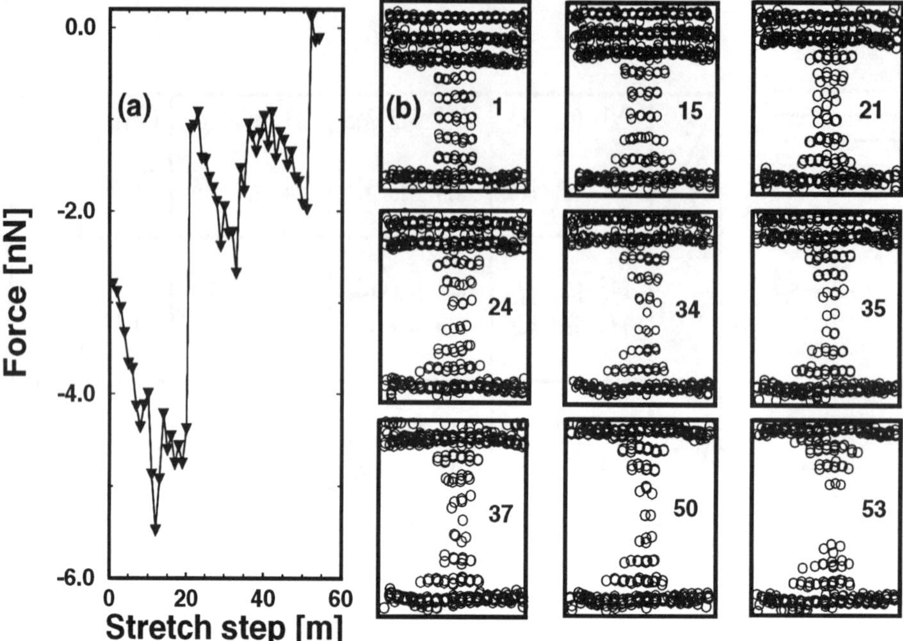

Figure 5. (a) The variation of the force $\mathbf{F}_z$(nN) with the strain or elongation along the z-axis applied by pulling the end layers of Cu (WN1) by $m\Delta l$ at $T = 300K$. (b) The side views of the nanowire showing the atomic positions at some relevant stretch values m.

subsequent stretches of Δl the nanowire is relaxed for $2500ps$. The tensile force $\mathbf{F_z}$ (which is the attractive force between the fixed top three layers and the rest of the wire) is obtained after averaging over the last 800fs, so that the force fluctuations are minimized.

4.2. MD RESULTS FOR Cu(001) NANOWIRE

In this section we examine the elongation and various physical events during the stretch of the nanowire, $WN1$ at $T = 300K$. The initial atomic configuration of the nanowire is described in Fig.4. The interlayer distance between the bulk Cu(001) layers is $c \sim 1.8\AA$ (approximately half of the lattice parameter, i.e $\sim a/2$). Initially, the layers of the wire have A-B-A-B sequence; the atoms of the B-layer face the hallow sites (the center of square unit cell) of the adjacent A-layers. We specify this stacking as the H-site registry. In Fig.5, we illustrate the variation of $\mathbf{F_z}$ as a function of stretch $m\Delta l$, (m, being an integer multiples of increment) and the side views of the structure at the neck region. The $\mathbf{F_z}(m)$ curve displays small fluctuations and also abrupt jumps. The magnitude of the average force increases with increasing m between two consecutive jumps. This is the "quasi" *elastic stage*[13]. Our results clearly indicate that a new abrupt jump of $\mathbf{F_z}(m)$ occurs whenever approximately $\sim 18-20$ increments are made in an elastic stage following the previous jump. Stated differently, *the tensile force makes a sudden jump and decreases significantly* whenever the length l of the nanowire is elongated by approximately interlayer distance, c. The jump of $\mathbf{F_z}(m)$ starts as the atomic structure of the neck becomes disordered; it lasts until a new layer with relatively smaller crossection is generated. This is the *yielding stage.* The formation of neck layers at fairly regular intervals was predicted earlier[5]. The overall behavior of $\mathbf{F_z}(m)$ is in agreement with the experimental results[3,12,13]. We now examine these in detail.

At the last increment before the first jump in $\mathbf{F_z}(m)$ ($m \sim 19$), the layer structure in the neck is destroyed, and becomes disordered; however it is recovered after a few increments, at $m \sim 24$, with the creation of a new layer (see Fig.5). At the end of this transformation, the crossection of the neck is reduced from 8 atoms to 5 atoms; whereas the crossection of neck layers adjacent to end layers ($n1$ and $n5$) is not altered. The local reduction of the crossection due to stretching causes $|\mathbf{F_z}|$ to reduce, and hence the interlayer spacings between M_3 and $n1$ (and similar spacing at the other end) to decrease. The layer structure of the neck is conserved until $m \simeq 33$. Beyond that point each increment of stretch by Δl causes one atom from the central layer to migrate and stay in the adjacent interlayer spacing which already became wide open due to pulling from $m = 25$ to $m = 33$. This way a new "layer" with two atoms is formed. As a result, the crossection of the connective neck is further reduced with three central layers formed by 5,2,3 atoms, respectively. Owing to the repulsive force induced between the new layer and adjacent ones, $|\mathbf{F_z}|$ decreases abruptly. The asymmetric distribution of atoms in the neck layers may be due to a dynamical effect and due to stretch of the wire from one end, as well. However, in the steps from m=35 to m=38, the asymmetry is lifted by a peculiar, transient event which may be relevant for transport properties: One of the two atoms in the neck layer created at $m = 35$, jumps back to the layer it emerges. During the following increments, the single atom neck is strengthen by the inclusion of one atom from another layer so that the necking becomes uniform by the layers including 4,2,4 atoms. Due to this exchange of atom, the conductance is expected to get a transient dip provided that its duration is long enough. In fact, such a dip has been observed[5]. The two-atom neck becomes stable until the break; but the atoms rotate in the $(xy)-$plane and they become slightly inclined in the z-direction. Such a fluctuation in configuration may give rise to changes in the conductance. It is also seen that the narrowest crossection of the neck prior to the break can have two atoms.

The above results point to the fact that the structural transformations followed by the abrupt change of $\mathbf{F_z}$ result in necking. An additional layer is

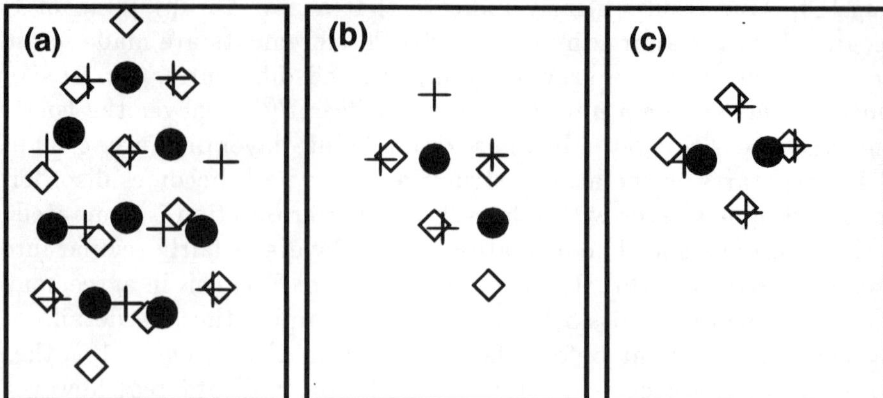

Figure 6. The top view of three layers at the neck showing atomic positions and their relative registry at *three* different levels of stretch. (a) Before the first yielding at $m = 15$ for layers $n2, n3$ and $n4$, with their atomic positions indicated by $+, \bullet$ and $\diamond$, respectively. (b) corresponds to $m = 38$, just after the *second* yielding, the structure keeps its registry; while (c) corresponds to $m = 46$, starting the formation of bundle structure. The position of atoms in the *third*, *fourth* and *fifth* layers are indicated by $+, \bullet$ and $\diamond$, respectively.

formed and the narrowest crossection decreases usually by more than one atom. In addition to these abrupt changes, we find another mechanism in necking that gives rise to relatively smaller and also slower changes in the crossection. This is a single atom process, in which individual atoms migrate from central layer towards the end layers. The tendency to minimize the surface area and hence to reduce surface energy is the main driving force for this type of necking.

Next we examine atomic rearrangements and structural changes in every atomic layer, which become gradually non-planar under tensile stress. The arrangements of atoms in the layers are shown in Fig.6. In our study we have found that upto $m = 12$, the Cu(001) structure in the neck layers exhibits very small and random deformation (with atomic displacements less than 10% and with no preferable direction), especially those atoms away from the boundary. As shown in Fig.6-(a), the original stacking and registry are kept with least deformation. Beyond a certain stretch, the neck atoms start to built up a structure which deviates from the A-B-A-B stacking sequence of the ideal Cu(001) planes.This deformation starts earlier in the layers $n2, n3$ and $n4$ at $m = 15$. Here we note its two aspects: First, the interatomic distances have slightly increased, especially those of central layers. Second, the $2D$ square lattice changes and then looks like a hexagonal lattice. Upon further elongation of the nanowire, the rest of the neck undergoes the same deformation. The tendency towards a $2D$ hexagonal-like lattice is mediated by the reduced interlayer interaction as a result of stretch; the atoms try to take relatively low energy configuration in the quasi independent layer behavior. As soon as a new neck layer is formed (at $m \sim 24$), the original features, especially the registry of the Cu(001) layers are recovered at both ends of the neck ($n1$,$n2$ and $n5$,$n6$ layers). These layers (being close to the relatively more stable M_3 and M_4 layers) are forced to return to the original structure. Whereas it is impossible to recover original A-B-A-B sequence of the Cu(001) planes and to match the new layer to the adjacent ones, the incommensurability and deviation from the bulk ordering continues until a new layer is formed. In the present case a new "layer" made of two atoms at the neck is formed at $m \sim 33-35$, but they do not have enough extension to be considered as a real atomic layer. An interesting situation owing to the reduced interlayer interaction and excessive tensile strain appears in the last stage of stretch: The H-site registry disappears eventually shortly before the break and hence the atoms of the adjacent neck layers tend to face the top (T)-sites. This way they are aligned to form atomic chains as shown in Fig.6. As a result, the crystal-like structure and original registry at the neck is replaced by a bundle of atomic chains. In some circumstances, a single atomic chain can form rather than the bundle. Earlier, *ab initio* calculations confirm that the T-site registry is only a local minimum occurring at relatively larger interlayer separation for the Al(111) and Al(001) slabs[28]. However, for some materials oriented in certain atomic planes, the break can take place before such a transition. The transition from the H-site registry to the T-site registry leading to a bundle of chain structure is interesting not only for the point of view of structural modification, but also for novel properties[27,30] (such as "giant" yield strength[12,13] and positive slope of conductance variations[7]).

4.3. MD-RESULTS FOR Cu(111) NANOWIRE

Except for some minor changes, the stretch of nanowire made from Cu(111) layers (as specified by $WN2$ in Fig.4) has overall features of the yielding mechanism at $T = 300K$, which are similar to the above Cu(001) orientation. In Fig.7, we show the force variation with stretch. The force curve depicts consecutive elastic and yielding stages. The elastic stages prior to the first and second structural transformation are rather linear with strain. This is not so in the third stage before the break, where either the structure becomes disordered or changes the registry from the H- to T-site. In fact, after few stretch steps, the magnitude of $\mathbf{F_z}$ drops suddenly due to onset of disorder without formation of a new layer. During the last stage of stretch, the strain is almost accommodated by the layers at the central part of the neck. The $2D$ hexagonal-like structure is quite robust for the layers adjacent or close to the end layers M_3 and M_4. Here the A-B-C stacking sequence

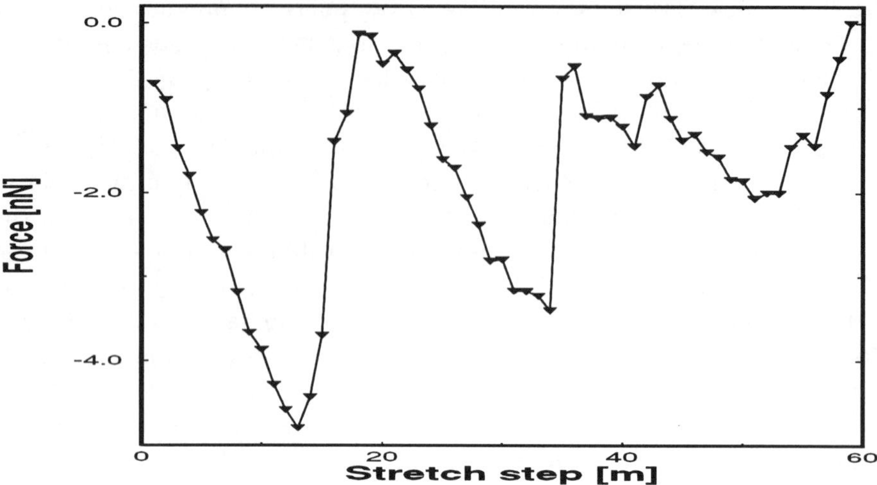

Figure 7. The variation of the tensile force $\mathbf{F}_z$ with the stretch along the z-axis for Cu(001) nanowire ($WN2$) at $T = 300K$.

and the H-site registry are forced by these end layers. In the central part of the neck, the distortion of the $2D$ hexagonal lattice structure becomes severe and the A-B-C sequence disappears, but the H-site registry is kept after the yielding stage. However, the atomic chains form as a result of the transition to the T-site registry in the last stage of stretch.

If the same Cu(111) wire is stretched at low temperature ($T = 150K$), it exhibits features, some of which are dramatically different from those at room temperature. For example, the same wire undergoes *five* yielding stages resulting in an elongation of $9.5\AA$; whereas at room temperature, it had only *three* yielding steps with $6\AA$ elongation. Apparently, while the necking is faster and sharper and hence, it is more trumpet-like at room temperature, the neck is longer at low temperature. We explain this with faster atomic motion towards the ends of the wire at room temperature. Another feature which does not exist at room temperature is a new type of structural transformation. The $\mathbf{F_z}(m)$ curve and some relevant side views of the wire in Fig.8 distinguish different types of yielding mechanisms. The first and second yielding stages are similar to the previous cases in which the structure undergoes layer-disordered-layer transformations during the generation of a new layer. At $m = 15$ a well-defined layer structure becomes broader and slightly disordered, and a new layer forms at $m = 16$. This yielding stage occurs more abruptly with least necking; the size of the layers decreases by 2-3 atoms. Similar process repeats at $m = 35$, and creates a new layer at $m = 36$. Again, in this process most of the layers are affected and reduced by two atoms with no significant necking effect. The hexagonal

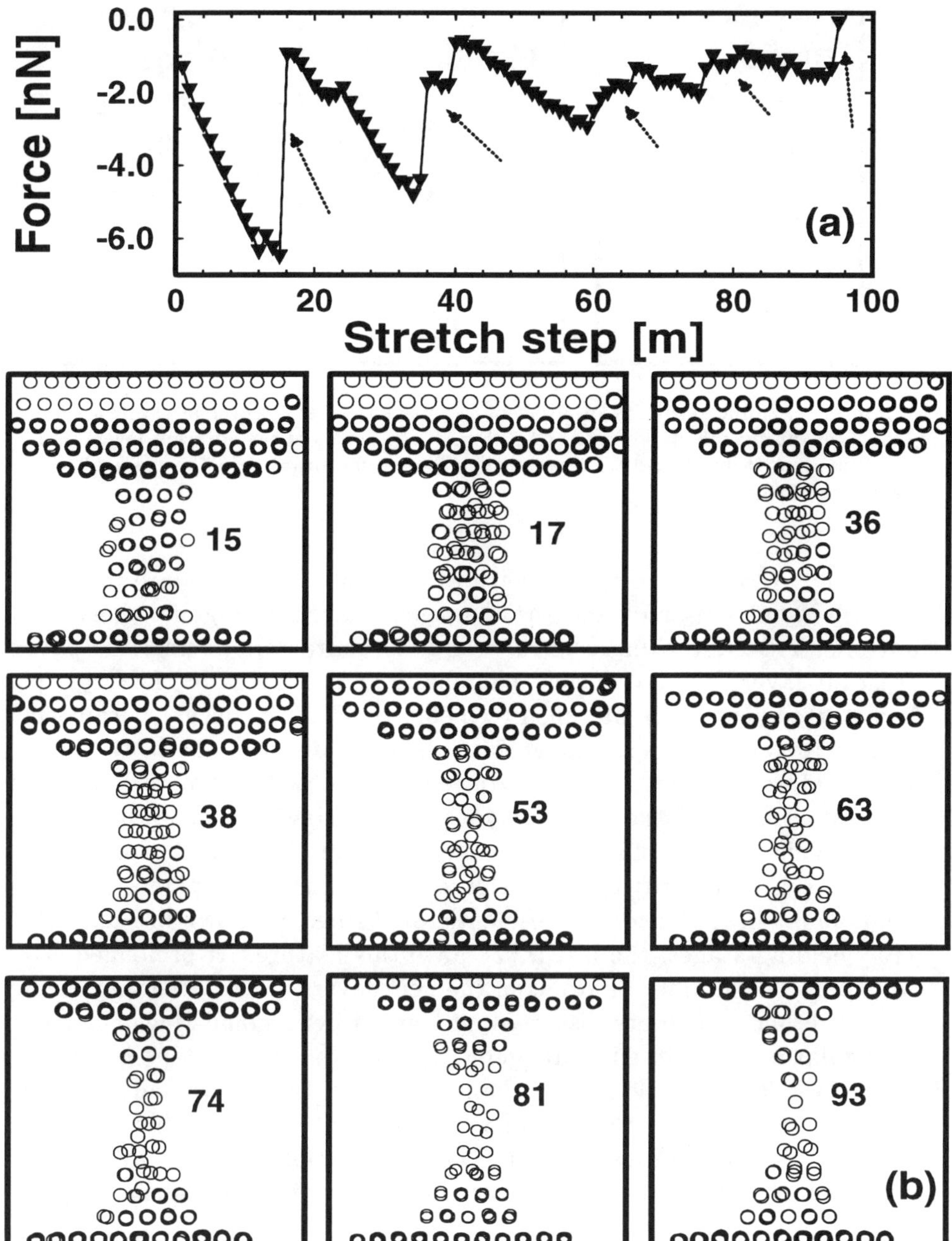

Figure 8. (a) Variation of the tensile Force $\mathbf{F}_z$ with the stretch for the Cu(111) nanowire ($WN2$) at $T = 150K$. We mark the yielding stages by arrows. (b) The side views of the structure at some relevant stretch steps m.

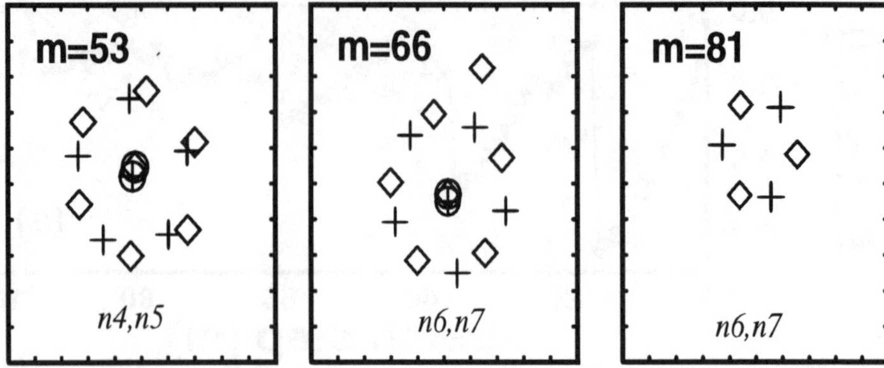

Figure 9. Atomic positions of adjacent layers at the neck center in the (xy)-plane for some relevant stretch steps described in Fig.8. The atoms of the adjacent layers indicated by n's are shown by $+$ and $\diamond$, respectively. The atoms situated in the space above these layers are circled.

rings and A-B-C stacking sequence of the Cu(111) planes are robust near the end layers (M_3 and M_4). At the central layers hexagonal rings persist, but the A-B-C stacking sequence and its characteristic registry between adjacent layers are destroyed. The shape of the layers in the $(xy)-$plane is distorted. For $m > 39$, an intermediate situation occurs within elastic stages: Hexagonal rings change into pentagonal ones; the latter becomes staggered to the rings in the adjacent layers. Moreover, one layer at a time (starting from the central layer) ejects a single atom to the interlayer space. This way a column (or chain) of atoms is formed along the axis of the neck. The projection of this chain in the $(xy)-$plane is almost a single point near the centers of pentagonal rings as shown in Fig.9. Out of *five* yielding stages, the last *three* have this feature. At prolonged stretch ($m > 75$), the number of atoms in the neck layers are reduced and the layer structure becomes more distorted; staggered pentagonal rings change into triangular ones again with staggered orientation. A few steps before break ($84 < m < 94$), a single atomic chain forms eventually. We understand that at low temperatures, the mobility of atoms are not suitable for massive changes, so the neck can accommodate the applied strain by these new intermediate structures which require least atomic rearrangement.

5. Mechanical properties of nanowires

To reveal physical phenomena underlying the "giant" Young's modulus observed in thin wires shortly before the break we carried out *ab initio* calculations by using SCF pseudopotential method within local density approximation. The effective force constant k_{eff}, that is defined as the mechanical force per elongation can be estimated from the slope of the curve $\mathbf{F_z}(m)$. For a contact, the effective force constant is given by $k_{eff} = R\gamma E/(1-\nu)^2$ (see Ref.12 and 29). Here R, γ, E and ν are respectively contact radius, a factor describing the exact distribution of stress, Young's modulus and Poisson's ratio. From $\mathbf{F_z}$ versus displacement curves measured by the combined STM and AFM, Agraït *et al.*[12] obtained $k_{eff} = 240N/m$ and $R \sim 3nm$. Using the above contact expression of k_{eff}, they estimated $E \sim 66-44$ GPa which is consistent with the macroscopic value of the Young's modulus of Au. On the other hand, Stalder and Dürig[13] found that the intrinsic yield strength of a neck with $R \sim 1nm$ (corresponding to a crossection $A \sim 300\AA^2$) between the Au(111) sample and Au tip is 8 GPa, which is one order of magnitude larger than the macroscopic value. This value approached to $20GPa$ shortly before the break. The discussions in the previous section point to the fact that in the last steps of the stretch, where the neck is reduced to $1-3$ atoms, either the bundle of atomic chains or single atomic chain is formed. This structure of the neck is rather different from the original structure derived from the atomic arrangement of the macroscopic bulk and is expected to lead dramatic changes in the mechanical properties. For example, the observed "giant" yield strength may originate from the unusual changes in the structure. By definition, the yield strength σ_{Y} is the stress, $\sigma = \mathbf{F_z}/A$ at the yield point where the linear relation between stress and strain starts to break. Since $\sigma_{\mathsf{Y}} = E\epsilon_{\mathsf{Y}}$ (ϵ_{Y} being the strain at the yield point), the observed "giant" yield strength may be due to high strain value and/or high Young's modulus attained by atomic chain. As a matter of fact, owing to the transition from the H-site registry to the T-site registry, a neck can attain large values[28,30] for ϵ_{Y}. Moreover, *ab initio* calculations that we discuss below demonstrate that E also increases in the neck consisting of atomic chain. Of course, the yield strength and Young's modulus are definitions, which are developed within the continuum description of matter. By adopting some of the variables (such as the crossection A) in these definitions we attempt to calculate E in the quantum limit.

The Young's modulus of an atomic chain can be expressed by $E = kb/A$, where k is the force constant and b is the interatomic distance. Apparently, the size and dimensionality of the chain go beyond the limits of continuum description, and the estimation of k requires *ab initio* quantum mechanical calculations. To this end, we carry out total energy calculations by using SCF pseudopotential method in momentum representation. Accordingly, we take into account the infinite Al-chain whereby we can use periodic boundary condition and pseudopotentials which converge easily for a simple metal. For example, by using the kinetic energy cut-off $|\mathbf{k}+\mathbf{G}| < 7.5Ry$, one can achieve accurate representation of the wave function and band energies. As pointed earlier[28,30], a short Al-chain can form in an Al-neck.

Moreover, the mechanical properties of a short Al-chain can be revealed from those of infinite chain. We use non-local and norm-conserving pseudopotentials[31] with Wigner exchange-correlation potential. We treat the $1D$ infinite atomic chain in a $3D$ supercell geometry in which the interchain distance is taken $7.5Å$. The minimum total energy is obtained for $b = 2.403Å$. The force constant k is obtained from the variation of total energy relative to interatomic distance b, i.e $k = -\partial^2 E_T(b)/\partial b^2$, and is found $164N/m$. There are ambiguities in the definition of crossection; here we used two schemes which yield similar values: i) $A = 0.38\pi\lambda_F^2$ which is the crossection necessary to open a ballistic conduction channel within free electron and infinite wall cylindrical potential approximation. ii) $A = \pi R^2$; here R is the radius at the value of calculated SCF potential, which is equal to E_F. For Al, we find $A = 15.38Å^2$ at $T = 300K$ from the first scheme, and $A \sim 14.0Å^2$ from the second scheme. Eventually, we calculate the Young's modulus of Al-chain, $E \sim 260\ GPa$ which is approximately *four* times larger than the bulk value. This implies that the mechanical properties of metal chain or bundle structure is different from those of bulk values, as well as from those of the neck or metal wires having large crossection, such as $A \sim 300Å^2$ as in the experiment[13]. The raise of E upon bundle or chain formation may have dominant contribution to the observed "giant" yield strength.

6. Discussion

Our calculations of the conductance for the idealized and circularly symmetric constriction fail to explain the experimental results. In view of this we investigated the effect of the atomic rearrangements induced by the tensile force. Our analysis based on the state of the art molecular dynamics simulations using the embedded atom potential show that the yielding and fracture mechanisms of a nanowire under uniaxial tensile force depend on atomic structure and temperature, and also on the pulling conditions. There are, however, features which are common to all nanowires under uniaxial strain. The important aspects of the present study are summarized:
1. Deformation (or elongation) in two different stages, that occur consecutively and repeat until fracture, are common to all nanowires treated in this study. In the first stage, the stored strain energy and average tensile force increase with increasing stretch, while the layer structure persists. Excluding the fluctuations (possibly due to displacement and relocation of atoms within the same layer or atom exchange between adjacent layers), the variation of $\mathbf{F_z}(m)$ is approximately linear. This stage was identified as elastic. Once the elongation of the nanowire reaches the interlayer separation at the end of the elastic stage, the structure becomes disordered; but after a

few increments of stretch, it is recovered with the formation of a new layer. In this yielding stage, $|\mathbf{F_Z}|$ decreases abruptly; the crossection of narrowest neck layers that determines the conductance, in particular the crossection of the layer formed at the end of yielding stage are reduced abruptly by a few atoms. When the neck becomes very narrow (having 3-4 atoms) the yielding is realized, however, by an atom jumping from one of the adjacent layers to interlayer space. On the other hand, it may occur through slips or the motion of dislocations when the wire is much wider. At low temperature, the yielding takes place in relatively shorter time-interval within one or two stretch steps. Owing to the limited mobility of atoms at low temperature, the character of elastic stage changes if the neck is long enough. In this situation, each layer ejects one atom to the adjacent interlayer space as it widens with the applied strain. The layers at the neck are made from the pentagon rings which become staggered in different layers. The interlayer atoms make a chain passing through the center of pentagon rings. In this new phase, the elastic and yielding stages are intermixed and elongation which is more than one interlayer distance can be accommodated.

2. In the initial steps of stretch, the layered structure and ordered $2D$ atomic arrangement within the layers are maintained. Upon increased uniaxial strain (increased m), the layers become wider and rougher due to the atoms departing from the atomic plane, the $2D$ lattice is distorted, and interatomic distances start to deviate from the bulk equilibrium value. In the elastic stage, one can still distinguish layer structure and some kind of order in the atomic arrangement within a layer. In particular, the hallow-site registry between layers are maintained if the layers contain enough number of atoms. The ordered atomic arrangement becomes disordered at low temperature towards the end of stretch.

3. The elastic and following yielding stages are reminiscent of the stick-slip motion. In the course of stretch, elastic and yielding stages repeat; the surface of the nanowire roughens and deviates strongly from circular symmetry. The narrowest part, which is only a layer thick, is connected to horn-like ends. Our results of atomic simulations point to the fact that neither adiabatic evolution of discrete electronic states, nor circular symmetry induced degeneracy can occur in the neck. Consequently any quantized sharp structure shall be smeared out by channel mixing and tunneling[9]. On the other hand, owing to the sudden reduction of crossection by a few atoms, some states contributiong to current carrying states are eliminated suddenly. This leads to sudden drop of G with stretch as described by dashed line in Fig3-b.

4. In addition to the stick-slip behavior under strain, atoms also migrate in a much slower process. The migration process speeds up with increasing temperature and its effect on necking becomes significant when a thin neck

has a canonical connection to the ends. Also a transient and permanent exchange of atoms between adjacent layers can take place. This may lead to transient dips on spikes in the $G(s)$ curve.
5. If the interlayer interaction is reduced as a result of extensive strain the $2D$ atomic structure (square-like lattice) is transformed into hexagonal rings. At prolonged stretch, just before the the break, the crossection of layers at the central part of the neck is reduced to $2-3$ atoms. In this case, the H-site registry may change to the T-site registry. This leads to the formation of bundle of atomic chains or single atomic chain. We consider this a dramatic change in the atomic structure of the wire and may have important implications. The positive slope of $G(s)$ curve at the last plateau of Al can be explained by the chain formation[30].
6. Our *ab initio* calculations of Young's modulus on the $1D$ metal chains indicate that the neck having chain structure may have yield strength much higher than the bulk value. This is in good agreement with experimental results[3,12,13], and also implies that the elastic properties of an atomic size neck deviate from those of the bulk defined in the continuum limit.

In conclusion, the simulations of atomic structure in nanowires under tensile stress indicate novel mechanisms of plastic deformation and elongation. When the restoring effect of the initial structure is weakened under the applied uniaxial force, the atomic arrangement deviate from the global minimum and can be stabilized in a different local minimum. The corresponding configuration exhibits a non-crystalline, short-range order and interesting mechanical and transport properties induced thereform.

Acknowledgements

We would like to thank Prof. M. S. Daw for providing embedded atom potential for Cu and Dr. E. Tekman, Prof. Ş. Erkoç and Dr. B. Tanatar for helpful discussions.

References

1. For extensive review see:(1991) J.L. Beeby(eds.), *Condensed Systems of Low Dimensionality*, NATO ASI Series B, Phys. Vol. **253**, Plenum Press; (1991) L. Esaki(eds.), *Highlights in Condensed Matter Physics and Future Prospects*, NATO ASI Series B, Phys. Vol. **285**, Plenum Press.
2. Gimzewski, J.K. and Möller, R. (1987) Transition From the Tunneling Regime to Point Contact Studied by Scanning Tunneling Microscopy, *Phys. Rev. B***36**, 1284-1287.
3. Landman, U., Luedtke, W. D., Burnham, N. A. and Colton, R. J. (1990) Atomistic Mechanisms and Dynamics of Adhesion, Nanoindentation and Fracture, *Science* **248**, 454-461.
4. Agraït, N., Rodrigo, J. G. and Vieiria, S. (1993) Conductance Steps and Quantization in Atomic-Sized Contacts, *Phys. Rev. B***47**, 12345-12348.

5. Pascual, J. I., Mèndez, J. , Gòmez-Herrero, J., Barò, A. M., Garcia, N. and Binh, V. T. (1993) Quantum Contact in Gold Nanostructures by Scanning Tunneling Microscopy, *Phys. Rev. Lett.* **71**, 1852-1855; Pascual, J.I., Mèndez, J., Gòmez-Herrero, J., Barò, A. M., Garcia, N., Landman, U., Luedtke, W.D., Bogachek, E.N. and Cheng H.P.(1995), Properties of Metallic Nanowires: From Conductance Quantization to Localization, *Science* **267**, 1793-1795.
6. Olesen, L. , Laegsgaard, E., Stensgaard, I., Besenbacher, F., Schiøtz, J., Stoltze, P., Jacobsen, K.W. and Nørskov, J. K. (1994) Quantization Conductance in an Atom-sized Point Contact, *Phys. Rev. Lett.* **72**, 2251-2254; (1995) Olesen et al. Reply, *Phys. Rev. Lett.* **74**, 2147-2147.
7. Krans, J.M., Müller, C.J., Yanson, I.K., Gowaert, Th.C.M., Hesper R. and Ruitenbeek, J.M. (1993) One Atom Point Contact, *Phys. Rev. B***48**, 14721-14724.
8. Ciraci, S. and Tekman, E. (1989) Theory of Transition from the Tunneling Regime to Point Contact in Scanning Tunneling Microscopy, *Phys. Rev. B***40**, 11969-11972; Ciraci, S. (1990) Tip Surface Interactions, in R.J. Behm, N. Garcia and H. Rohrer(eds.), *Scanning Tunneling Microscopy and Related Methods*, Kluwer Academic Publishers, Volume **184**, pp.113-141.
9. Tekman, E. and Ciraci, S. (1991) Theoretical Study of Transport through a Quantum Point Contact, *Phys. Rev. B***43**, 7145-7169.
10. Todorov, T.N. and Sutton, A.P. (1993) Jumps in Electronic Conductance due to Mechanical Instabilities, *Phys. Rev. Lett.* **70**, 2138-2141.
11. Krans, J.M. Müller, C.J., Van der Post, N., Postama, F.R., Sutton, A.P., Todorov, T.N. and Ruitenbeek, J.M. (9995) Comments on Quantized Conductance in an Atom-sized Point Contact, *Phys. Rev. Lett.* **74**, 2146-2146.
12. Agraït, N., Rubio, G and Vieiria, S. (1995) Plastic Deformation of Nanometer-Scale Gold Connective Necks, *Phys. Rev. Lett.* **74**, 3995-3998; Rubio, G., Agraït, N. and Vieiria, S. (1996) Atomic-sized Metallic Contacts, *Phys. Rev. Lett.* **76**, 2302-2305.
13. Stalder, A. and Dürig, U. (1996) Study of Yielding Mechanics in Nanometer-sized Au Contacts, *App. Phys. Lett.* **68**, 637-639.
14. Landauer, R. (1957) Spatial Variation of Currents and Fields due to Localized scatterers in Metallic Conduction, *IBM J. Res. Develop.***1**, 223-231.
15. Sharvin, Yu.V. (1965) A Possible Method for Studying Fermi Surfaces, *Zh. Eksp. Teor. Fiz.* **48**, 984-985; [*Sov. Phys.-JETP* **21**, 655-656].
16. Van Wees, B.J., Van Houten, H., Beenakker, C.W.J., Williamson, J.G, Kouwenhoven, L.P., van der Marel, D. and Foxon, C.T.(1988) Quantized Conductance of Point Contacts in a Two-Dimebsional Electron Gas, *Phys. Rev. Lett.* **60**, 848-850.
17. Wharam, D.A., Thorton, T.J., Newbury, R., Pepper, M., Ahmed, M., Frost, J.E.F, Peacock, D.G., Ritchie, D. A. and Jones, G.A.C. (1988) One-Dimensional Transport and the Quantization of the Ballistic Regime, *J. Phys. C***21**, L209-L214.
18. Büttiker, M., Imry, Y., Landauer, R. and Pinhas, S. (1985) Generalized Many Channel Conductance Formula with Application to Small Rings, *Phys. Rev. B***31**, 6207-6215; Imry, Y. (1986) Physics of Mesoscopic Systems, in G. Grinstein and G. Mazenko(eds.), *Directions in Condensed matter Physics*,World Scientific Publisher, Singapore, pp.101-163.
19. Beenakker, C.W.J and van Houten, H. (1991) Quantum Transport in Semiconductor Nanostructure, in H. Ehrenreich and D. Turnbull(eds.), *Solid State Phys.***44**, Academic Press,Inc., pp.1-170.
20. Lang, N.D. (1987) Resistance of a One-Atom Contact in the Scanning Tunneling Microscope, *Phys. Rev. B***36**, 8173-8176.
21. Ferrer, J., Martin-Rodero, A. and Flores, F. (1988) Contact Resistance in the Scaaning Tunneling Microscope at very small Distances, *Phys. Rev. B***38**, 10113-10115.
22. Daw, M.S. and Baskes, M.I. (1984) Embedded-atom Method: Derivation and Application to Impurities, Surfaces and Other Defects in Metals, *Phys. Rev. B***29**, 6443-6453.

23. Foiles, S.M., Baskes, M.I. and Daw, M.S. (1986) Embedded-atom Method functions for the FCC Metals Cu, Ag, Au, Ni, Pd, Pt and their Alloys, *Phys. Rev. B*33, 7983-7991.
24. Daw, M.S. (1989) Model of Metallic Cohesion: The Embedded-atom Method, *Phys. Rev. B*39, 7441-7452.
25. Lynden-bell, R.M. (1994) Computer Simulations of Fracture at the atomic level, *Science* **263**, 1704-1705.
26. Bratkovsky, A.M, Sutton, A.P. and Todorov, T.N. (1995) Conditions for Conductance Quantization in Realistic Models of Atomic-scale Metallic Contacts, *Phys. Rev. B*52, 5036-5051.
27. Mehrez, H., Ciraci, S. and Erkoç, Ş. (1997) Yielding and Fracture Mechanisms of Nanowires, *Phys. Rev. B* (to be published).
28. Ciraci, S., Baratoff, A. and Batra, I.P. (1990) Site-dependent Electronic Effects, Forces and Deformation in Scanning Tunneling Microscopy, *Phys Rev. B*42, 7618-7621, for Al(111) surface; and Ciraci, S., Tekman, E., Baratoff, A. and Batra I.P. (1992) Theoretical Study of Short- and Long- Range Forces and Atom Transfer in Scanning Tunneling Microscopy, *Phys. Rev. B*46, 10411-10422, for Al(001) surface.
29. Johnson, K.L. (1986) *Contact Mechanics*, Cambridge University Press, Cambridge.
30. Mehrez, H., Ciraci, S., Buldum, A. and Batra, I.P. (1997) Conductance through a Single Atom, *Phys. Rev B*55.
31. Bachelet, G.B., Hamann, D.R. and Schlüter, M. (1982) Pseudopotentials that work from H to Pu, *Phys. Rev. B*26, 4199-4228.

ELECTRONIC PROPERTIES OF QUANTUM DOTS AND ARTIFICIAL ATOMS

JEAN-PIERRE LEBURTON AND SATYADEV NAGARAJA
Department of Electrical and Computer Engineering
and Beckman Institute
University of Illinois at Urbana-Champaign
Urbana, IL 61801

1. Introduction

Three-dimensionally (3D) confined carriers in quantum dots are interesting systems because of their resemblance to atoms [1]. Early studies of quantum dots were motivated by the observation of single-electron charging in granular metallic islands containing a "small" number of conduction electrons (N$\sim$100-1000) surrounded by an insulator characterized by a small capacitance C [2,3]. In metallic dots, quantum confinement is relatively weak and the large effective mass of conduction electrons makes the energy spectrum quasi- continuum with negligible separation between electron states, even at low temperatures, $\Delta E << kT$. Hence, the addition of an electron to the island requires a charging energy $e^2/2C$ from a supply voltage source to overcome the electrostatic repulsion or "Coulomb-blockade" from the electrons present in the dot; in these systems the influence of energy quantization is negligible [4].

Advances in patterning and nanofabrication techniques have made possible the realization of semiconductor quantum dots with precise geometries and characteristic sizes comparable to the de Broglie wavelength of charge carriers [5]. These quantum dots are realized in various configurations by combining heterostructures and electrostatic confinement resulting from charged metal electrodes patterned on the semiconductor surfaces. In 3D confined III-V compound semiconductors, the small effective mass of conduction electrons results in an energy spectrum of discrete bound states with energy separation comparable to, or even larger than the charging energy $e^2/2C$. The ability to vary the electrostatic potential over large

G. Abstreiter et al. (eds.), Optical Spectroscopy of Low Dimensional Semiconductors, 235–256.

voltage ranges allows for fine tuning of the quantum dot charge of just a few electrons ($N \sim 1 - 10$) [6].

In semiconductor quantum dots, electron filling of discrete energy levels by taking into account the Coulombic interaction between particles is reminiscent of atomic structures with shell orbitals whose electron occupation is determined by specific selection rules. However, while atomic orbitals are to a large extent spherically symmetric, and the potential strength is quantized in integer numbers of the elementary charge, "artificial atoms" can be designed to depart strongly from the 3D spherical symmetry of the central Coulombic potential and the discreteness of the nuclear charge. In this context the physics of few electron quantum dots offers new opportunities to investigate fundamental concepts such as electron-electron interaction in arbitrary 3D confining potentials and elementary excitations of carriers from equilibrium [7]. Aside from the investigation of basic phenomena, and because of the ability to precisely tailor their electronic structures, "artificial atoms" are also important structures for applications in high functionality nanoscale electronic and photonic devices such as ultra-small memories [2] or high gain performance lasers [8,9].

2. The Many-Body Hamiltonian of Artificial Atoms

The electronic spectrum of N-electron quantum dots are computed by considering the many-body Hamiltonian

$$\hat{H} = \sum_i \hat{H}_{oi} + \sum_{i \neq j} \hat{H}_{ij} \tag{1}$$

where $\hat{H}_{oi}$ is the single-particle Hamiltonian of the ith electron and

$$\hat{H}_{ij} = e^2/\epsilon \mid \vec{r}_i - \vec{r}_j \mid \tag{2}$$

is the interaction Hamiltonian describing the Coulomb interaction between carriers. Here ϵ is the dielectric constant of the material, and the summation in the second term of Eq. 1 is carried over all $i \neq j$ to avoid the interaction of carriers with themselves. Quite generally the Hamiltonian (1) is used for solving the Schrödinger Equation for the many-particle energies and wavefunctions

$$E = E_N(1, 2, 3, ..., N) \tag{3}$$

$$\Psi = \Psi_N(\vec{r}_1, \vec{r}_2, \vec{r}_3, ..., \vec{r}_N) \tag{4}$$

which, given the many-body interaction (2), can only be solved exactly for N=2. In this section we shall describe a natural approach toward the solution of this problem for a general number N of electrons by considering successive approximations.

2.1. SINGLE-PARTICLE HAMILTONIAN AND SHELL STRUCTURES

We start by considering a system of independent and 3D confined electrons in the conduction band of a quantum dot. By neglecting the interaction $\hat{H}_{ij}$, the Hamiltonian (1) is reduced to a summation of single particle Hamiltonians, each of the same form

$$\hat{H}_{oi} = \hat{H}_i = \frac{\hat{p}_x^2 + \hat{p}_y^2 + \hat{p}_z^2}{2m^*} + \hat{V}(\vec{r}) \tag{5}$$

Here we assume electrons can be described with an effective mass m*; $\hat{p}_x, \hat{p}_y, \hat{p}_z$ are the components of the electron momentum and $\hat{V}(\vec{r})$ is the external potential producing carrier confinement in the quantum dot. We will assume that confinement of electrons is achieved by a double heterostructure quantum well along the y-direction, and by an electrostatic potential in the x-z plane (Fig. 1). The latter confinement results usually from dopant atoms in neighboring semiconductor layers, and from the fringing fields of the metal electrodes deposited on the semiconductor surface. This configuration is most commonly achieved in planar quantum dots and vertical quantum dots, and results, in a first approximation, in a 2D parabolic potential in the x-z plane (Fig. 1.b). Confinement due to heterostructures along the y-direction is generally strong (~10 nm) with energy separation of the order of 30-100 meV, while the x-z planar confinement is much weaker with energy separation of the order of 1 meV over larger distance ($> 100nm$).

In a first approximation, the external potential is separable into two terms,

$$\hat{V}(\vec{r}) = \hat{V}_1(x,z) + \hat{V}_2(y) \tag{6}$$

which results in the energy spectrum $E_{\nu,n_x,n_z} = E_\nu + E_{n_x,n_z}$ with corresponding wavefunctions $\Psi_\nu(y)\Psi_{n_x,n_z}(x,z)$ where $E_\nu(E_{n_x,n_z})$ is the spectrum resulting from the y-potential (x-z potential). Hence, each value of the ν-quantum number gives a series of x-z energy levels. At low temperature, given the large separation between the E_ν energy states, only the first levels of the lowest series $\nu = 0$ are occupied by electrons. If one further assumes that the $\hat{V}_1(x,z)$ potential is cylindrically symmetric, the ($\nu = 0$) energy spectrum is written as [7]

$$E_{0,n_x,n_z} = E_{0,m,l} = E_0 + m\hbar\omega \tag{7}$$

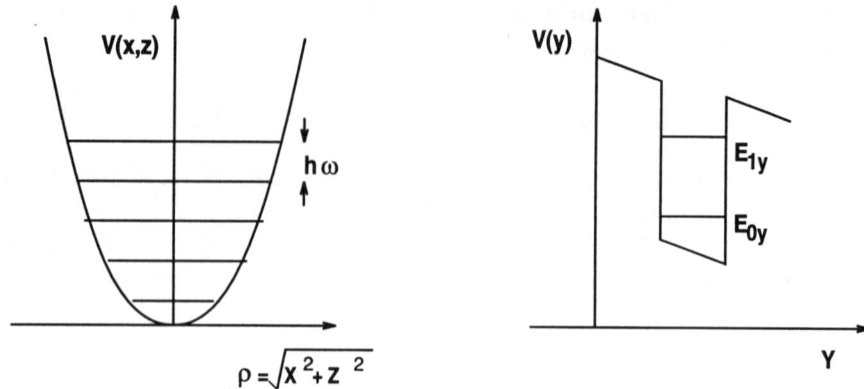

Figure 1. Schematic representation of a) a 2D parabolic potential with cylindrical symmetry in the x-z plane showing equally spaced energy levels and b) the square potential with the first two quantized levels in the y-direction: $E_{1y} - E_{oy} >> \hbar\omega$

where ω is frequency of the cylindrical parabolic potential. Here each m-level is $2m$-times degenerate with the factor 2 accounting for the spin degeneracy. The number m (=1,2,3....) is the radial quantum number and the number $l(= 0, \pm 1 \pm 2, ...)$ is the angular momentum quantum number. Hence the 2D cylindrical parabolic potential results in 2D orbitals s,p,d,f,... supporting 2, 4, 6, 8.... electrons which give rise to shell structures filled with 2, 6, 12, 20,.... particles, thereby creating a sequence of numbers which can be regarded as the 3D analogues of "magic numbers" in atomic physics [10,11].

In the absence of cylindrical or square symmetry, the parabolic potential are characterized by two different frequencies, ω_x and ω_z which lift the azimuthal degeneracy on the l-number of the 2D artificial atoms. Therefore electronic states are spin-degenerate only, and determine a sequence of shell filling numbers 2, 4, 6, 8,..... of period or increment 2. Only when the ratio ω_x/ω_z is commensurable, does the sequence of filling numbers deviates from the period 2. This provides a new sequence of "magic numbers" for particular combinations of the n_x and n_z quantum numbers in the case of accidental degeneracy.

Another important class of 3D confined systems are quantum dots obtained by self-assembled or self-organized Stranski-Krastanov (SK) epitaxial growth of lattice-mismatched semiconductors which result in the formation of strained induced nanoscale islands of materials. InAs and InGaAs islands on GaAs have been obtained with this technique in well-controlled size and density [9,12,13]. For these materials, shapes vary between semi-spherical and pyramidal form, and size is so small that these quantum dots contain only one fully 3D quantized level of conduction electrons. In this paper we restrict our analysis to parabolic shape quantum dots which contain a large number of quantum states for conduction electrons.

2.2. HARTREE-FOCK APPROXIMATION AND HUND'S RULES

The natural extension of the atomic model for independent 3D confined electrons is the consideration of the Coulomb interaction between particles in the Hartree-Fock (HF) approximation. The HF scheme has the advantage of conserving the single particle picture for the many-body state of the system by representing the total wavefunction as a product of single particle wavefunctions in a Slater determinant which obeys Fermi statistics. The main consequence of the HF approximation for the Coulomb interaction between particles is a correction of two-terms to the single particle energies derived from the $\hat{H}_o$ Hamiltonian [14];

$$E_i = E_{\nu ml} + \frac{e^2}{\epsilon} \sum_{j \neq i} \int d\vec{r}_i, d\vec{r}_j \frac{| \Psi_i(\vec{r}_i) |^2 | \Psi_j(\vec{r}_j) |^2}{| \vec{r}_i - \vec{r}_j |} - \frac{e^2}{\epsilon} \sum_{j \neq i} \int d\vec{r}_i d\vec{r}_j \frac{\Psi_j(\vec{r}_i) \Psi_i^*(\vec{r}_i) \Psi_i(\vec{r}_j) \Psi_j^*(\vec{r}_j)}{| \vec{r}_i - \vec{r}_j |} \tag{8}$$

where the second term with the summation over all occupied states j different from the state i, and irrespective of their spins, is the Hartree energy accounting for the classical repulsion between electrons. The last term is the attractive exchange interaction which occurs between carriers with parallel spins. In this scheme, the wavefunctions $\Psi_i(\vec{r}_i)$ satisfy the N-HF integro-differential equations where the Coulomb interaction terms depend upon all the other single particle wavefunctions of the occupied states. The HF equations are therefore non-linear and must be solved self-consistently for all wavefunctions of occupied states.

One of the important consequences of the HF approximation for inter-electron interaction (2) is the prediction of spin effects in the shell filling of artificial atoms similar to Hund's rules in atomic physics [11]. These effects are illustrated in the charging energy of a few electron quantum dot with cylindrical parabolic potential achieved in planar or vertical quantum structures [5]. In Fig 2.a we show schematically the Coulomb staircase resulting from charging a quantum dot with a few electrons as a function of energy or the voltage between the metal gate electrode and the semiconductor substrate. The relative step sizes of the staircase represent the amount of energy needed to put an additional electron in the dot. The vertical arrows on each steps represent the spin of each individual electron on the successive orbitals during the charging process. The filling of the first shell

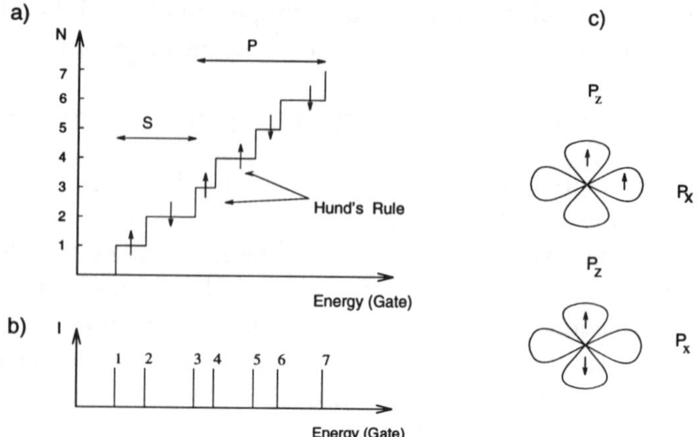

Figure 2. a) Coulomb staircase as a function of the charging energy with the spin states of each electron. N is the number of electrons, and the horizontal two head arrows indicate the occupation of the s- and p-orbitals in the dot. b) electron current through the dot versus the charging energy. c) 2D p-orbitals illustrating the two possible occupations of two electrons with parallel (top diagram) and anti-paprallel spins (bottom diagram).

(s-orbital with 2 electrons) consists of one electron with spin up followed by one electron with spin down. The step size of the spin-up electron measures the charging energy needed to overcome the Coulomb repulsion against the spin down electron, which is just the Hartree energy between the two particles. The larger step size of the second (spin-down) electron is due to the fact that the charging of the third electron requires the Coulomb blockade energy augmented by the energy to access the next quantized level which is the first p-orbital. The latter process starts the filling of the second shell with the third electron on either one of the degenerate $l = \pm 1$ orbitals of either spins (here we choose the $l = -1$ and the spin-up). At this stage, the configuration with the fourth electron on the $l = 1$ orbital with a parallel spin becomes more favorable because the resulting attractive exchange energy between the two electrons minimizes the Hartree energy between orbitals of different quantum numbers (Fig. 2.c). This is the reason why the third step is smaller than the first and the second steps, requiring less energy and manifesting Hund's rule in electron filling of the 2D artificial atom. The fourth step is long because the addition of the fifth electron on either of the p-orbitals with $l = \pm 1$ must correspond to a spin-down electron which undergoes a repulsion from the two other p-electrons without benefiting from the exchange with them since its spin is anti-parallel. Fig. 2.b shows the current peaks resulting from the single- electron charging of the quantum dot which is obtained by differentiating the Coulomb staircase. Current characteristics with similar structure have recently been observed in gated double barrier $GaAs/AlGaAs/InGaAs/AlGaAs/GaAs$

vertical quantum dot tunnel devices which revealed the shell structure for a cylindrical parabolic potential as well as spin effects obeying Hund's rule in the charging of the dot [11].

Hence the HF approximation provides a reasonable picture of the contribution of electron-electron interaction and spin effects in the spectrum of quantum dots. However, it is well known from atomic physics and theoretical condensed matter physics that this approximation suffers from two important drawbacks in that it neglects electron correlation and overestimates the exchange energy [15]. Moreover, it leads to a tedious solution of the self-consistent problem when involving a large number of electrons, and makes it difficult to apply to quantum devices the accounting of the environment of heterolayers, doping, and boundary conditions at the semiconductor interfaces with metal electrodes

3. Full Scale Simulation of Quantum Dots

Beyond the HF approximation, the density functional theory is extremely powerful in predicting with relative accuracy the electronic properties of atomic systems, and even of large ensemble of atoms. It is therefore well suited for computing the electronic spectra of quantum dots and artificial atoms in the nanostructure environment [16-18]. Moreover, like the HF approximation, it has the advantage of retaining the single-particle picture for the wavefunctions of the system.

3.1. THE LOCAL DENSITY APPROXIMATION (LDA) HAMILTONIAN

In the envelope function-effective mass approximation the Hamiltonian reads

$$\hat{H} = -\frac{\hbar^2}{2}\nabla\left[\frac{1}{m^*(\vec{r})}\nabla\right] + E_c(\vec{r}) + \mu_{xc}(n). \tag{9}$$

The conduction-band edge energy $E_c(\vec{r})$ is given by $E_c(\vec{r}) = -e\phi(\vec{r}) + \Delta E_c$, where ΔE_c is the band offset between different materials. The electrostatic potential $\phi(\vec{r}) = \phi_H(\vec{r}) + \phi_{ions}(\vec{r})$ where $\phi_H(\vec{r})$ is the Hartree potential equivalent to the second term in Eq. 2 and ϕ_{ions} is the potential resulting from ionized donors and acceptors present in the nanostructure, is the solution of the 3D Poisson equation,

$$\nabla\left[\epsilon(\vec{r})\nabla\phi(\vec{r})\right] = e\left[n(\vec{r}) - N_D^+(\vec{r}) + N_A^-(\vec{r})\right], \tag{10}$$

which is solved self-consistently with the stationary Schrödinger equation (9). In the above equations, we allow for the position dependence of the effective mass m^* and the dielectric constant ϵ; $n(\vec{r})$ is the electron concentration, and $N_D^+(\vec{r})$ and $N_A^+(\vec{r})$ are the ionized donor and acceptor concentrations, respectively. In n-type devices we neglect the charge due to holes concentration. Exchange and correlation are self-consistently incorporated into the model using the Kohn-Sham density-functional method [15]. This approach accounts for the high-order electron-electron interactions by adding a correction to the Hartree potential in $\hat{H}$:

$$\mu_{xc}(n) = \frac{d}{dn}[n\epsilon_{xc}(n)], \tag{11}$$

where $\epsilon_{xc}(n)$ is the exchange-correlation energy per electron derived from the local electron density (LDA) which can be expressed in terms of the Slater exchange energy and the Perdew-Zunger parametrization of the correlation potential [19]. The effect of exchange and correlation is predominant in the localized quantum-dot region, where the conduction band is altered by as much as 1 meV [18].

In simulating quantum devices, it is often useful to partition the device into regions of different confinements. Hence the eigenenergies and eigenstates are computed numerically in each region based upon the degree of quantization and the 3D electrostatic potential arising from the Poisson equation. The wavefunctions are then assembled using the appropriate occupation probabilities to generate a 3D charge density throughout the device. If the quantum device contains 2D gas regions, the adiabatic approximation can be used by assuming the Schrödinger equation to be separable into vertical (y) and transverse components (x,z). This approximation assumes that the vertical component of the wavefunction responds instantaneously to variations of the transverse potential. Since the longitudinal wave functions generally exhibit a propagating nature, they are cast as plane waves to enhance the performance of our code. In quasi-2D leads, therefore, the quantum-mechanical problem breaks down into solving the 1D Schrödinger equation in slices down the x-z plane of the 2D regions. The use of plane waves in 2D regions results in an incomplete treatment of evanescent leakage and reflection in regions where propagating states impinge on barriers. The plane-wave approximation could be justified, however, when in the calculation of charge densities the counting of occupied propagating states is of interest rather than rigorously calculated quasi-bound eigenenergies.

3.2. THE ITERATIVE EXTRACTION ORTHOGONALIZATION METHOD (IEOM)

The use of conventional numerical eigenvalue solvers in the generation of multiple solutions of the Schrödinger equations over a large grid is prohibitively expensive from a computational standpoint. The execution time of these methods scale as N_G^3 where N_G is the number of grid points. Furthermore, conventional methods generate N_G eigenvalues, a majority of which are unnecessary and therefore lead to inefficient use of computer time. To overcome these difficulties, we have employed the iterative extraction-orthogonalization method (IEOM) for solving the general eigenvalue problem which is given by

$$\hat{\Lambda} \mid \alpha\rangle = \lambda_\alpha \mid \alpha\rangle, \tag{12}$$

where $\mid \alpha\rangle$ is an eigenstate of the operator $\hat{\Lambda}$, and λ_α is the eigenvalue of $\mid \alpha\rangle$. The IEOM was originally developed by Kosloff and Tal-Ezer [20] and has found use in the generation of initial states in time-dependent problems [21-23].

The theory behind the IEOM involves the creation of a function operator $f(\hat{\Lambda})$ that effectively extracts the lowest eigenvalue basis state $\mid 0\rangle$ (i.e., the ground state) from an initial guess vector $\mid \psi_0\rangle$ comprised of a mixture of basis states. Operating on $\mid \psi_0\rangle$ with $f(\hat{\Lambda})$ gives

$$\begin{aligned}(f\hat{\Lambda}) \mid \psi_0\rangle &= \sum_m \mid \alpha\rangle f(\lambda_\alpha)\langle\alpha \mid \psi_0\rangle \\ &= \mid 0\rangle f(\lambda_0)\langle 0 \mid \psi_0\rangle + \sum_{m=1} \mid \alpha\rangle f(\lambda_\alpha)\langle\alpha \mid \psi_0\rangle\end{aligned} \tag{13}$$

Clearly, if $f(\lambda_\alpha)$ decreases with increasing λ_α the first term in Eq. (13) will dominate and the new basis composition will favor the ground state. Successive applications of $f(\hat{\Lambda})$ further extract the ground state such that after some number of iterations N_i, the high-order contributions become vanishingly small:

$$f(\hat{\Lambda})^{N_i} \mid \psi_0\rangle \simeq \mid 0\rangle f(\lambda_0)^{N_i}\langle 0 \mid \psi_0\rangle \tag{14}$$

A constraint on the extraction capability of $f(\hat{\Lambda})$ is that it must have a well-behaved Taylor expansion to cast it into a computationally tractable form. In addition, physical eigenvalue problems typically require more eigenstates than just the lowest state, so a mechanism must be used to maintain the spectrum of higher (excited) eigenstates. This is accomplished by creating a set of N_E initial guess states ψ_n and applying the Gram-Schmidt orthogonalization algorithm

$$\mid \psi_n'\rangle = \mid \psi_n\rangle - \sum_{m=1}^{n-1} \frac{\mid \psi_m'\rangle\langle\psi_m' \mid \psi_n\rangle}{\langle\psi_m' \mid \psi_m'\rangle} \tag{15}$$

after each iteration of Eq. (13). The net effect of the Gram-Schmidt orthogonalization is to eliminate all basis components with indices $m = 1, 2, ..., n-1$ from ψ_n, such that a unique basis state $| n\rangle$ is extracted for each ψ_n. The application of multiple extraction and orthogonalization iterations thereby converts a set of initial states $| \psi_n\rangle$ into a set of solution states $| n\rangle$.

The chief advantage of the IEOM is the ability to generate an arbitrarily small number of eigenstates N_E. As a result, the IEOM scales as $N_E^2 N_G$ as opposed to N_G^3 for conventional eigenvalue methods. The iterative extraction-orthogonalization method therefore maps particularly well into the simulation of nanostructures exhibiting a small number of occupied states and a large number of mesh points [18].

3.3. BOUNDARY CONDITIONS

Boundary conditions for the Schrödinger equation are imposed by assuming vanishing wave functions around the perimeter of each quantized region. Special consideration, however, is required for the boundary conditions around 0D regions, particularly in the direction of current flow. Since the localized quantum-dot states hybridize into propagating states in the lead (2D) regions, the constraint of a vanishing wave function at the 0D-2D boundary artificially alters the weakly localized eigenenergies. To avoid this problem, we impose the longitudinal boundary conditions well into the lead regions and let the localized states acquire propagating components. This, however, generates an additional problem since the parity of the quasilocalized 0D states is now determined in part by the propagating portion as well as the localized portion. The Gram-Schmidt calculation would then orthogonalize the quasilocalized states based on their overall parity (localized and propagating) and tend to collapse them into the resonant linewidth of the ground state. This problem could be avoided by creating an interior boundary for the integrations in the Gram-Schmidt step that treats the parity only in the localized region and allows us to retain an orthogonal set of localized 0D states.

The boundary conditions imposed on the potential reflect the confinement geometry of the structure. Potentials along exposed surfaces are modeled by Dirichlet conditions which convey Fermi-level pinning. The Schottky-barrier heights ϕ_s are strongly influenced by surface chemistry; so we use experimental data of Grant et al. [24] for GaAs and Best [25] for $Al_{0.3}Ga_{0.7}As$. Gated surfaces are treated by modifying the Schottky barriers by $\phi_s - V_G$, where V_G is the gate bias. Unexposed boundaries are assumed to be under flat-band conditions and are modeled by Neumann boundary conditions.

3.4. CHARGE DENSITIES AND EQUILIBRIUM STATISTICS

Once the wave functions are obtained in each region exhibiting quantum confinement, they are assembled to create a local charge density. As mentioned above, the wave functions are cast into quasibound forms with free-electron behavior in the unconfined directions where applicable. The charge densities therefore take on the form

$$n(\vec{r}) = \sum_{m}^{N_E^{0D}} \mid \psi_m(x,y,z) \mid^2 f_0(E_m) \ \ (0D) \tag{16}$$

$$= \sum_{m}^{N_E^{2D}} \mid \psi_m(y) \mid^2 F_0[E_m(x,z)] \ \ (2D) \tag{17}$$

In the above expressions, $F_0(E)$ is the Fermi integral of order 0 and $f_0(E_m)$ is the distribution function for a quantum dot which is derived from the grand canonical ensemble. If the dot is in strong contact with the leads through tunneling, the distribution is well approximated by the Fermi-Dirac form. If, on the other hand, the dot wave functions exhibit a high degree of localization, only an integer number of electrons N can occupy the dot and this constraint is imposed on the Gibbs' distribution such that [26,27]

$$f_0(N \mid E_m) = \frac{1}{Z(N)} \sum_{\{n_i\}} exp[-\beta[\sum_i n_i E_i - \frac{1}{2} E_H\{n_i\} - N\mu]]\delta_{m,1} \tag{18}$$

where the partition function Z(N) is given by

$$Z(N) = \sum_{\{n_i\}} exp[-\beta[\sum_i n_i E_i - \frac{1}{2} E_H\{n_i\} - N\mu]] \tag{19}$$

$\frac{1}{2}E_H$ is half the Hartree energy which subtracted from the total energy avoids from double counting of the electron-electron interaction. In the above expressions, μ is the chemical potential in the leads, and the summations are carried out over all possible occupation configurations $\{n_i\}$ with the constraint that $\sum_i n_i = N$. The lack of diffusive contact between the quantum dot and the rest of the structure implies that μ is well defined only in the leads and contacts. For carriers that do not exhibit confinement (holes) and regions that are not governed by quantum mechanics a semiclassical Thomas-Fermi model is used to generate the charge density.

Under nondiffusive conditions, the equilibrium number of electrons in the quantum dot N_{eq} will generally be determined by the configuration that minimizes the free energy $F(N) = -kTln[(Z(N)]$.

4. Results

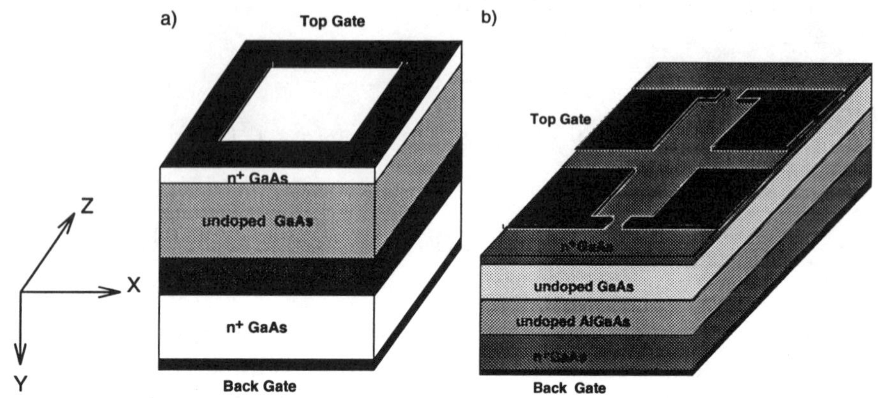

Figure 3. Square (a) and quad gate (b) quantum dot devices with layer structure

Figure 3 shows a schematic representation of two planar quantum dot devices fabricated by growing an inverted GaAs/$Al_xGa_{1-x}As$ heterostructure which confines the electrons to a 2D gas at the interface. The layer dimensions used in the device consist of a 22.5-nm undoped $Al_{0.3}Ga_{0.7}As$ barrier, a 125-nm undoped layer of GaAs, and a 18-nm-thick GaAs cap layer. The doping of the cap layer was uniformly distributed with a concentration $N_D = 5 \times 10^{18}$ cm^{-3} to bring the conduction band just above the Fermi energy at the GaAs-cap layer/undoped GaAs interface [1,18]. The inverted heterostructure is grown on an n^+ GaAs substrate, and charge control is achieved by modulating the bias V_{back} on a backgate. In our model, we assume a negligible potential drop over the conductive substrate and apply V_{back} directly to the bottom of the $Al_{0.3}Ga_{0.7}As$ layer. We first consider a square gate quantum dot device with a 250×250 nm^2 open area in the x-z plane on the top and a 65nm wide metal gate (Fig. 3.a). This structure is idealized because the gate is completely closed and does not allow for electron tunneling from the 2D gas to charge the dot. However it is well suited for studying the electronic spectrum and dot filling without considering the influence of splitting the gate on the electron confinement.

The confining potential in the square device is displayed in Fig. 4, which shows the conduction-band edge for slices taken through the device active region. A potential-energy minimum occurs at the interface (y = 0.143μm) between the undoped GaAs and $Al_{0.3}Ga_{0.7}As$ spacer layer

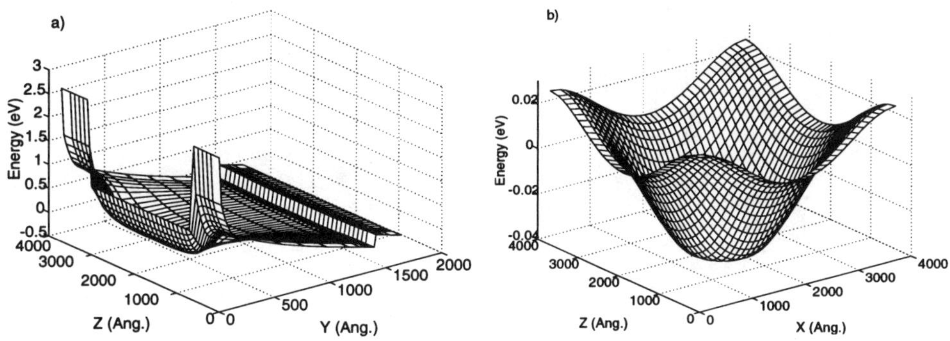

Figure 4. Conduction band-edge profile: a) A vertical slice in the y-z plane taken at the center of the dot shows the lateral confining potential arising at the GaAs/AlGaAs heterojunction and between gate along the z-direction. b) A horizontal slice in the x-z plane taken at the heterojunction shows the effect of confinement in the corners of the square gate.

[Fig. 4(a)]. The confinement in this vertical direction (y) is the strongest, and therefore provides the largest contribution to the eigenenergy spectrum ($\Delta E_y \simeq 30 - 40$meV). The transverse confinement arising from the square gate at y $=$ 0μm is evident in Fig. 4(a), which reveals the square-gate separation of 0.25 μm in the z-direction. The strength of the planar (x,z) confinement is considerably weaker and barely noticeable; therefore it plays the most significant role in determining the quasi-equilibrium properties of the device. This fact has often been used to justify two-dimensional approximations of quantized structures. Figure 4(b) shows the conduction-band edge in a horizontal plane taken at the undoped GaAs/$Al_{0.3}Ga_{0.7}As$ interface (y $= 0.143\mu$m). The square shape of the gate is evident by the high value of the potential in the four corners while in the center of the dot, the potential is roughly parabolic and assumes a cylindrical symmetry. However as we shall see later on, the cylindrical symmetry and the parabolic dispersion hold only for no more than two electrons in the dot since the ground state is cylindrically symmetric. For a larger number of electrons the effect of the Coulomb interaction between carriers tends to recover the square symmetry and flattens the potential in the center of the dot.

Figure 5 shows the first four localized 3D wave functions that reside in the quantum dot. The labeling scheme we use reflects the number of nodes in each spatial direction. For instance, state (002) has no nodes in the x- and the y- directions, and two nodes in the z-direction. Given the square symmetry of the gate, the (100)- and the (001)- states are degenerate and form the first p-orbital with 4 electrons. The (002)-state is the next higher state which is equivalent to the (200) state by a 90° rotation of the wavefunction. By symmetry considerations one may think that this state

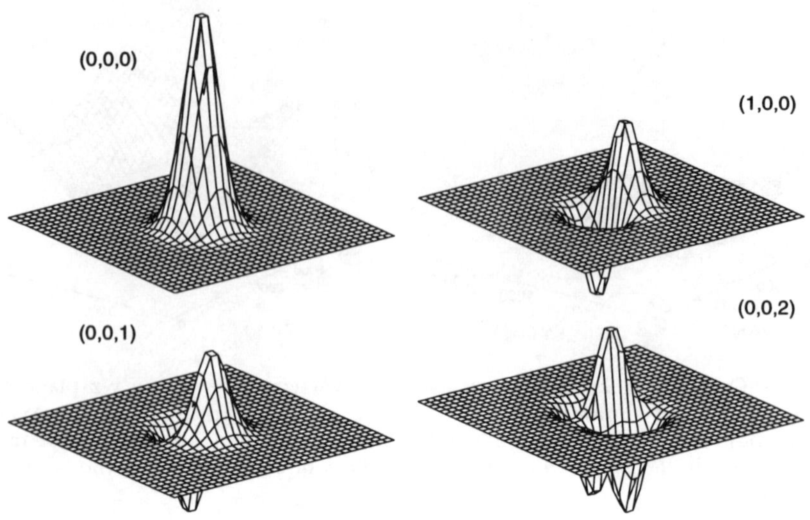

Figure 5. The four lowest wavefunctions at the hetero-interface. The (100) and (001)-states are degenerate.

is also degenerate with the (101)-state. However this is only correct if the confining potential is strictly parabolic in the x- and z-variables, which is hardly the case with the Coulomb interaction between carriers which induces some anharmonicity in the potential.

Figure 6.a shows the potential profile in the dot along the x-direction for increasing values of the back gate voltage V_{back}. Here the zero of the vertical scale corresponds to the position of the Fermi level. Notice the depth of the potential well ($40meV$) which takes into account the energy resulting from the y-confinement, unseparable from the other degrees of freedom in our model computation of quantized states. Each curve corresponds to the potential calculated just after the addition of an electron to the dot, starting from the first electron with the highest curve to the thirteenth electron with the lowest curve. It is seen that the potential energy drops as expected when V_{back} increases to charge the dot with electrons. However the potential also flattens out with the increasing number of electrons, thereby deviating significantly from the initial parabolic shape and reflecting the Coulomb interaction between particles. The remarkable feature here is the grouping of the potential curves in series of tight potential "bunches" which are indicative of the shell structure in the quantum dot. Indeed a closer look reveals the successive "bunching" of the first two curves for the s-orbital, then the next four curves for the p-orbitals indicating the filling of the 4-electron shell (N=6) and then the six next curves for the filling of the 6-electron shell (N=12) and so on. The evolution of the single-particle energy levels as a function of the backgate voltage is illustrated in Fig. 6.b. which shows a staircase variation of the first ten levels. It is seen that for low gate voltage ($V_{back} < 0.98V$) the levels are well separated and drop rapidly with V_{back}. However, when the first s-level crosses the Fermi level, it "sticks" to this level for a voltage range which corresponds to the charging of the first two electrons in the dot. This effect produces a change of slope in the upper levels which remains well separated from the first level owing to the electrostatic confinement in the dot. Meanwhile, the third and fourth levels which are respectively threefold and fourfold degenerate, and are responsible for the 6- and 8-electron orbitals, split each into two new levels. This effect is due to the Coulomb interaction between carriers which induces some anharmonicity in the confining potential and lifts the degeneracy of the (101)-state with the (200)- and (002)-states, on one hand, and the degeneracy of the (201)- and (102)-states with the (003)- and (300)-states on the other hand. However, because of the conservation of the square symmetry the (002)- and (200)-, the (300)- and (003)-, and the (201)- and (102)-states remain degenerate, respectively.

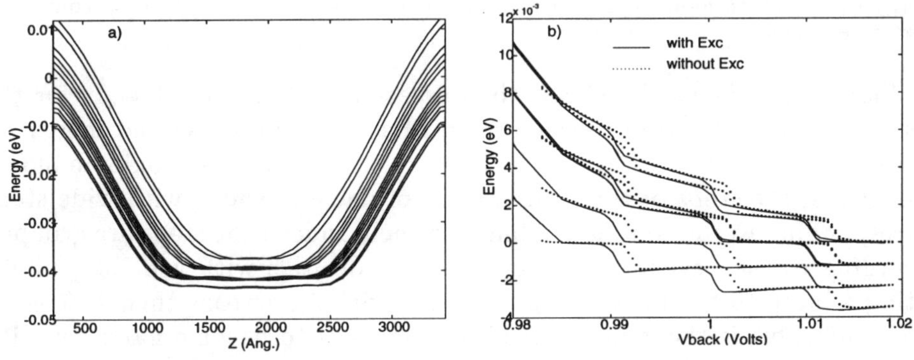

Figure 6. a) A slice of the conduction band edge along the x-direction for various values of N between 1 to 13. The shell structure of the dot energies is manifested in the bunching of the conduction band bottom for various Ns. b) Electronic energy spectrum as a function of the back gate voltage V_{back}. The lowering of the levels due to the exchange interaction is seen in the shifting of the levels to lower voltages with E_{xc}.

When the first s-orbitals is filled, it drops suddenly below the Fermi level, leaving room for the next p-orbitals to be filled with four electrons. During this process, the degenerate (100)-and (001)-states remain on the Fermi level for the charging of four electrons over a longer range of backgate voltage than for the charging of the first level, as seen in the diagram. In the meantime, the split upper levels reconverge because the potential becomes flatter and wider with the addition of electrons in the dot and the many-body interaction. This effect is also noticeable in the decreasing separation between levels below the Fermi energy as a function of V_{back}. Hence the process of shell filling is repeated for each successive orbital. Also visible on Fig. 6.b. is the influence of electron exchange-correlation which shifts all the single particle levels to lower energy because of the attractive nature of this many-body interaction.

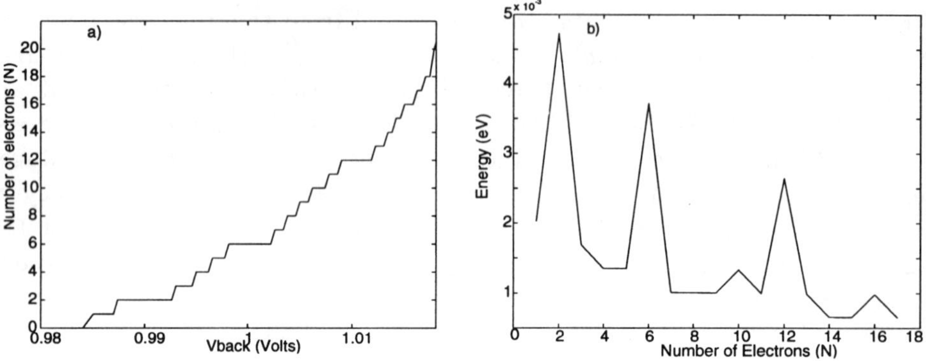

Figure 7. a) Coulomb staricase for the square dot structure. b) Variation of the addition energy with electron number. The peaks correspond to the additional energy required to start a new shell.

Figure 7.a. shows the Coulomb staircase as a function of V_{back} for the first eighteen electrons in the square dot. Each step indicates the energy required to add an electron to the dot. Unlike in conventional Coulomb blockade effects, the steps of the staircase are of unequal width with wide steps corresponding to the complete filling of a shell. Indeed the steps are grouped according to the degenerated single particle orbitals with wide steps indicating a jump to the next orbital, starting with 2 electrons then 4, 6, etc... Also noticeable is the general trend for the steps to become narrower with high gate voltage which results from an increase in the dot capacitance, as well as a reduction in confinement which decreases the energy separation between levels as the dot widens. These effects are illustrated in Fig.7.b. which shows the addition energy $e^2/C_s(N) = \mu(N+1) - \mu(N)$ to add an electron to the dot which is exactly the inverse of the dot self-capacitance $C_s(N)$ [10]. Each peak indicates the filling of a shell and corresponds to the wide steps in the Coulomb staircase of Fig. 7.a. Between the peaks, the addition energy remains small with the overall decreasing trend at high V_{back}. However two additional peaks at N=10 and N=16 electrons appear in the diagram, which indeed correspond to wider steps within the orbital grouping in the Coulomb staircase [28]. These anomalously large addition energies are caused by the splitting of the degenerate d- and f-orbitals due to anharmonicity induced by many body interaction in the confining potential as mentioned in our comments of Fig. 6. Indeed a detailed analysis shows that the (101)-state in the d-orbitals is slightly higher in energy than the (200)- and (002)-states bearing the tenth electron. Similarly the (201)- and (102)-states in the f- orbitals are higher than the (300)- and the (003)-states occupied by the sixteenth electron. These general features in the addition energy spectrum e.g. high peaks for the shell filling and the N=16 peak, have indeed been observed experimentally by Tarucha et al., although in a vertical quantum dot with circular cross section [11]. It is worth mentioning that the experimental data shows more peaks than the theory because of spin effects and Hund's rule in the shell filling. At this stage, our model cannot account for these effects because the LDA does not distinguish between spin states in the calculation of the exchange and correlation. Extention of the LDA into the more advanced Local Spin Density Approximation (LSDA) capable of resolving the spin of each single particle is more suited for this purpose [15]. Let us also point out that the circular geometry of the experimental dot weakens the effect of anharmonicity and that experimental data show a minimum instead of a peak at N=10 in the addition energy diagram, which may be due to spin effects offsetting the influence of anharmonicity in the potential.

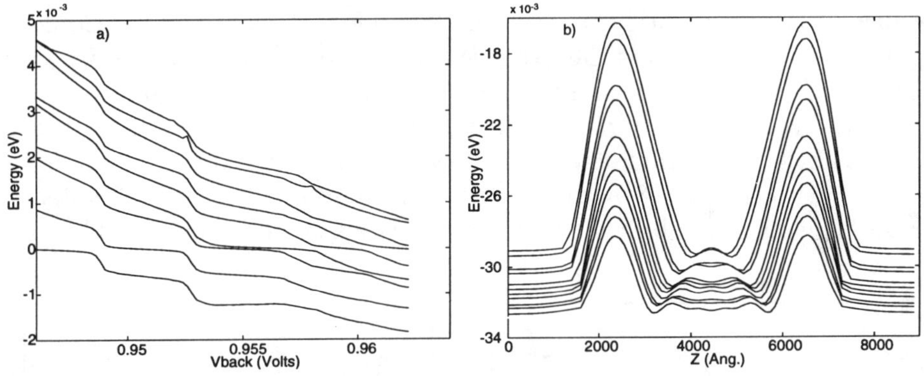

Figure 8. a) Electron energy spectrum as a function of back gate voltage for the quad-gate structure. b) Slice of the conduction band-edge along the z-direction for different values of N between 1 and 11 showing the bunching of the conduction band edge for 5th, 6th 7th and 8th electron occupation under "Coulomb degeneracy".

In Fig. 8.a., we show the energy spectrum of the quad gate quantum dot of Fig. 3.b. In addition to the rectangular shape of the dot, this configuration has also split gates along the x- and z-directions, which are suited for electron charging by tunnel injection from the adjacent 2D electron gas. The rectangular open area on the top is 230×408 nm^2 wide, the two 40 nm wide stubs at the edges of the large side of the rectangle (z-direction) are separated by 70 nm, and the opening between the gates in the middle is 80 nm wide. This new gate geometry breaks the square symmetry of the former dot and lifts the degeneracy between single particle levels. For an empty dot, the succession of the first lower levels is (000), (001), (002), (100), (003), (101), (004), (005), etc.... with similar profile as shown in Fig. 3. The variation of the energy levels as a function of gate bias shows a trend qualitatively similar to the square dot except that each curve now represents a spin-degenerate level only which reduces the shell structure to a simple superposition of doubly degenerate states. Because the ratio between the sides of the rectangle are incommensurable, accidental geometrical degeneracy mentioned in section 2.1 is absent from the spectrum. However careful analysis of the energy spectrum shows several interesting features such as a convergence between the (002)- and the (100)-states, and several level anti-crossings at high gate bias i.e. at V_{back}=0.958V and 0.961V. The former effect is due to the presence of openings in the gate which makes the variation of the (100)-level more rapid than the variation of the (002)-level with back gate bias. As a result of the convergence these levels would cross on the Fermi level when the lower (002) state is filling with electrons. However, the level crossing cannot take place if the (002) is only partially filled with one electron because that would imply that the (100) would be immediately filled with two electrons, but this is impossible because of the many body interaction. Therefore, because of the Coulomb repulsion between carriers the two levels appear to be degenerate during the charging of the dot. Indeed Fig. 8.b. shows the potential profile variation with gate bias. By comparison with Fig. 6.a. for the square gate, it is seen that the shell "bunching" effect has now disappeared, except for the filling the 5th, 6th, 7th and 8th electrons which shows a "bunch" of potential curves. This "Coulomb degeneracy" effect is also evident on the Coulomb

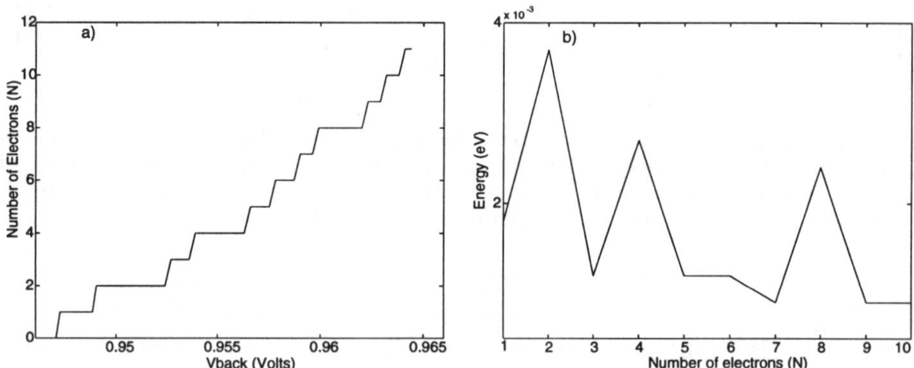

Figure 9. a) Coulomb staircase for the quad-gate structure b) Variation of the addition energy with electron number showing the low charging effect during Coulomb degeneracy.

staircase of fig 9.a. for V_{back} varying between 0.954V and 0.959V. Indeed, aside from the regular series of short and long steps reflecting the successive charging of separated and doubly degenerate states, the Coulomb staircase shows a grouping of steps with the same width for N=5,6 and 7 electrons. Similar features are seen in the addition energy diagram of Fig. 9.b. which shows an alternance of peaks and valleys of decreasing amplitude except for N=5,6, and 7 which illustrate the effect of "Coulomb degeneracy " for the (002)- and (100)-states. Finally let us mention that the level anti-crossing seen at V_G= 0.958V and 0.961V are of the same origin as the "Coulomb degeneracy" except that the level convergence occurs when the lower level is completely filled with its two electrons. Therefore it drops below the Fermi Energy before the second level begins to charge. It should be pointed out that these effects disappear if the split gates are closed along the z-direction since it is the presence of the openings which is responsible for the different variations of confinement along the x- and z-directions and the level convergence with gate bias.

5. Conclusions and Outlook

The investigation of quantum dots and artificial atoms requires comprehensive self-consistent theoretical tools for describing realistically the respective influence of the external confining potential and the many-body interaction between carriers on the electronic properties. We have shown that the LDA coupled to efficient eigenvalue solvers such as the IEOM is so far the most powerful technique for treating the quantum mechanical properties of 3D confined systems in arbitrary confinement. The consideration of quantum dots with two different geometries has illustrated the capability of the LDA-IEOM scheme to describe within a comprehensive model, encompassing the whole device, the electronic properties of few electron confined systems from the formation of shell structures in highly symmetric configuration to such fine features as the variation of the charging energy caused by anharmonicity in the electrostatic potential or the "Coulomb degeneracy" due to many-body interaction in rectangular dots. However, comparison with experimental data has shown the limitation of the LDA and its inability to treat spin-related effects such as Hund's rule in the shell formation. The main reason resides in the formulation of the LDA in terms of single-particle spin- degenerate energy levels as expressed by Eq. 9., while a more realistic model should resolve the spin of each particle. In principle, the HF approximation can treat this issue but it has only qualitative predictive capability. Therefore, extension of the LDA into a formalism treating single-spin states such as the LSDA appears to be the most realistic approach to address this issue. This new technique appears also particularly important in the description of the electronic properties of self-assembled dots for electrons but mainly for hole states which can accommodate several single- particle levels in the dots.

Acknowledgements: The authors are indebted to R. Martin, P. von Allmen, L. Fonseca, I.H. Lee, P. Matagne, V.Y. Thean, Y.H. Kim, S. Tarucha and H. Tamura for stimulating discussions, and to K. Hess for providing a preprint of his manuscript. They also acknowledge the technical assistance of Ms Marian "Miq" Hanna. This work was supported by NSF Grant No NSF-ECS 95-09751 and ARO Grant No DAAH04-95-1-0190.

6. References

1. Kastner, M.A. (1993) Artificial Atoms,*Physics Today* **46**, 24-31.

2. Averin, D.V., and Likarev, K.K. (1991), Single Electronics: A correlated transfer of single electrons and Cooper Pairs in systems of small tunnel junction, in B.L. Altshuler, P.A. Lee, and R.A. Webb (eds.), *Mesoscopic Phenomena in Solids*, Elsevier, Amsterdam, pp. 173-271.

3. Devoret, M.H. and Grabert, H., Introduction to Single Charge Tunneling NATO ASI B294 in Grabert, H. and Devoret, M.H. (eds.), *Single Charge Tunneling: Coulomb Blockade Phenomena in Nanostructures*, Plenum Press, pp. 1-19.

4. Korotkov, A. Coulomb Blockade and Digital Single-Electron Device to be published in Jortner, J. and Ratner, M.A. (eds.), *Molecular Electronics*, Blackwell, Oxford.

5. Meirav, V. and Foxman, E.B. (1995) Single-Electron Phenomena in Semiconductors, *Semicond. Sci. Technol.* **10**, 255-284.

6. Ashoori, R.C., Störmer, H.L., Weiner, J.S., Pfeiffer, L.N., Baldwin, K.W., and West, K.W. (1992) N-Electron Ground State Energies of a Quantum Dot in a Magnetic Field, *Phys. Rev. Lett.* **71**, 613-616.

7. Johnson, N.F. (1995) Quantum Dots: Few Body, Low Dimensional Systems, *J. Phys. Condens. Matter* **7**, 965-988.

8. Arakawa, Y. and Yariv, A. (1986) Quantum Well Lasers-Gain, Spectra, Dynamics, *IEEE Journal of Quantum Electronics* **22**, 1887-1899.

9. Ledentsov, N.N., Grundmann, M., Kirstaedter, J., Schmidt, O., Heitz, R., Böhrer, J., Bimberg, D., Ustinov, V.M., Shchukin, V.A., Egorov, A.Yu., Zhukov, A.E., Zaitsev, S., Kop'ev, P.S., Alferov, Zh.I., Ruvimov, S.S., Kosogov, A.O., Werner, P., Gösele, U., and Heydenreich, J. (1986) Ordered Arrays of Quantum Dots: Formation, Electronic

Spectra, Relaxation Phenomena, Lasing, *Solid State Electronics* **40**, 785-798.

10. Maccuci, M., Hess, K., and Iafrate, G.J. (1995) Simulation of Electronic Properties and Capacitance of Quantum Dots, *J. Appl. Phys.* **77**, 3267-3276.

11. Tarucha, S., Austing, D.G., Honda, T., VanDerHage, R.J., and Kouwenhoven, L.P. (1996) Shell Filling and Spin Effects in a Few Electron Quantum Dot, *Phys. Rev. Lett.* **77**, 3613-3616.

12. Leonard, D., Pond, K., and Petroff, P.M. (1994) Critical Layer Thickness for Self-Assembled InAs Island on GaAs, *Phys. Rev. B* **50**, 11687-11692.

13. Miller, M.S., Malm, J.O., Pistol, M.E., Jeppesen, S., Kowalski, B., Georgsson, K., and Samuelson, L. (1996) Stacking InAS Islands on GaAs Layers: Strongly Modulated One-Dimensional Electronic Systems, *Appl. Phys. Lett* **80**, 3360-3364.

14. Madelung, O. (1978) *Introduction to Solid State Physics*, Springer Series in Solid-State Sciences **2**, Springer-Verlag, Berlin.

15. Jones, R.O. and Gunnardson, O. (1989) The Density Functional Formalism, its Applications and Prospect, *Reviews of Modern Physics* **61**, 689-746.

16. Kumar, A., Laux, S.E., and Stern, F. (1990) Electron States in a GaAs Quantum Dot in a Magnetic Field, *Phys. Rev. B* **42**, 5166-5175.

17. Stopa, M. (1993) Coulomb Oscillation Amplitudes and Semiconductor Quantum-Dot Self-Consistent Level Structure, *Phys. Rev. B* **48**, 18340-18343.

18. Jovanovic, D. and Leburton, J.P. (1994) Self-Consistent Analysis of Single Electron Charging Effects in Quantum-Dot Nanostructures, *Phys. Rev. B* **49**, 7474-7483.

19. Perdew, J.P. and Zunger, A. (1981) Self-Consistent Correction to Density-Functional Approximations for Many-Electron Systems, *Phys. Rev. B* **23**, 5048-5079.

20. Kosloff, R. and Tal-Ezer, H. (1986) A Direct Relaxation Method for Calculating Eigenfunctions and Eigenvalues of the Schrödinger Equa-

tion on a Grid, *Chem. Phys. Lett.* **127**, 223-230.

21. Feit, M.D., Fleck, Jr., J.A., and Steiger, A. (1982) Solution of the Schrödinger Equation by a Spectral Method, *J. Comp. Phys.* **47**, 412-433.

22. Degani, M.H. (1991) Stark Ladders in Strongly Coupled GaAs-AlAs Superlattices, *Appl. Phys. Lett.* **53**, 57-59.

23. Bigelow, J. and Leburton, J. P. (1994) Self-Consistent Simulation of Quantum Transport in Dual Gate Field Effect Transistors, *J. Appl. Phys.* **76**, 2887-2892.

24. Grant, R.W., Waldrop, J.R., Kowalczyk, S.P., and Kraut, E.A. (1981) Correlation of GaAs Surface Chemistry and Interface Fermi-Level Position: A Single Defect Model Interpretation Grant, *J. Vac. Sci. Techn.* **19**, 477-480.

25. Best, J.S. (1979) The Schottky-Garrier Height of an on $-Ga_{1-x}Al_xAs$ as a function AlAs content, *Appl. Phys. Lett.* **34**, 522-524.

26. Beenakker, C.W.J. (1991) Theory of coulomb-blockade oscillations in the conductance of a quantum dot, *Phys. Rev. B* **44**, 1646-1656.

27. Averin, D.V., Korotkov, A.N., and Likharev, K.K. (1991) Theory of single-electron charging of quantum wells and dots, *Phys. Rev. B* **44**, 6199-6211.

28. A similar effect is also predicted by Macucci, M., Hess, K., and Iafrate, G.J. Numerical Simulation of Shell Filling Effects in Circular Dots, to be published.

SELF–ORDERING OF NANOSTRUCTURES ON SEMICONDUCTOR SURFACES

V.A. SHCHUKIN[*,†], N.N. LEDENTSOV[†], M. GRUNDMANN, and D. BIMBERG
Technische Universität Berlin, Hardenbergstraße 36, Berlin D–10623, Germany
[†] *On leave from A.F. Ioffe Physical Technical Institute St. Petersburg 194021, Russia*

A review is given of theoretical concepts and experimental results on self–ordering phenomena on semiconductor surfaces and on formation mechanisms of self–ordered nanostructures. Thermodynamic theory is reviewed for various classes of ordered semiconductor nanostructures, namely periodically faceted surfaces, systems of surface stress domains, and ordered arrays of three–dimensional coherently strained islands. All these structures are described as equilibrium structures of elastic domains. Despite the fact that the driving forces of the instability of a homogeneous phase are different in each case, the common driving force for the long–range ordering of the inhomogeneous phase is the elastic interaction. Kinetic theory of the formation of multiple–sheet structures of quantum dots is reviewed which includes both equilibrium ordering of coherently strained islands in the first sheet and the strain–driven kinetic stacking of islands in the next sheets or strain–driven shape transformation effects induced by complex growth modes. Experimental situation in fabrication technologies of both single–sheet and multiple–sheet ordered arrays of quantum dots is analyzed.

G. Abstreiter et al. (eds.), Optical Spectroscopy of Low Dimensional Semiconductors, 257–302.

I. INTRODUCTION

Spontaneous formation of periodically ordered domain structures in solids with the periodicity much larger than the lattice parameter is a general phenomenon which covers a large variety of different domain structures. This process, also called self–ordering, can occur if the homogeneous state of the system is thermodynamically unstable, and the system undergoes a phase transition into an inhomogeneous state. In an inhomogeneous state, a coordinate–dependent order parameter is, in general case, the source of a long–range field (electric, magnetic, strain). Therefore, a multi–domain state of the solid is energetically more favorable than the single–domain state, since the former provides compensation of the long–range field at large distances outside the domain structure. The long–range field is responsible for the periodic ordering of the equilibrium domain structure which meets the conditions of the total Helmholtz free energy minimum. The free energy may be written as a sum of three distinct contributions:

$$F_{total} = F_{domains} + F_{boundaries} + E_{long-range} \,. \tag{1}$$

Here $F_{domains}$ is the free energy of domains themselves, $F_{boundaries}$ is the free energy of domain boundaries, $E_{long-range}$ is the energy of a long–range field, which includes interaction between domains.

Conventional classification of domain structures is based upon the physical nature of the order parameter. Here we intend to consider the problem from another point of view and emphasize the nature of a long–range field associated with the domain structure. In this sense, an elastic strain field is of major importance since it exists nearly in all domain structures. The reason is that any phase transformation in solids is accompanied, as a rule, by crystal lattice rearrangement. In most cases, neighboring domains have different lattice parameters of unit cells of the bulk or of the surface crystal structure, whatever is the order parameter itself. For the case of ferroelectric or ferromagnetic domain structures this phenomenon is well known as electrostriction or magnetostriction, respectively [1].

The focus of the present review is given on domain structures where the elastic strain field is *the only long–range field* or at least where the major part of the energy of a long–range field is the elastic strain energy. Such domain structures are often called elastic domain structures. Whereas ferroelectric or ferromagnetic domain structures are for a long time dealt with in textbooks (see, e. g., Ref. [1]), elastic domain structures became only recently the subject of monographs. To the best of our knowledge, the first systematic theoretical approach to elastic domain structures *in bulk materials* is given in the monograph by Khachaturyan [2]. Both composition–modulated structures in phase–separating metal alloys and domain structures induced by martensite transformation are considered in Ref. [2], and a review of experimental data is presented. The theory of elastic domain structures in *epitaxial films of macroscopic thickness* has been developed by Roitburd [3] and by Bruinsma and Zangwill [4] for martensite–type domains and by Ipatova, Malyshkin, and Shchukin [5,6] for composition–modulated structures in alloys.

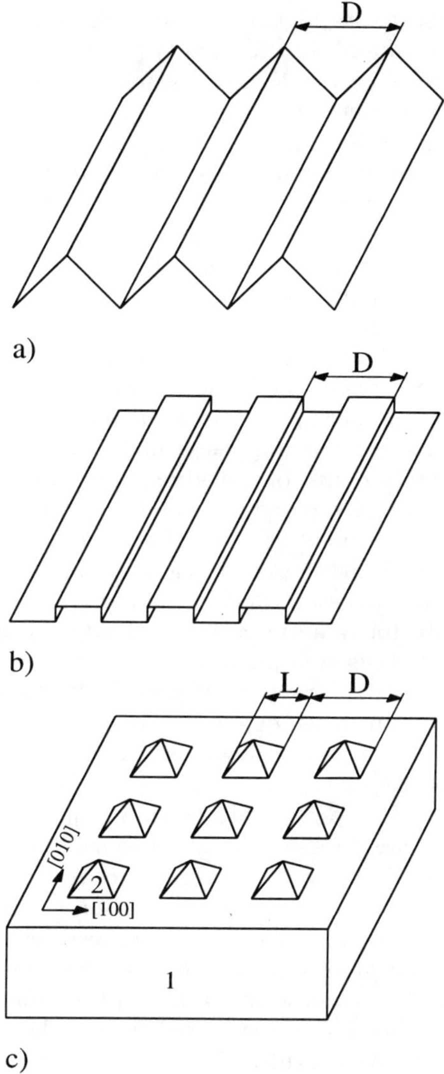

FIG. 1. Various classes of spontaneously ordered nanostructures. a) Periodically faceted surfaces; b) Surface domain structures; c) Ordered arrays of coherently strained islands (2) on a lattice–mismatched substrate (1).

In recent years, due to its high importance for fabrication of semiconductor heterostructures with reduced dimensionality (quantum wires and quantum dots), self–ordering phenomena *on crystal surfaces* became a subject of intense experimental and theoretical studies. In the present review article we will discuss *three classes* of self–ordered surface structures displayed in Fig. 1. These are periodically faceted surfaces of pure crystals, surface domain structures (e. g., ordered arrays of monolayer–height islands in heterophase systems at sub–monolayer coverage), and ordered arrays of three–dimensional coherently strained islands on lattice–mismatched substrates. Since typical linear dimensions of all these structures are of nanometer length scale, they are commonly termed nanostructures. We will demonstrate below that for all these types of nanostructures there exists a long–range strain field created at domain boundaries. This enables us to use a general approach and to consider all these nanostructures as *equilibrium structures of elastic domains*. In comparing with experimental data, particularly with crystal growth experiments, we will emphasize under which conditions and to which extent the surface structure are controlled by equilibrium thermodynamics.

Self–ordered nanostructures based on *semiconductor materials* are of particular interest not only from the point of view of self–ordering phenomena themselves but also as objects exhibiting unique optical properties and exciting device applications. Semiconductor nanostructures of interest are those where inclusions of a narrow–gap semiconductor material are embedded into the matrix of a wide–gap semiconductor material thus creating a confinement potential for electrons, or for holes, or for both. Periodic structure of such inclusions may create a superlattice comprising of quantum wells, quantum wires, or quantum dots. Self–ordering phenomena in semiconductors are now the basis of a novel fabrication technology of ordered arrays of quantum wires and dots. Advantages which self–ordered nanostructures have with respect to other types of nanostructures, particularly in applications for lasers, are discussed, e. g., in Refs. [7,8].

The three classes of self–ordered nanostructures presented in Fig. 1 are the content of the next three Sections, II — IV. In Section II the problem of the equilibrium crystal shape in relation to surface faceting is reviewed. A detailed discussion is presented on the concept of intrinsic surface stress of the solid since it is the common driving force for the ordering of all types of nanostructures in question. Energetics of periodic faceting is considered, where the elastic relaxation energy due to crystal edges always provides the existence of the optimum period of the faceted structure. Periodic arrays of step bunches on vicinal surfaces are discussed as a particular class of periodically faceted surfaces. The physical mechanism is discussed which favors cluster growth in the GaAs/AlAs system with corrugated interfaces and thus gives the possibility for direct fabrication of quantum wires and quantum–wire superlattices.

In Section III the ordering of surface domain structures is considered. Although the geometry of the structures is very different from that of faceted surfaces, the basic energetics is the same, and both the ordering of domains in size and the periodic lateral arrangement always occur.

Section IV is devoted to ordering phenomena in arrays of three–dimensional coherently strained islands on a lattice–mismatched substrate. The formation of 3D coherently strained islands and the onset of misfit dislocations are alternative mechanism of the stress relaxation. The possibility of formation of 3D islands depends on the interplay between the elastic relaxation energy, the surface energy of planar and tilted surfaces, and on the energy of dislocated interface. The complexity of this system is due to two sources of the long–range strain field, one being the lattice mismatch, and another being the surface stress. The total energy is calculated, and a phase diagram of the system is constructed which contains both a parameter region corresponding to the ordering in a *two–dimensional array of three–dimensional islands*, and a parameter region corresponding to Ostwald ripening of islands and to the formation of large dislocated clusters.

In Section V, the formation of *vertically* ordered array of islands in the semiconductor matrix is discussed, where the formation mechanism involves *both strain–driven equilibrium ordering* and *strain–driven growth kinetics.*

II. SURFACE FACETING

A. Equilibrium crystal shape: Two distinct formulations of the problem

The phenomenon of equilibrium faceting plays the key role in determining the equilibrium crystal shape (ECS). We emphasize here two distinct problems related to the ECS which correspond to two different physical situations. *The first problem is the ECS of the single crystal* which dates back at least to Wulff's paper of 1901 [9], further developings of the ECS theory can be found in Refs. [10–12]. The exact thermodynamic formulation may be found, e. g., in the review by Rottman and Wortis [13] who discussed the shape of a single solid inclusion of a fixed volume ω which is in equilibrium with the liquid or with the gas phase. The surface free energy of the inclusion is the integral over the surface of ω,

$$F_{SURF}(T,\omega) = \oint_{\partial\omega} \gamma(\hat{\mathbf{m}};T)dA\,. \tag{2}$$

Here $\gamma(\hat{\mathbf{m}};T)$ is the surface free energy per unit surface area dependent on the orientation $\hat{\mathbf{m}}$ of the surface element dA relative to crystal axes. The thermodynamics of the ECS states that a macroscopic inclusion of a fixed volume

$$V(\omega) = \int_{\omega} dV \tag{3}$$

takes at equilibrium that shape which minimizes the surface free energy (2) subject to the constraint (3).

The brief summary of the ECS theory may be given as follows [13]. The orientational dependence of the surface free energy $\gamma(\hat{\mathbf{m}};T)$ is expected to have cusps in symmetry directions leading to facets in the crystal shape at sufficiently low temperatures. These cusps represent discontinuities in the angular derivatives of the surface free energies, the discontinuities being associated with the free energy of the steps on a given facet. With the temperature increase, step free energies decrease, cusps blunt, and corresponding facets shrink. Given facet finally disappears at the roughening transition temperature T_R of the corresponding infinite planar surface, T_R being different for different symmetry directions. Above each T_R the corresponding region of ECS becomes smoothly rounded.

References to experimental data on equilibrium crystal shapes of micrometer–scale metal clusters is presented in Ref. [13]. TEM measurements of the equilibrium shape of voids in Si enabled to reveal the orientational dependence of the surface free energy of Si [14].

Statistical mechanics is the tool for microscopic evaluation of the orientational dependence of the surface free energy $\gamma(\hat{\mathbf{m}};T)$ and for the determining of the ECS, an overview of theoretical results may be found in Ref. [13].

The important issue of the ECS theory is that there exist surface orientations which are not present in the crystal shape at a given temperature. At $T = 0$ only several high–symmetry surface orientations are present in the ECS, and all others are passive in the sense that they do not contribute to the ECS. With the temperature increase the domain of passive orientations shrinks as the crystal becomes more rounded.

The importance of this issue becomes even more visible as one considers *the second problem of the ECS.* This concerns the crystal which is in equilibrium with the liquid or with the gas phase and has the volume fixed and *all surfaces fixed but the top one.* This formulation of the problem is relevant to experimental situation where only the top crystal surface is studied, e. g., to thermal annealing of the crystal or to growth interruptions introduced in crystal growth experiments.

The top crystal surface is not fixed and is allowed to rearrange into a hill–and–valley structure. The question which arises here is: when can the free energy of a plane surface be lowered by rearranging the atoms into hills and valleys. Since hills and valleys addressed have the size large compared to the lattice parameter, new tilt facets can be defined, and the free energy of the hill–and–valley structure can be written as a surface integral over tilt facets:

$$F_{SURF} = \int \frac{\gamma(\hat{\mathbf{m}};T)}{(\hat{\mathbf{m}}\cdot\hat{\mathbf{n}})} dA \,. \tag{4}$$

Here $\hat{\mathbf{m}}$ is the coordinate–dependent unit vector locally normal to the surface at each point, and $\hat{\mathbf{n}}$ is the constant unit vector normal to the initially planar surface. The scalar product $(\hat{\mathbf{m}}\cdot\hat{\mathbf{n}})$ in the denominator of the integrand means that the surface element dA of a tilt facet is normalized per unit projected area of the nominally planar surface. Fixed side surfaces of the crystal imply that the average normal to the top surface coincides with the normal to the nominally planar surface, i. e.

$$\frac{1}{A}\int \hat{\mathbf{m}} dA = \hat{\mathbf{n}} \,, \tag{5}$$

A being the total area of the nominally planar surface.

The theorem proved exactly by Herring [10] reads: "If a given macroscopic surface of a crystal does not coincide in orientation with some portion of the boundary of the equilibrium shape, there will always exist a hill–and–valley structure which has a lower free energy than a flat surface, while if the given surface does occur in the equilibrium shape, no hill–and–valley structure can be more stable."

In the case where the planar surface is unstable, the resulting hill–and–valley structure is determined by the minimum of the surface free energy (4) subject to the constraint (5). This minimization will yield the orientation of tilt facets as well as the fraction of the nominal planar surface onto which each facet is projected. Microscopic theory based on statistical mechanics can yield the orientational dependence of the surface free energy $\gamma(\hat{\mathbf{m}};T)$ and thus allow to determine the ECS. Recent developings in this area have been made by Williams *et al.* [15] for surfaces vicinal to Si(111) and by Mukherjee *et al.* [16] for surfaces vicinal to Si(001).

The theory formulated by Eqs.(4,5) does not give, however, the linear scale of the equilibrium faceted structure, and, therefore, the formation of periodic structures has not been discussed in Refs. [9–13]. The latter problem requires additional concepts of the *intrinsic surface stress* and of *capillarity effects on solid surfaces* which are addressed in the two next subsections.

B. Intrinsic surface stress of a solid

Since atoms in the surface layer of any material are in different environment than in the bulk, the surface layer energetically favors a lattice parameter different from the bulk value in the directions parallel to the surface. Being adjusted to the bulk lattice parameter, the surface layer is intrinsically stretched or compressed. Therefore the surface is characterized by *intrinsic surface stress.*

The intrinsic surface stress of a solid is an analogue to the surface tension of a liquid. However, there exists a fundamental difference in thermodynamic properties of a liquid surface and of a solid surface pointed out long ago by Gibbs [17], an explanation may be found also in Refs. [18,19]. *The basic reason is that any liquid is incompressible.* When a liquid film is stretched atoms or molecules move out from the bulk to form new surface which is structurally identical to the existing surface. Thus, an attempt to stretch a liquid film by 1% (say, by means of a thought experiment) results in 1% increase of the number of surface atoms or molecules, whereas the spacing between surface atoms remains unchanged. Thus the processes of the creation and deformation of a liquid surface are identical and are described by *a single parameter* γ, the energy required to create unit area of the surface. However, when a crystal surface is stretched, the distance between atoms increases and the nature of the surface itself changes. This process is quite different from the creation of a new surface by the cutting of bonds. The energy to create unit area of the surface of a given orientation is characterized by the *scalar quantity* γ *termed the surface energy,* but the energy change due to deformation of the crystal surface is described by the *intrinsic surface stress tensor* $\tau_{\alpha\beta}$. The concept of the intrinsic surface stress tensor has been proposed in Ref. [17] and discussed later in Refs. [18,20,21]. The linear in strain change of the surface energy may be written as the following integral over the surface

$$\int \tau_{\alpha\beta}(\hat{\mathbf{m}})\varepsilon_{\alpha\beta}dA\,, \tag{6}$$

where the intrinsic surface stress tensor $\tau_{\alpha\beta}$ has non–vanishing components only in the surface plane, α, $\beta = 1, 2$ [18]. The principle values of the intrinsic surface stress tensor can be either positive (tensile) or negative (compressive). A tensile surface stress is associated with the surface which favors contraction while a compressive surface stress favors expansion.

Values of the intrinsic surface stress for solids are known either from first–principles calculations, or from comparison with indirect experimental data on parameters of surface domain structures (see also the next Section). No direct experimental method for determining the intrinsic surface stress of a solid has been proposed so far. For most of solid surfaces, the intrinsic surface stress is tensile (see, e. g., Refs. [19,22]), whereas the compressive surface stress is known for Si(001) surface in the direction perpendicular to dimers [23]. Unpublished results on (2×4)–GaAs(001) surface also indicate compressive surface stress perpendicular to dimers [24]. The order of magnitude of τ both for tensile and compressive surface stress is 100 meVÅ^{-2}.

At this point it is worthwhile to note the following. Since the liquid surface is characterized by the single quantity γ which is both the energy for creation of unit surface area (i. e. the surface energy), and the quantity responsible to capillarity effects of the Laplace–pressure type, the term “surface tension” is widely used for the surface energy γ. Mainly for historical reasons, the usage of this term has been extended to surfaces and interfaces of solids. However, the usage of “surface tension” for solids turns out to be totally ambiguous since it may produce a confusion between the surface energy γ and the intrinsic surface stress $\tau_{\alpha\beta}$. Therefore we intend to distinguish every time the “surface energy” and the “intrinsic surface stress” and do not use the term “surface tension” for solids.

The change of the surface energy due to strain (6) indicates the interconnection which exists between surface effects and strain–related phenomena. Within the framework of the linear theory of elasticity where the bulk strain energy is a quadratic function of the strain, it is possible to expand the surface energy up to second–order terms in strain. The thermodynamics of solid surfaces and interfaces up to second–order terms in strain has been studied by Andreev and Kosevich [25], further developings have been done by Nozières and Wolf [26]. The dependence of the surface or interface free energy on strain may be written as the following expansion:

$$\begin{aligned}\gamma(\hat{\mathbf{m}}; \varepsilon_{\alpha\beta}) = \quad & \gamma_0(\hat{\mathbf{m}}) + \tau_{\alpha\beta}(\hat{\mathbf{m}})\varepsilon_{\alpha\beta} \\ & + \frac{1}{2}S_{\alpha\beta\varphi\psi}(\hat{\mathbf{m}})\varepsilon_{\alpha\beta}\varepsilon_{\varphi\psi} + \frac{1}{2}h_{\alpha\beta i}(\hat{\mathbf{m}})\varepsilon_{\alpha\beta}\sigma_{ij}m_j \,. \end{aligned} \tag{7}$$

Here $\hat{\mathbf{m}}$ is the normal to the surface or interface, Greek characters label two–dimensional indices in the surface plane whereas Latin indices are three–dimensional ones. Quadratic coefficients $S_{\alpha\beta\varphi\psi}$ and $h_{\alpha\beta i}$ have the meaning of *surface excess elastic moduli* and can be either positive or negative. σ_{ij} is the bulk elastic stress tensor, and the fourth term in Eq.(7) exists on solid–solid interfaces and vanishes on stress–free surfaces where the bulk stress tensor obey the boundary conditions $\sigma_{ij}m_j = 0$.

Following the conventional approach of the elasticity theory [27] we use here and everywhere below all surface quantities defined *per unit area of the unde-*

formed surface. The dependence of the surface free energy on the strain (7) can be interpreted as the *strain–induced renormalization* of the surface free energy.

C. Capillarity phenomena on solid surfaces

Apart from the difference between a solid surface and a liquid surface emphasized in the previous subsection, there exists the basic similarity between these two types of surfaces. Both the liquid and the solid surfaces are intrinsically stressed. Therefore solid surfaces exhibit capillarity (or surface stress–induced) phenomena similar to the familiar Laplace capillarity pressure near a curved liquid surface, and a strain field is generated at a curved surface of a solid. The existence of surface stress–induced strain field was pointed out by Marchenko and Parshin [18] who considered the elastic energy of a solid including both bulk and surface contributions:

$$E_{elastic} = \frac{1}{2}\int \lambda_{ijlm}\varepsilon_{ij}\varepsilon_{lm}dV + \int \tau_{\alpha\beta}(\hat{\mathbf{m}})\varepsilon_{\alpha\beta}dA\,. \qquad (8)$$

Since the components of the tensor $\tau_{\alpha\beta}$ depend on the surface orientation $\hat{\mathbf{m}}$ (and even the orientation of the axes where the tensor components do not vanish changes with $\hat{\mathbf{m}}$), there appears the divergence of the surface stress tensor which gives the effective elastic force applied to the crystal

$$F_i = \frac{\partial \tau_{i\beta}}{\partial r_\beta}\,. \qquad (9)$$

The force F_i creates the elastic strain field in the crystal.

The ultimate case of a curved solid surface is a sharp edge between neighboring facets. Fig. 2 depicts the force balance on the curved surface of a liquid and at the crystal edge and thus demonstrates the similarity between capillarity effects in liquids and solids. The surface tension on a curved liquid surface results in an excess pressure below the curved surface, and, similarly, the intrinsic surface stress of crystal surfaces causes an effective force applied to the edge of the crystal. This force generates the strain field which affects significantly the energetics of a faceted surfaces and promotes the formation of a periodic structure of facets.

D. Periodically faceted surfaces

If a planar crystal surface is unstable and breaks up into a system of tilt facets, the conservation of the average orientation of the normal to the surface (5) implies the coexistence of alternating facets (Fig. 1a). At the intersection of neighboring facets there appear either sharp crystal edges or narrow rounded parts of the surface. Both these types of intersections may be described as linear defects at the surface. These linear defects give a short–range contribution into the surface free energy, and a long–range contribution due to elastic strain energy.

It has been shown by Andreev [28] that if the order parameter related to the phase transition at the surface is linearly coupled to the strain field, i. e. if linear striction effects exist, they favor formation of a periodic structure which period can be macroscopically large. For the faceting phase transition, the possibility of the formation of a faceted surface with a macroscopic period has been pointed out in Ref. [29].

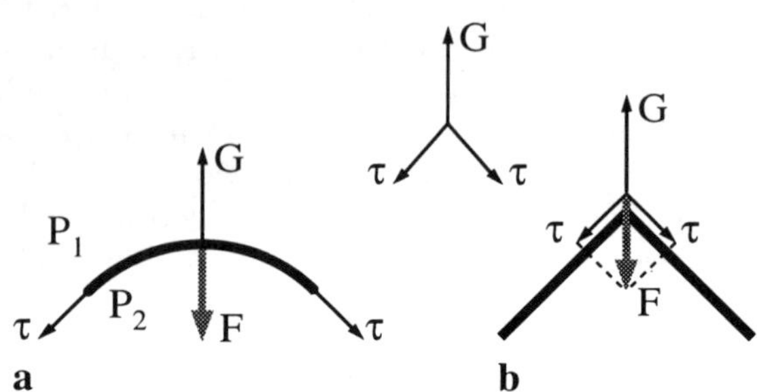

FIG. 2. Capillarity effects on liquid and on solid surfaces. a) The balance of forces acting on an element of the curved surface layer of a liquid. Forces caused by the surface tension τ are balanced by the force $\mathbf{G}$ acting from the bulk of the liquid. According to the third Newton's law, the opposite force $\mathbf{F} = -\mathbf{G}$ is acting from the surface layer on the bulk of the liquid. This force results in the excess Laplace pressure $\Delta P = P_2 - P_1$ below the curved surface of the liquid, P_1 and P_2 being the values of the pressure below and above the curved liquid surface. b) The balance of forces acting on a crystal edge. Forces caused by the intrinsic surface stress τ are balanced by the force $\mathbf{G}$ acting from the bulk of the crystal. According to the third Newton's law, the opposite force $\mathbf{F} = -\mathbf{G}$ is acting from the surface layer on the bulk of the crystal resulting in the strain field. The inset in the middle depicts the force balance similar for a liquid and for a crystal. (All forces in Figure 2 are defined per unit length in the direction perpendicular to the Figure.)

The theory of a periodically faceted surface has been developed by Marchenko [30], and the period of the equilibrium structure has been found. Following Ref. [30], we consider the faceted surface with a one–dimensional periodic saw–tooth profile depicted in Fig. 3. The free energy per unit projected area equals

$$F = F_{surf} + E_{edges} + \Delta E_{elastic} \,. \tag{10}$$

Here F_{surf} is the free energy of tilted facets, E_{edges} is the short–range energy of the edges, and $\Delta E_{elastic}$ is the elastic energy due to the discontinuity of the surface stress tensor τ_{ij} at the crystal edges.

The free energy of tilted facets per unit projected area depends only on orientation of facets, $E_{surf} = \gamma(\varphi)\sec(\varphi)$ and does not depend on the period of the structure D. The short–range energy of the edges per unit projected area equals

$$E_{edges} = \frac{\eta}{D} \equiv \frac{\eta^+(\varphi) + \eta^-(\varphi)}{D} \,, \tag{11}$$

where $\eta^+(\varphi)$ is the short–range energy of the convex edge per unit length of the edge, and $\eta^-(\varphi)$ denotes the same energy for the concave edge.

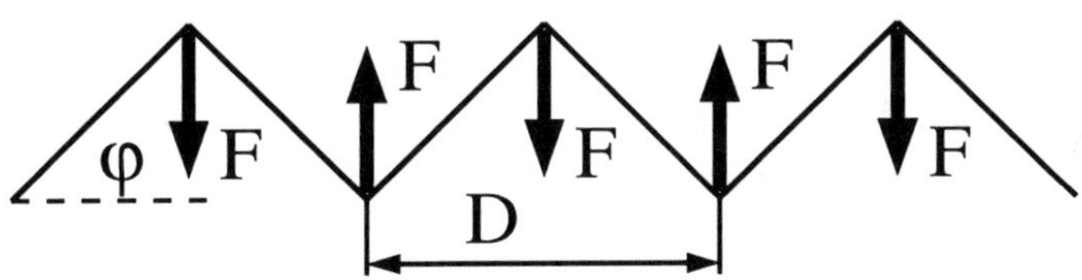

FIG. 3. Saw–tooth profile of the faceted surface. Effective forces of alternating sign are applied to neighboring edges.

Discussing the elastic strain energy, we note the following. First, $E_{elastic}$ is given by the general formulae (8) where the energy is zero in the absence of strain, and contains both linear and quadratic terms as a function of strain. Therefore, the elastic strain energy in the equilibrium is *negative* which corresponds to the relaxation of the surface stress at crystal edges. This negative elastic energy will be termed below the *elastic relaxation energy due to crystal edges.* Second, effective elastic forces acting at the edges are displayed in Fig. 3. *Force monopoles* are acting at each edge, and forces applied to neighboring edges are balanced, so that the total force applied to the system vanishes. Elastic strains generated by linear crystal edges propagate into the crystal over the distance of the order of D and decay at larger distances from the surface. Since the strain field is generated at linear defects at the surface, namely at the linear crystal edges, the elastic relaxation energy depends logarithmically on the period D of the structure:

$$\Delta E_{elastic} = -\frac{\widetilde{C}(\varphi)F^2}{YD}\ln\left(\frac{D}{a}\right) = -\frac{C(\varphi)\tau^2}{YD}\ln\left(\frac{D}{a}\right). \tag{12}$$

Here τ is the characteristic value of the intrinsic surface stress tensor, Y is the Young's modulus, a is the lattice parameter, $C(\varphi)$ is the geometric factor which accounts particular symmetry of the tensor τ_{ij}, elastic anisotropy of the crystal, etc.

By substituting Eq.(11) and Eq.(12) into Eq.(10), one obtains the following expression for the free energy of the faceted surface per unit projected area:

$$F = \frac{\gamma(\varphi)}{\cos\varphi} + \frac{\eta(\varphi)}{D} - \frac{C(\varphi)\tau^2}{YD}\ln\left(\frac{D}{a}\right). \tag{13}$$

The dependence of the free energy versus the period of the faceted surface D is displayed in Fig. 4. Due to the logarithmic dependence of the elastic relaxation energy on the period D, there always exists an optimum period of faceting D_{opt} equal to

$$D_{opt} = a \exp \left[\frac{\eta Y}{C(\varphi)\tau^2} + 1 \right] . \tag{14}$$

All material parameters which enter the exponential in Eq.(14) have typical atomic values. Therefore the combination which appears as the argument of the exponential, is of the order of 1. Since the exponential function is steep, the argument which can eventually be equal to, say, 3 or more, results in a macroscopic period D which exceeds the lattice parameter a at least by an order of magnitude. In this case the macroscopic approach is justified.

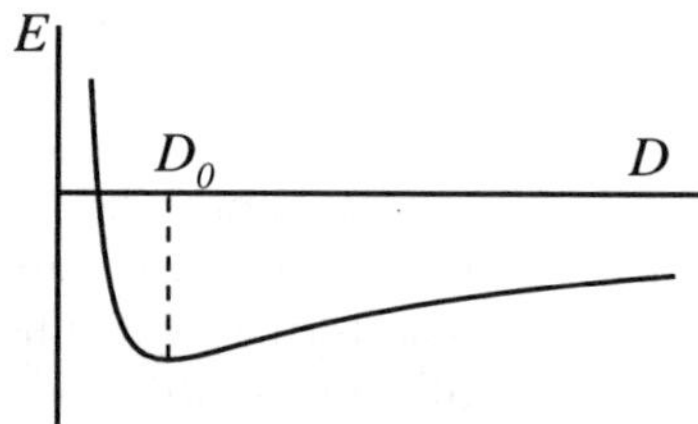

FIG. 4. The energy of a periodically faceted surface versus the period D. There always exists an optimum period of faceting D_{opt} due to the logarithmic dependence of the elastic relaxation energy on the period D.

It should be noted that the free energy of facets, i. e. the first term in Eqs.(10,13), contains the entropy contribution which describes the dependence of the faceting itself and of the facet orientation on temperature. However, there exists another entropy contribution to F associated with possible deviations of the structure from a perfect periodic structure (the configuration entropy). For periodically faceted surfaces, the configuration entropy has not been considered in literature so far. Therefore the present discussion refers, strictly speaking, to the case $T = 0$. Below we will omit everywhere the entropy contribution to F and will discuss only the total energy of the system.

Periodic faceting with macroscopic period have been observed on various surfaces of different materials, e. g., on surfaces vicinal to Si(111) [15,31], on surfaces vicinal to GaAs(001) [32–34], on surfaces vicinal to Pt(100) [35], on high–index surfaces of GaAs [36] and Si [37], on low–index surface TaC(110) [38]. The formation of facets has been also observed on Ir(110) [39] although no periodicity has been revealed. All above cited experimental observations of faceting refer to conditions of thermal annealing.

E. Periodic arrays of macroscopic step bunches

An important particular example of surface faceting is the faceting of a vicinal surface. A homogeneous vicinal surface of a crystal consists of planar terraces with low Miller indices, neighboring terraces being separated by equidistant monoatomic or monomolecular steps. The phenomenon of step bunching is known for a long time as one of possible kinetic instabilities of the crystal growth on vicinal surfaces (see, e. g., Ref. [40]). Recent experiments on surfaces vicinal to Si(111) [15] or GaAs(100) [32–34], have revealed step bunching which appears after the annealing of a cleaved surface or during growth interruption. This indicates an *equilibrium step bunching.* The height of step bunches was found to be homogeneous throughout the sample [15,32–34]. The observation of equal–width terraces (or equidistant step bunches) was reported for GaAs(100) [32–34].

The height of step bunches observed in Refs. [15,32,34] (7 — 15 monolayers) allows to consider them as macroscopic step bunches forming new facets (Fig. 5). Another example of faceting of a vicinal surface is its breaking into two vicinal surfaces with different concentration of steps [15]

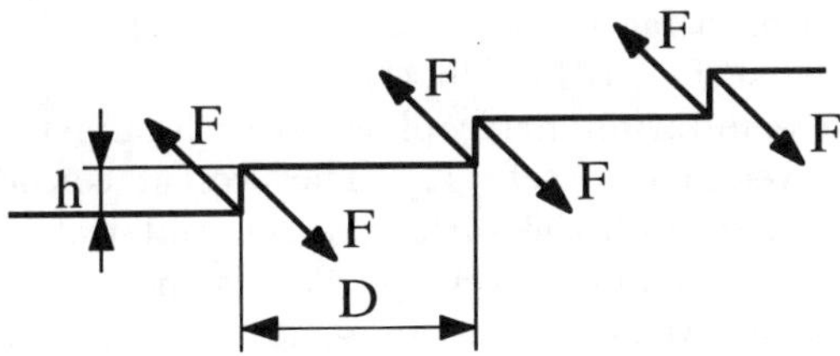

FIG. 5. A periodic array of macroscopic step bunches resulting from the faceting of a vicinal surface. Force monopoles are acting at the edges of the structure. The elastic interaction between the two edges of one step bunch is the monopole–monopole interaction, whereas the interaction between different step bunches is the dipole–dipole one.

Fig. 5 depicts the periodic array of step bunches of a macroscopic height. The discontinuity of the intrinsic surface stress tensor τ_{ij} at the edges results in effective force monopoles acting at the edges. Thus, the elastic interaction between the two edges of the same step bunch is the monopole–monopole interaction. Force monopoles acting at the two edges of a same bunch compensate each other, and the strain field due to a single step bunch equals at large distances to a strain field of an elastic dipole. Therefore, the elastic interaction between step bunches is the dipole–dipole interaction which decreases with the separation L as L^{-2}, similar to the elastic interaction between steps of a microscopic height [18]. The total energy per unit surface area equals (where dipole–dipole interaction and higher terms are truncated)

$$E = \gamma_0 + \gamma_1 \frac{h}{D} + \frac{C_1}{D} - \frac{C_2}{D} \ln\left(\frac{h}{a}\right) . \tag{15}$$

Since the average orientation of the faceted surface in Fig. 5 coincides with the orientation of initially homogeneous vicinal surface, there exists a relation between the height of macroscopic step bunches h and the period D of the structure, $h = D\varphi$, where φ is the miscut angle of the initially homogeneous vicinal surface. By using this relation, it is possible to write Eq.(15) in the form similar to the Eq.(13), $E = \gamma_0 + \gamma_1\varphi + \varphi\left[C_1 H^{-1} - C_2 H^{-1}\ln(H/a)\right]$. Due to the logarithmic dependence of the elastic relaxation energy on the height of the step bunch, there always exists an optimum equilibrium height of step bunches.

F. Formation of quantum wires on faceted surfaces

Periodic surface faceting gives a possibility for direct fabrication of ordered arrays of quantum wires. The growth of the deposited material 2 on the faceted surface of material 1 allows, in principle, the fabrication of quantum wires provided the growth proceeds in grooves of the faceted substrate. Here we focus on the heteroepitaxial growth in the system GaAs/AlAs where the two materials are nearly lattice–matched.

Two examples of the heteroepitaxial growth *on faceted surfaces* in GaAs/AlAs system are the growth on a faceted vicinal surface to the (001) surface [32] and the growth on a faceted surface (311) [36,41].

The theory of quasi–equilibrium heterophase growth on periodically corrugated substrates has been developed in Ref. [42]. The present consideration is focused on the practical case where both the surface of the material 1 and the surface of the material 2 are unstable against faceting. This is the case both for vicinals to (001) surfaces of GaAs and AlAs [32], on one hand, and for (311) surfaces of GaAs and AlAs [36], on the other hand. For this case, there are several possibilities for the morphology of the heterophase system depicted in Fig. 6.

The total energy of the heterophase system equals [42]

$$E = E_{surf} + E_{interface} + E_{edges} + \Delta E_{elastic}\,. \tag{16}$$

Here, in addition to the three contributions to the energy of the faceted surface (see Eq.(10)), one more contribution enters which is the interface energy. The comparison of total energies for several distinct types of the heterophase structures depicted in Fig. 6 has been carried out in Ref. [42]. It yields the following conclusions.

The selection between two possible growth modes is determined by the fact whether the deposited material *wets or does not wet* the substrate. If the deposited material wets the substrate, then the homogeneous coverage of the periodically corrugated substrate occurs (Fig. 6a). The example is the growth of AlAs on periodically corrugated vicinal surface of GaAs(001) 3^0–off towards $[1\bar{1}0]$ [32].

If the deposited material does not wet the substrate, then isolated clusters of the deposited material appear on periodically corrugated substrate (Fig. 6b). This situation is likely to be realized for the growth of GaAs on vicinal surface of AlAs(001) 3^0–off towards $[1\bar{1}0]$ [32], and for both GaAs/AlAs and AlAs/GaAs heterophase growth on (311)A surface [36,41].

In the case of inhomogeneous cluster coverage the periodic surface corrugation is restored after the deposition of first several monolayers. Then the hills of the top surface of the heterophase system appear over the valleys of the substrate, and *vice versa*, and a continuous layer of the deposited material with periodically modulated thickness is formed (Fig. 6d). Thus, the formation of clusters allows direct fabrication of quantum wires and quantum wire–superlattices in heterophase semiconductor systems.

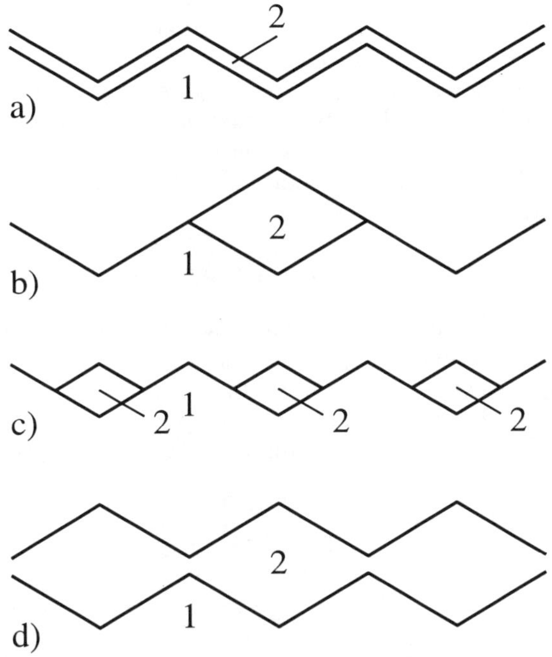

FIG. 6. Distinct structures of the heterophase system. a) — Homogeneous coverage; b) — System with separated "thick" clusters; c) — System of "thin" clusters; d) — The heterophase system at high coverage, where the periodic surface corrugation is restored, and the hills of the top surface of the heterophase system appear over the valleys of the substrate, and *vice versa*. The heterophase system contains a continuous layer of material 2 with periodically modulated thickness.

Since any periodically faceted surface is a structure of elastic domains, its geometrical parameters (e. g., the period) can be tuned in a controlled way by applying an external stress. Detailed consideration may be found in Ref. [43].

To conclude the present Section, we emphasize that the spontaneous periodic faceting of semiconductor surfaces and the cluster growth in grooves gives the possibility for direct fabrication of isolated quantum wires, of quantum wire–superlattices, and of quantum well–superlattices with modulated thickness of quantum wells.

III. SURFACE STRUCTURES OF PLANAR DOMAINS

Another class of self–ordered nanostructures is associated with periodically ordered structures of planar surface domains. Surface domain structures occur if different phases can coexist on the surface, e. g., phases of (2×1) and (1×2) surface reconstruction of Si(001), monolayer–height islands in heterophase systems, etc. Then neighboring domains possess different values of the intrinsic surface stress tensor τ_{ij} which gives rise to the elastic relaxation. The existence of surface domain structures governed by the discontinuity of τ_{ij} has been predicted in 1981 by Marchenko [44]. Although the geometry of these structures is very different from that of periodically faceted surfaces, the energetics is basically the same. Fig. 7 depicts force monopoles which are acting at domain boundaries. These force monopoles lead to the elastic relaxation. The total energy of the domain structure per unit surface area equals

$$E = E_{surf} + E_{boundaries} + \Delta E_{elastic} \,. \tag{17}$$

The surface energy E_{surf} does not depend on the period of the structure D, the energy of domain boundaries equals $E_{boundaries} = C_1 \eta D^{-1}$, and the elastic relaxation energy equals $\Delta E_{elastic} = -\, C_2 (\Delta\tau)^2 Y^{-1} D^{-1} \ln(D/a)$. Due to the logarithmic dependence of the elastic relaxation energy on the period of the structure D, the total energy (17) always has a minimum at a certain optimum period

$$D_{opt} = a \, \exp\left[\frac{C_1 \eta Y}{C_2 (\Delta\tau)^2} + 1\right] . \tag{18}$$

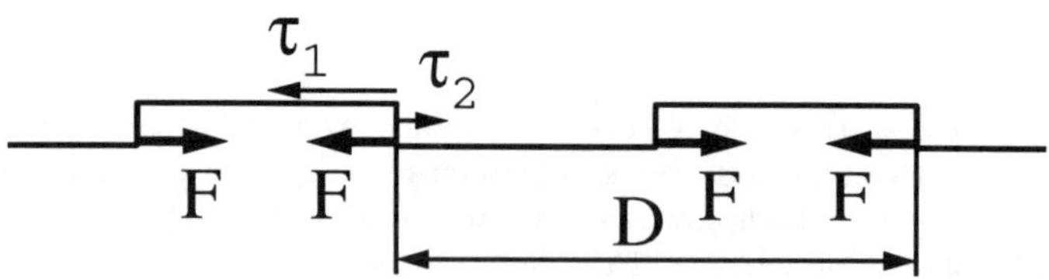

FIG. 7. Effective forces applied to domain boundaries of a system of planar surface domains.

The most well–known system exhibiting surface stress domain structures is the Si(001) surface and corresponding vicinal surfaces. Coexisting domains are domains of (2×1) and (1×2) surface reconstruction separated by single–height atomic steps. Such a domain structure was first observed by Men *et al.* [45], and the explanation was given by Alerhand *et al.* [46]. The comparison of the measured period of the domain structure with the theoretical formulae (18) allows to extract the value of $\Delta\tau$ from experiment (provided the short–range energy of single–height atomic steps is known) and to compare $(\Delta\tau)$ with results of first–principles calculations. A detailed discussion may be found, e. g., in Ref. [47]. Force monopoles applied to single–height atomic steps lead to a variety of structures of vicinal surfaces as a function of miscut angle. The review of both experimental and theoretical works on these structures may be found, e. g., in Ref. [16].

Another class of surface–stress domain structures include monolayer–height islands formed in heterophase systems *at submonolayer coverage.* Periodic domains of strips have been observed in O/Cu(110)–system [48], in InAs/GaAs(001)–system [49].

It should be noted that force monopoles acting at opposite boundaries of a given domain are balanced, and the interaction between domains at large distances is the dipole–dipole one. Similar to elastic domain structures are domain structures known for systems with dipole–dipole electrostatic interactions (Langmuir monolayers at the water/air interface), or for systems with dipole–dipole magnetic interactions in ferromagnetic films or in planar–confined ferrofluid/water mixtures in magnetic fields. Corresponding references may be found in paper by Kwok–On Ng and Vanderbilt [50].

Although the scaling behavior of the energy is similar for all systems mentioned, the particular pattern of the domain structure depends strongly on the orientational dependence of the long–range interaction and of the energies of domain boundaries. For isotropic interaction, there exists the transition from 1D array of stripes to 2D hexagonal array of disks [50,51], disks being more favorable if the coverage of the surface is close to 0 or to 1. Anisotropic interaction which is always the case for elastic interaction, and anisotropic domain boundaries, which is always the case for crystal surfaces, favor striped domains over disk–shaped domains.

In Ref. [52], a relation was established between the size of a single surface domain (for low domain concentration) and the minimum separation between them at intermediate concentration. In the particular case of monolayer–height islands, this result means that there exists an optimum island size for a dilute array of islands. The existence of the optimum size of islands implies that two–dimensional strained islands in the heterophase system do not undergo Ostwald ripening.

Similar results are valid for a dilute array of strained islands of a macroscopic height for the case where islands have a planar top surface, and the height of the islands is *kinetically limited* to a value considerably smaller than the lateral size [53]. The global geometry of such islands is similar to the geometry of two–dimensional islands. The important issue of Ref. [53] is that there exists an optimum size in the system of planar strained islands, and no Ostwald ripening occurs.

In the next Section, another situation will be considered which occurs for essentially three–dimensional strained islands (e. g., pyramid–shaped islands) where two different regimes are possible, one being the regime where an optimum size of islands exists, and another being the regime of ripening.

IV. ORDERED ARRAYS OF THREE–DIMENSIONAL COHERENTLY STRAINED ISLANDS

A. General morphology of lattice–mismatched systems

There are three well known modes of the heteroepitaxial growth: Frank–van der Merwe, Volmer–Weber, and Stranski–Krastanow. They may be described as layer–by–layer growth (2D), island growth (3D), and layer–by–layer plus islands (Fig. 8). The particular growth mode for a given system depends on the interface energies and on the lattice mismatch.

In lattice–matched systems, the growth mode is governed by the interface and surface energies only. If the sum of the epilayer surface energy γ_2 and of the interface energy γ_{12} is lower than the energy of the substrate surface, $\gamma_2+\gamma_{12} < \gamma_1$, i. e., if the deposited material *wets* the substrate, the FvdM mode occurs. A change in $\gamma_2 + \gamma_{12}$ alone may drive a transition from FvdM to VW growth mode. For a strained epilayer with small interface energy, initial growth may occur layer–by–layer, but a thicker layer has a large strain energy and can lower its energy by forming isolated islands in which strain is relaxed. Thus SK growth mode occurs.

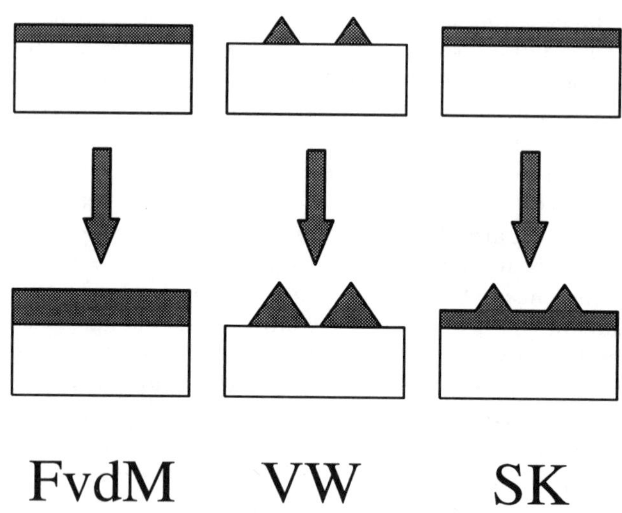

FIG. 8. Schematic diagrams of the three possible growth modes: Frank–van der Merwe, Volmer–Weber, and Stranski–Krastanow.

It was traditionally believed that islands formed in SK growth are dislocated. However, experiments on InAs/GaAs(001) [54], and on Ge/Si(001) [55,56] have demonstrated the formation of *three-dimensional coherently strained islands.*

The relaxation of the elastic energy due to the formation of coherently strained islands is related to the Asaro–Tiller–Grinfield instability [57–60] of a strained layer against a long-wavelength corrugation of the surface. To illustrate the physical mechanism of the elastic relaxation, it is convenient to consider a strongly pronounced corrugation. Among examples of such strongly pronounced corrugation, there are islands [61], troughs [61], surface cusps [62], and cracks [63]. The formation of troughs, cusps, cracks can occur in a strained epitaxial film of a certain macroscopic thickness under annealing. At the same time, for the first stages of the heteroepitaxial growth on a substrate, the formation of islands seems to be the only coherent mechanism of the elastic relaxation.

Fig. 9 demonstrates two islands of a different shape. A flat island with a small height–to–width ratio is practically non–relaxed whereas a hypothetical island having a shape of a vertical bar with a large height–to–width ratio is relaxed almost completely. Thus the elastic relaxation depends strongly on the island shape. For a given shape, the elastic relaxation energy is proportional to the volume of the island.

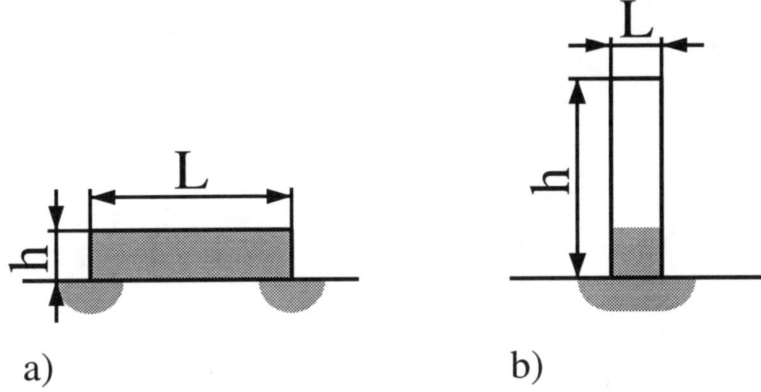

FIG. 9. Effect of the island shape on the volume elastic relaxation of coherently strained island. The grey area is that with a large strain energy density. a) The island with the height–to–width ratio $h/L \ll 1$ is not relaxed. b) The island with the height–to–width ratio $h/L \gg 1$ is nearly totally relaxed.

Thus the volume elastic relaxation of coherently strained islands is an alternative mechanism of relaxation which competes with the formation of dislocations. The theory developed by Vanderbilt and Wickham [61] compares the two mechanisms of elastic relaxation and yields the phase diagram of the lattice–mismatched system where all possible morphologies are present, i. e. uniform films, dislocated islands, and coherent islands (Fig. 10). The formation of an island from an uniform film is accompanied, first, by a relaxation of the elastic energy, $\Delta E^V_{elastic} < 0$ and, second, by a change of the surface area, $\Delta A > 0$. The corresponding change of the surface energy is then caused by the formation of side facets of the islands and by the disappearance of certain areas of a planar surface. It is usually believed that the change of the surface energy caused by the formation of islands is *positive*, $\Delta E_{surf} > 0$. It was shown in Ref. [61] that the morphology of the mismatched system is determined by the relation between the ΔE_{surf} and the energy of the dislocated interface E_{inter}^{disl}. The ratio of these two energies, denoted $\Gamma = E_{inter}^{disl}/\Delta E_{surf}$, is the control parameter which governs the morphological phase diagram of Fig. 10.

If ΔE_{surf} is positive and large, or the energy of dislocated interface is relatively small, the corresponding value Γ on the phase diagram of Fig. 10 is smaller than Γ_0. Then formation of coherently islands is not favorable. With the increase of the amount of the deposited material, a transition occurs from a uniform film to dislocated islands, and coherently strained islands are not formed.

If ΔE_{surf} is positive and small, or the energy of dislocated interface is relatively large, the corresponding value Γ on the phase diagram of Fig. 10 is larger than Γ_0. With the increase of the amount of the deposited material, a transition from a uniform film to *coherent islands* occurs. Further deposition or growth interruption may cause the onset of dislocations in the islands. The theory of Ref. [61] deals with islands having the shape of elongated prisms ("ridges"). The theory of Ref. [64] yields the existence of all the same morphologies, i. e., uniform films, coherent islands, and dislocated islands in the case of pyramid–shaped islands.

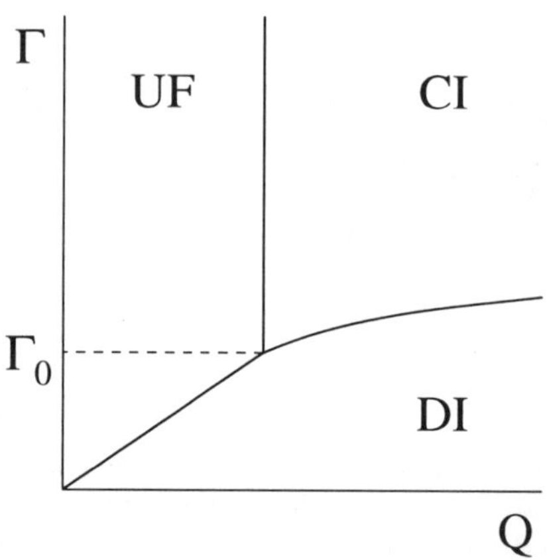

FIG. 10. Phase diagram by Vanderbilt and Wickham [61] showing the preferred morphology as a function of the amount of deposited material Q (horizontal axis) and of the quantity $\Gamma = \Delta E_{surf}/E_{inter}^{disl}$, where ΔE_{surf} is the change of the surface energy due to island formation, and E_{inter}^{disl} is the energy of a dislocated interface. Labels UF, CI, and DI refer to "uniform film", "coherent island ", and "dislocated island", respectively.

Thus, when coherent islands are formed, the common belief is that they should undergo Ostwald ripening in order to reduce the overall surface of the islands and thus to reduce the total surface energy of the system. The process of ripening implies the growth of large islands at the expense of the evaporation of small islands. Thus islands, undergoing ripening should become dislocated.

In the case $\Delta E_{surf} < 0$, there is no driving force to decrease the total surface area of the system, and ripening becomes unfavorable.

Experimental studies of coherent islands of InGaAs/GaAs(001) and InAs/GaAs(001) systems have revealed surprisingly narrow size distribution of the islands [65–68] which does not follow from SK growth mode itself. Besides that, the authors of Ref. [67–71] have reported that coherent islands of InAs form, under certain conditions, a quasi–periodic square lattice on the GaAs(001) surface. The periodicity and the island size do not change with time upon growth interruptions. Experimental results of Ref. [67–69,71] indicate the existence of a new class of self–ordered nanostructures, namely, ordered arrays of coherently strained *three–dimensional islands.*

To reveal the driving force responsible for ordering of three–dimensional coherently strained islands, we consider the energetics of these arrays.

B. Energetics of a dilute array of islands

We study the energetics of an array of three–dimensional coherent strained islands under the constraint of the fixed amount of the deposited material Q assembled in all islands, where Q is defined in numbers of monolayers. We treat both the substrate and the deposited material as *elastically anisotropic cubic media* with equal elastic moduli λ_{ijlm}, and the lattice mismatch between the two materials being equal $\epsilon_0 = \Delta a/a$, where a is the lattice spacing. The energy of the uniformly strained film of the thickness (Qa) on the (001)–substrate is given by: $E_{EL}^{(0)} = \lambda\epsilon_0^2 A(Qa)$, where the elastic modulus λ equals $(c_{11}+2c_{12})(c_{11}-c_{12})c_{11}^{-1}$, c_{11} and c_{12} are elastic moduli in the Voigt notation, and A is the surface area.

The energy of the heterophase lattice–mismatched system with islands is equal:

$$\widetilde{E} = \widetilde{E}_{elastic} + \widetilde{E}_{surf} + \widetilde{E}_{edges}\,. \tag{19}$$

Here $\widetilde{E}_{elastic} = (1/2)\int \lambda_{ijlm}\left(\varepsilon_{ij}(\mathbf{r}) - \varepsilon_0\delta_{ij}\vartheta(\mathbf{r})\right)\left(\varepsilon_{lm}(\mathbf{r}) - \varepsilon_0\delta_{lm}\vartheta(\mathbf{r})\right)dV$, where $\varepsilon_{ij}(\mathbf{r})$ is the strain tensor, $\vartheta(\mathbf{r}) = 1$ in the deposited material, and $\vartheta(\mathbf{r}) = 0$ in the substrate. The surface energy per unit area γ is *renormalized* in the strain field: $\gamma(\varepsilon_{\alpha\beta}) = \gamma_0 + \tau_{\alpha\beta}(\varepsilon_{\alpha\beta} - \varepsilon_0\delta_{\alpha\beta}) + \frac{1}{2}S_{\alpha\beta\mu\nu}(\varepsilon_{\alpha\beta} - \varepsilon_0\delta_{\alpha\beta})(\varepsilon_{\mu\nu} - \varepsilon_0\delta_{\mu\nu}) + \ldots$, where $\tau_{\alpha\beta}$ is the intrinsic surface stress tensor, $S_{\alpha\beta\mu\nu}$ is the tensor of the "surface excess elastic moduli" [77], and α, β, μ, ν are 2D indices in the local facet plane. The total *renormalized* surface energy of the heterophase system equals:

$$\widetilde{E}_{surf} = \int \Big[\gamma_0(\hat{\mathbf{m}}) + \tau_{\alpha\beta}(\hat{\mathbf{m}})\left(\varepsilon_{\alpha\beta}(\mathbf{r}) - \varepsilon_0\delta_{\alpha\beta}\right) + \frac{1}{2}S_{\alpha\beta\mu\nu}(\hat{\mathbf{m}})\left(\varepsilon_{\alpha\beta}(\mathbf{r}) - \varepsilon_0\delta_{\alpha\beta}\right)\left(\varepsilon_{\mu\nu}(\mathbf{r}) - \varepsilon_0\delta_{\mu\nu}\right)\Big](\hat{\mathbf{m}}\cdot\hat{\mathbf{n}})^{-1}\,dA\,. \tag{20}$$

Here $\hat{\mathbf{m}} = \hat{\mathbf{m}}(\mathbf{r})$ is the local normal to the facet, $\hat{\mathbf{n}} = (0,0,1)$ is the normal to the flat surface, and the integration in Eq.(20) is carried out over the reference flat surface. The surface energy $\gamma_0(\hat{\mathbf{m}})$ usually has cusped local minima for low–index facets [10]. We consider here the situation where the quantity $\gamma_0(\hat{\mathbf{m}})\,(\hat{\mathbf{m}}\cdot\hat{\mathbf{n}})^{-1}$ has, besides the *cusped* absolute minimum for the (001)–surface, also *cusped local minima* for 4 equivalent facets: (k0l), (0kl), ($\bar{\text{k}}$0l), and (0$\bar{\text{k}}$l). Then, under a range of conditions, the island will be bounded by (k0l), (0kl), ($\bar{\text{k}}$0l), and (0$\bar{\text{k}}$l) facets, e. g., will have the shape of a pyramid or of a prism, which are shown in the inset of Fig. 11. The third term in Eq.(19) is the short–range energy of edges.

For lattice–mismatched systems with edges, the total strain field is the sum of two contributions, one due to the lattice mismatch, and another due to the discontinuity of the intrinsic surface stress tensor τ_{ij} at the edges. By evaluating the change of the energy of the heterophase system due to the formation of a single island of a characteristic size L, one obtains [74]:

$$\widetilde{E}(L) = \left[-f_1(\varphi_0)\lambda\varepsilon_0^2 L^3 + (\Delta\Gamma)L^2 - \frac{f_2(\varphi_0)\tau^2}{\lambda} L \ln\left(\frac{L}{2\pi a}\right) + f_3(\varphi_0)\eta L\right] . \quad (21)$$

Here $\varphi_0 = \tan^{-1}(k/l)$ is the tilt angle of island facets. The first term in Eq.(21) is the energy of the volume elastic relaxation $\widetilde{\Delta E}^{V}_{elastic}$, it is always negative. The second term is the change of the renormalized surface energy of the system due to the island formation, $\widetilde{\Delta E}^{renorm}_{surf} = (\Delta\Gamma)L^2$, where $(\Delta\Gamma) = \gamma_0(\varphi_0)\sec\varphi_0 - \gamma_0(0) - g_1(\varphi_0)\tau\varepsilon_0 - g_2(\varphi_0)S\varepsilon_0^2$. The bare difference of the surface energies of the tilted facet and the planar surface, $\gamma_0(\varphi_0)\sec\varphi_0 - \gamma_0(0) > 0$, which indicates that unstrained planar surface is stable against faceting. Due to renormalization terms, $(\Delta\Gamma)$ can be *of either sign.* The third term in Eq.(21) is the contribution of the edges of the island to the elastic relaxation energy, $\widetilde{\Delta E}^{edges}_{elastic} \sim -L\ln L$, it is always negative. The fourth term in Eq.(21) is the short range energy of edges, where η is the characteristic energy per unit length of the edge. Coefficients f_1, f_2, f_3 depend only on the tilt angle of facets φ_0.

C. Ordering of islands in shape

For a dilute system of islands, where the average distance between islands is large compared to the island size L, the equilibration of the island shape by atomic migration on the island is faster than material exchange between islands. Then for any given volume of an island, there exists an equilibrium shape. For sufficiently large islands, the first two terms in the island energy (21), $\widetilde{\Delta E}^{V}_{elastic}$ and $\widetilde{\Delta E}^{renorm}_{surf}$ are the two dominant ones.

We focus on the situation here where the surface free energy of tilted facets of the deposited material per unit projected area, $\gamma_0(\hat{\mathbf{m}})\,(\hat{\mathbf{m}}\cdot\hat{\mathbf{n}})^{-1}$, has *cusped local minima* for 4 equivalent facets: (k0l), (0kl), ($\bar{\text{k}}$0l), and (0$\bar{\text{k}}$l). It yield cusped minima in $\widetilde{\Delta E}^{renorm}_{surf}$ which fix the tilt angle of island facets for a certain interval of island volumes (or sizes). Then, we consider islands bounded by four facets with the same tilt angle relative to the substrate, i. e., by facets (k0l), (0kl), ($\bar{k}$0l), and (0$\bar{k}$l).

Comparison of pyramids and prisms is presented in Fig. 11 where the volume elastic relaxation energy versus the tilt angle of facets is displayed for a square–based pyramid and for an infinitely elongated prism. For the prism, the component of the strain field in the direction along the prism remains equal to ε_0 throughout the entire volume of the island, whereas for the pyramid, all components of the strain undergo relaxation and decrease with the height. Thus, Fig. 11 demonstrates that the volume elastic relaxation is more efficient for pyramids ("quantum dots") than for infinitely elongated prisms ("quantum wires"), which explains the preferred pyramid–like shape of the islands.

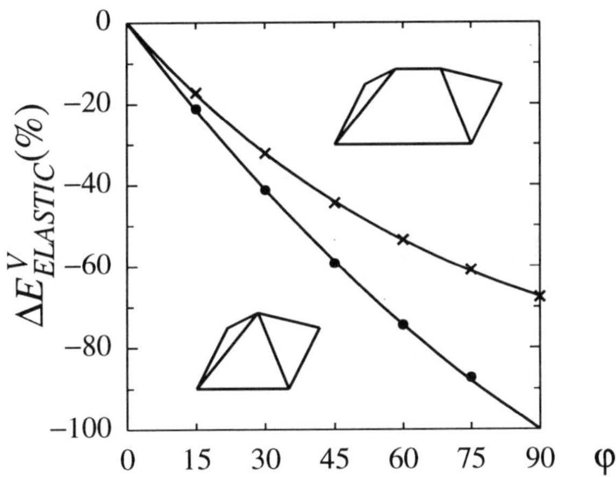

FIG. 11. The energy of the volume elastic relaxation of 3D coherently strained islands versus the tilt angle of island facets for pyramids (lower curve) and for elongated prisms (upper curve). Shaded circles and crosses represent results of the calculations by the finite element method, which are fitted by solid curves.

D. Ordering of islands in size versus Ostwald ripening

The equilibrium in the array of islands can be reached by the material exchange between islands which occurs via surface migration. For a *dilute* system of islands, the elastic interaction between islands via the strained substrate may be neglected. Then the energy of the array is the sum of contributions of single islands (21). Here we consider the dilute array of identical pyramid–shaped islands with the base length L.

The equilibrium meets the condition of the total energy minimum under the constraint of the fixed amount of the overall deposited material. Then, instead minimization of the total energy, it is possible to use an equivalent procedure and to minimize the energy *per one atom in the island* $E(L)$. Dividing $\widetilde{E}(L)$ from

Eq.(21) by the volume of a single island $(1/6)\tan\varphi_0 L^3$ and multiplying by the atomic volume Ω, one obtains [74]:

$$E(L) = 6\cot\varphi_0\Omega\left[-f_1(\varphi_0)\lambda\varepsilon_0^2 + \frac{\Delta\Gamma}{L} - \frac{f_2(\varphi_0)\tau^2}{\lambda L^2}\ln\left(\frac{L}{2\pi a}\right) + \frac{f_3(\varphi_0)\eta}{L^2}\right]. \quad (22)$$

It is worth noting that the volume elastic relaxation energy $E^V_{elastic}$ (the first term in Eq.(22)) does not depend on the island size L. To search the minima of $E(L)$ from Eq.(22) we introducing the characteristic length

$$L_0 = 2\pi a \exp\left[\frac{f_3(\varphi_0)\eta\lambda}{f_2(\varphi_0)\tau^2} + \frac{1}{2}\right], \quad (23)$$

and the characteristic energy per one atom,

$$E_0 = \frac{1}{2}\frac{\Omega f_2(\varphi_0)\tau^2}{\lambda L_0^2}. \quad (24)$$

Then we may write the sum all L–dependent terms in $E(L)$ as follows:

$$E'(L) = E_0\left[-2\left(\frac{L_0}{L}\right)^2\ln\left(\frac{e^{1/2}L}{L_0}\right) + \frac{2\alpha}{e^{1/2}}\left(\frac{L_0}{L}\right)\right]. \quad (25)$$

The function $E'(L)$ is governed by the control parameter

$$\alpha = \frac{e^{1/2}\lambda L_0}{f_2(\varphi_0)\tau^2}(\Delta\Gamma), \quad (26)$$

which is the ratio of the renormalized surface energy and of the contribution of the edges to the elastic relaxation energy, $|\Delta E^{edges}_{elastic}|$. The energy of the dilute array of islands per one atom versus the size of the island L is displayed in Fig. 12 for different values of α. If $\alpha \leq 1$, there exists an optimum size of islands L_{opt}, corresponding to the absolute minimum of the energy, $\min E'(L) \equiv E(L_{opt}) < 0$. On the other hand, the ripening of islands would correspond to $L \to \infty$ where the energy $E'(L) \to 0$. It means that the array of identical islands of the optimum size L_{opt} is a *stable array*, and *islands do not undergo ripening*. If $1 < \alpha < 2\,e^{-1/2} \approx 1.2$, then there exists only a local minimum of the energy, corresponding to a metastable array where $E'(L') > 0$. If $\alpha \geq 1.2$, the local minimum in the energy $E'(L)$ disappears. For the both latter cases where $\alpha > 1$, there exists the thermodynamic tendency to ripening. The energy minimum corresponds then to one large cluster where all deposited material is collected.

If $\Delta\Gamma < 0$ (and $\alpha < 0$), the formation of a 3D island, besides the decrease of the strain energy due to the relaxation, leads also to the *decrease of the renormalized surface energy.*

For the InAs pyramid with the (101)–type side facets over the InAs wetting layer deposited on the GaAs(001) surface, the evaluation of $\Delta\Gamma$ yields:

$$\Delta\Gamma = \quad 1.41\,\gamma_{InAs}^{(101)} + \gamma_{interface} - \gamma_{WL}^{(001)} \tag{27a}$$

$$- \left[0.72\,\tau_{\mu\mu}^{(101)} + 0.40\,\tau_{\nu\nu}^{(101)} + 0.15\left(\tau_{\zeta\zeta}^{(001)} + \tau_{\eta\eta}^{(001)}\right)\right]\varepsilon_0 \tag{27b}$$

$$+ \left[0.22\,S_{\mu\mu\mu\mu}^{(101)} + 0.08\,S_{\nu\nu\nu\nu}^{(101)} + 0.25\,S_{\mu\mu\nu\nu}^{(101)} + 0.10\,S_{\mu\nu\mu\nu}^{(101)}\right. \tag{27c}$$

$$\left. + 0.01\left(S_{\zeta\zeta\zeta\zeta}^{(001)} + S_{\eta\eta\eta\eta}^{(001)}\right) + 0.03\,S_{\zeta\zeta\eta\eta}^{(001)} + 0.01\,S_{\zeta\eta\zeta\eta}^{(001)}\right]\varepsilon_0^2\,, \tag{27d}$$

where the axes μ, ν, ζ, η are defined in Fig. 13. The change of the surface energy due to the formation of a pyramid contains contributions due to the appearance of tilted $< 101 >$–facets of InAs (the first term in (27a)), the appearance of the InAs/GaAs interface under the pyramid (the second term in (27a)), the disappearance of the L^2–area of the wetting layer (the third term in (27a)), linear renormalization terms (27b), and quadratic renormalization terms (27c,27d).

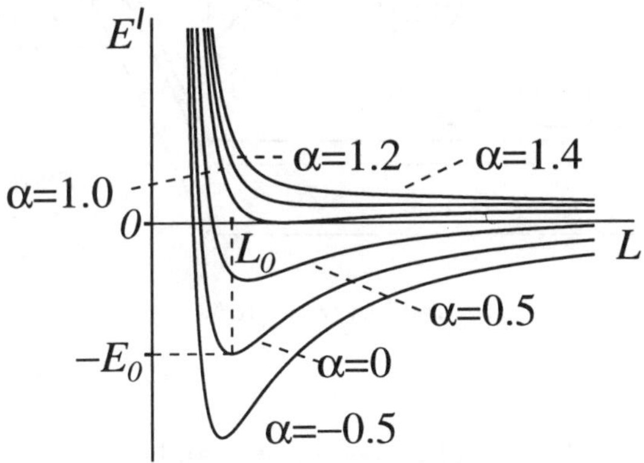

FIG. 12. The energy of the dilute array of 3D coherently strained islands per one atom versus the size of the island. The parameter α is the ratio of the renormalized surface energy of the island, ΔE_{surf}^{renorm}, and of the contribution of the edges to the elastic relaxation energy, $\left|\Delta E_{elastic}^{edges}\right|$.

To calculate $\Delta\Gamma$ from Eq.(27), we use *ab initio* values of materials parameters [24,86]: $\gamma_{InAs}^{(101)} = 41\text{meV}\text{Å}^{-2}$, $\gamma_{WL}^{(001)} = 54.7\text{meV}\text{Å}^{-2}$, $\gamma_{interface} = 2.1\text{meV}\text{Å}^{-2}$, $\tau_{\mu\mu}^{(101)} = 26\text{meV}\text{Å}^{-2}$, $\tau_{\nu\nu}^{(101)} = 54\text{meV}\text{Å}^{-2}$. For $(\tau_{\zeta\zeta}^{(001)} + \tau_{\eta\eta}^{(001)})$ we take its value known for GaAs(001), $13\text{meV}\text{Å}^{-2}$. Quadratic coefficients S are known to reach extremely high values for some metals [75–77]. The estimation of S for InAs(101) from [24] yields large negative values for this facet, $S \sim -1670\text{meV}\text{Å}^{-2}$. The substituting of these parameters into Eq.(27) gives us $\Delta\Gamma \approx +1\text{meV}\text{Å}^{-2}$.

In Ref. [86], it was shown that the optimum island shape is in many cases a *truncated pyramid*. By considering truncated pyramids with the same square base and $< 101 >$–type facets, and using the surface energy of the top InAs(001)–facet $\gamma_{InAs}^{(001)} = 49\text{meV}\text{Å}^{-2}$, we obtain that $\Delta\Gamma \approx 0$ if the full pyramid is truncated at

0.65 of its height, and $\Delta\Gamma \approx -1\text{meV}\text{Å}^{-2}$, if the pyramid is truncated at the level 0.5. It should be noted that the decrease of the surface energy in InAs/GaAs(001) system ($\Delta\Gamma < 0$) may occur in spite of the fact, that (001) surface of the bulk InAs is stable against faceting. The reason is that the appearance of the tilted facets of InAs competes with the disappearance of the certain area of the *wetting layer* which has larger surface energy than InAs(001). The above analysis clearly confirms, that the formation of a 3D island in mismatched system, in spite of the increase of the surface area, may lead to the *decrease of the surface energy* of the system and provide the *ordering of 3D islands in size.*

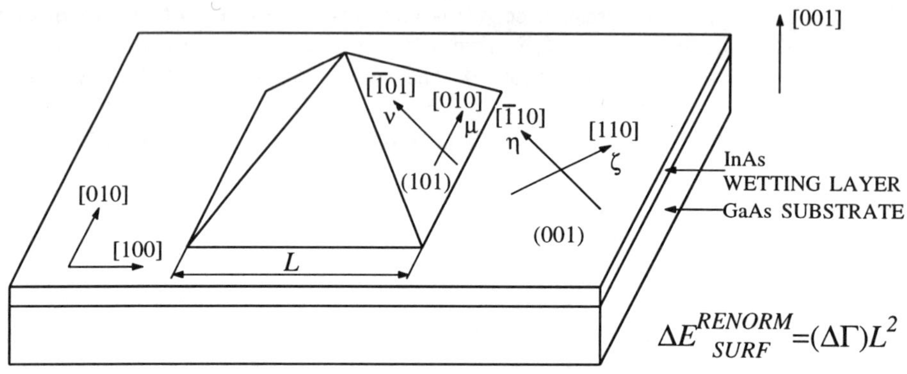

FIG. 13. Geometry of the InAs pyramid over the InAs wetting layer deposited on the GaAs(001) surface.

E. Lateral ordering of islands

For a *dense* system of islands, elastic interaction between islands *via the strained substrate* is essential. Interacting islands form a system of elastic domains where the strain energy minimum corresponds to a *periodic domain structure* [2,5,44,51,52]. In the approximation of small tilt angle of facets φ_0, the sum of the energy per unit surface area for isolated islands, $\Delta E^{(1)}_{elastic}$, and of the interaction energy per unit surface area, $E_{interaction}$, equals $\Delta E^{(1)}_{elastic} + E_{interaction} = (Qa)\tilde{\lambda}\epsilon_0^2\varphi_0 \times (-g_i(q))$. Here $\tilde{\lambda} = (c_{11} + 2c_{12})^2 (c_{11} - c_{12})^3 c_{11}^{-3} (c_{11} + c_{12})^{-1}$, q is the fraction of the surface covered by islands, and plots $-g_i(q)$ for different arrays are given in Fig. 14.

Among different 2D arrays of pyramid–shaped islands on the (001)–surface of the elastically anisotropic cubic medium, the minimum energy corresponds to the 2D square lattice of islands with primitive lattice vectors along the "soft" directions [100] and [010]. There exist two factors which favor the square lattice. First is the cubic anisotropy of elastic moduli of the medium, and second is the square shape of the base of a single island.

The approximation of small tilt angles used in deriving $\Delta E^{V}_{elastic} + E_{interaction}$ in Fig. 14, gives significant elastic relaxation energy already for small φ_0, in agreement with Ref. [61]. For large values of φ_0, this approximation overestimates the elastic relaxation energy. However, the difference between the exact and approximate values of $\Delta E^V_{elastic}$ is small even for the tilt angle $\varphi_0 = 45^0$. The approximation of small tilt angles yields the energy of the volume elastic relaxation equal to (-64%) of the strain energy of a flat film, whereas the exact calculations Calculations by the finite element method gives the elastic relaxation energy of a pyramid equal to (-60%) of the strain energy of a flat film.

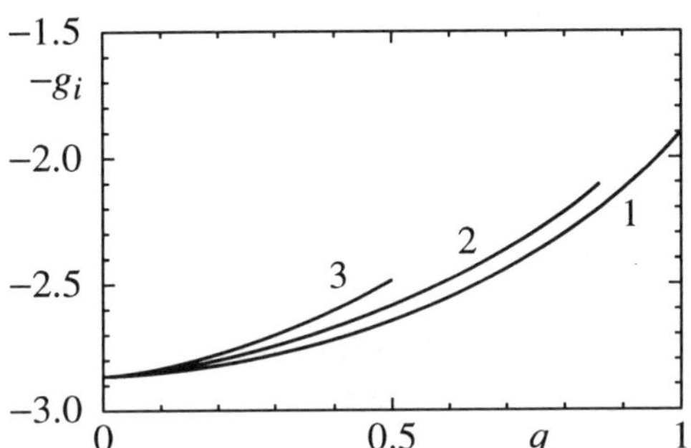

FIG. 14. The energy per unit area $\Delta E^{(1)}_{elastic} + E_{interaction} = (Qa)\widetilde{\lambda}\epsilon_0^2\varphi_0 \times (-g_i(q))$ for different arrays of coherently strained islands versus the fraction q of the surface covered by islands. (1) – 2D square lattice of pyramids with primitive lattice vectors $(1,0,0)$ and $(0,1,0)$. (2) – 2D hexagonal lattice of pyramids with primitive lattice vectors $(-\frac{1}{2}, -\frac{\sqrt{3}}{2}, 0)$ and $(1,0,0)$. (3) – 2D "checkerboard" square lattice of pyramids with primitive lattice vectors $(1,-1,0)$ and $(1,1,0)$. Curves 2 and 3 terminate at maximum possible coverages for given arrays.

The close similarity between the approximation of the small tilt angles and the exact numerical solution remains for arrays of interacting islands, too. Evaluation of the elastic energy of interacting arrays of islands by the finite element method for the tilt angle $\varphi_0 = 45^0$ shows that the cubic anisotropy of elastic moduli favors in this case also 2D *square* lattice of islands with primitive lattice vectors along the "soft" directions [100] and [010].

F. Phase diagram of 2D array of islands

The main part of the interaction energy is the energy of dipole–dipole elastic repulsion between islands, $E_{interaction} = (Qa)\,6\cot\varphi_0 f_5(\varphi_0)\lambda\epsilon_0^2 q^{3/2}$. Thus, the curve $-g_1(q)$ in Fig. 14 in the region $0 \le q \le 0.5$ can be approximated by $-g_1(q) = -2.87 + 0.61q^{3/2}$ with the accuracy 0.01. To express q in terms of Q and L, we equate the volume of the pyramid and the volume of the initially uniform film per unit period $D \times D$ of the square lattice: $(1/6)L^3\tan\varphi_0 = D^2(Qa)$. Hence, $q \equiv L^2/D^2 = 6Qa\cot\varphi_0/L$. Now, adding $E_{interaction} \sim q^{3/2} \sim L^{-3/2}$ to the energy of Eq.(22), we get the energy of the array of interacting islands. To consider in detail L–dependent terms in the energy per one atom $E'(L) = E_{dilute}(L) +$

$E_{interaction}(L)$, we introduce the characteristic length L_0 from Eq.(23) and the characteristic energy E_0 from Eq.(24). Then the sum of all L-dependent terms in $E(L)$ may be written as the function of the dimensionless length L/L_0:

$$E'(L) = E_0\left[-2\left(\frac{L_0}{L}\right)^2 \ln\left(\frac{e^{1/2}L}{L_0}\right) + \frac{4\beta}{e^{3/4}}\left(\frac{L_0}{L}\right)^{3/2} + \frac{2\alpha}{e^{1/2}}\left(\frac{L_0}{L}\right)\right]. \quad (28)$$

This function is governed by two control parameters, where α is defined in Eq.(26), and

$$\beta = (Qa)^{3/2} \times \frac{e^{3/4} f_5(\varphi_0)(6\cot\varphi_0)^{3/2}(\lambda\epsilon_0)^2 L_0^{1/2}}{2 f_3(\varphi_0)\tau^2}. \quad (29)$$

The parameter β is the ratio $E_{inter}/|\Delta E^{edges}_{elastic}|$, it increases with the amount of the deposited material as $Q^{3/2}$.

Searching the minima of the energy $E'(L)$ from Eq.(28) for different α and β, we obtain the phase diagram of Fig. 15. For the region 1 of the phase diagram, there exists an optimum size of islands L_{opt}, corresponding to the absolute minimum of the energy, $\min E'(L) \equiv E'(L_{opt}) < 0$. On the other hand, the ripening of islands would correspond to $L \to \infty$ where the energy $E'(L) \to 0$. It means that the 2D periodic square lattice of islands of the optimum size L_{opt} is a *stable array*, and *islands do not undergo ripening.* For region 2 of the phase diagram, there exists only a local minimum of the energy, corresponding to a metastable array where $E'(L') > 0$. For region 3, the local minimum in the energy $E'(L)$ disappears. In both regions 2 and 3, there exists the thermodynamic tendency to ripening. If the system initially corresponds to the region 1, and the amount of the deposited material Q increases, then the point in the phase diagram moves to regions 2 and 3, and islands undergo ripening.

If $\alpha \le 0$, there exists the absolute minimum of the energy $E'(L)$ for an arbitrary value of β, and $\min E' \equiv E'(L_{opt}) < 0$. Besides the absolute minimum of $E'(L)$, there may also exist a local minimum at $L = L'$, the energy of a corresponding metastable state $E'(L') < 0$ in region 4; $E'(L') > 0$ in region 5; and no metastable state exists in region 6.

To estimate characteristic values of α and β, we substitute $\tau \approx 100\text{meV}/\text{Å}^2$, $\lambda \approx 500\text{meV}/\text{Å}^3$, $L_0 \approx 100\text{Å}$, $a = 2.8\text{Å}$ to Eq.(29). This gives $\beta \approx 1$ for $|\epsilon_0| \approx 1\%$ and $Q = 0.5$ monolayers. Therefore, if $\alpha > 0$, the array of islands may correspond to region 1 of the phase diagram only for low coverage Q. On the other hand, the typical difference in the surface energy between facets of different orientation is of the order of $\approx$ 5–10meV/Å^2 [14]. It follows then from Eq.(26) that the parameter α for a strongly mismatched system may become negative due to mismatch–induced renormalization of the surface energy of island facets. If $\alpha < 0$, then the increase of $|\epsilon_0|$, e. g., by the increase of x for the heterophase system $In_xGa_{1-x}As/GaAs(001)$, results in the decrease of L_{opt}. This agrees with experimental data of Ref. [67].

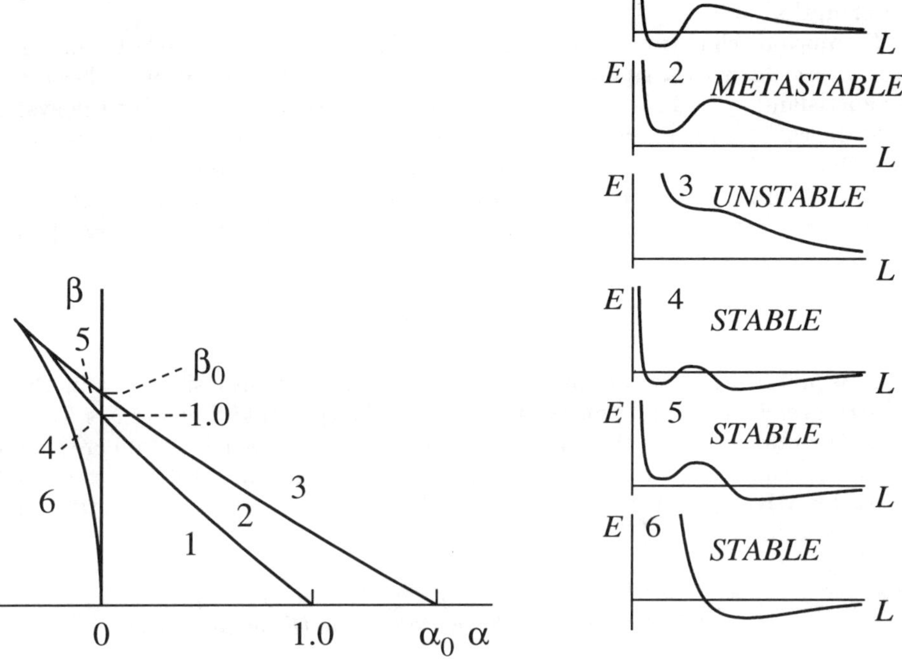

FIG. 15. Schematic phase diagram of the stability of a square lattice of coherently strained islands in the plane of control parameters $\alpha - \beta$. Here α and β are defined in Eqs. (26,29). On the right side of the Figure, the energy of the interacting array of islands versus the size of the islands is plotted, plots correspond to different regions of the phase diagram. In regions 1, 4, 5, and 6, there exist stable arrays of islands which do not undergo ripening. In regions 2 and 3, all arrays of islands undergo ripening. $\alpha_0 = 2\,e^{-1/2} \approx 1.213$; $\beta_0 = (4/3)\,e^{-1/4} \approx 1.038$.

G. Ordering–to–ripening phase transition

The key difference of the arrays of three–dimensional coherently strained islands from periodically faceted surfaces and from periodic structures of surface domains is the existence of both the ordering regime and the ripening regime, where the possible transition between these two regimes is governed by the surface energy of island facets. The remarkable fact is that such phase transition driven by the As–pressure has been observed experimentally [7].

The MBE–growth of 4 monolayers of InAs for $T = 480^0$ C and As–pressure $P_{As}^{(0)} = (1.5 — 3) \times 10^{-6}$ torr results in an array of 140–Å pyramid–shaped quantum dots arranged in a 2D square lattice with principal axes along [100] and [010] directions. This array is stable and does not undergo ripening upon growth interruption.

Significant changes of MBE growth conditions ($T = 480^0$ C and $P_{As} = 10^{-5}$ torr) lead to a ripening regime which results in formation of large macroscopic islands [7]. This indicates that kinetics is sufficient to drive the system to a state with lower energy, i. e. to a stable array for the ordering regime or to macroscopic clusters for the ripening regime. The shift from the ordering regime to the ripening regime may be caused by the As–pressure induced change of the surface reconstruction and the consequent change of the surface energy [79].

H. Discussion

It is noteworthy to discuss in detail the experimental situation in the self–ordering of InAs islands on GaAs surface. The present discussion is focused on experimental evidences that the self–ordering phenomena in question are goverened rather by thermodynamics than by kinetics. The experimental results refered are those for MBE growth of InAs on GaAs(001) [67–69,71,72], a more detailed review may be found in Ref. [7].

1. Effect of growth interruption on the island morphology

The study of the morphology of the heterophase InAs/GaAs(001) system by the plan–view transmission electron microscopy and by the photoluminescence spectroscopy has revealed the following. After the critical average thickness of InAs deposition is reached (1.6 — 1.7 monolayers (ML)), morphological transition to three–dimensional InAs islands occur. The deposition of 2ML InAs results in the formation of islands (dots) which are small, mostly do not show well resolved crystalline shape and exhibit large size dispersion. Increase to 4ML of the amount of InAs deposited results in a dense array of well–developed dots of a size ~140 Å. The sides of the square–shaped base of the dot are parallel to either [100] or [010] directions.

The introducing of *growth interruption* results in dramatic changes of the heterophase morphology with respect to *as–grown samples.* Growth interruptions of 40 s (10 s) are sufficient to let the dots reach the size ~140 Å for 2.5 ML (3 ML) InAs deposition. With very long growth interruption (100 s) one can even force 2 ML dots to reach the same size. The above cited data allow to conclude that the size of InAs dots ~140 Å is *the equilibrium size* which can be attained by growth interruption.

2. Effect of As pressure on the island morphology

As concerns the stability of the array of InAs dots, experimental data of Ref. [7] allow to reveal the following. The growth at the temperature $T = 480^0$ C and standard MBE arsenic pressure ($P^0_{As} \approx 2.0 \cdot 10^{-6}$ torr) results in equilibrium dots of high density ($5 \cdot 10^{10}$cm^{-2}). The dot array is stable when the As pressure is changed by ≈ 50 % around this value. An increase in As pressure by a factor of 3 ($3P^0_{As}$) results in dramatic change of the morphology. The size of dots reduces, and a high concentration of large ($\sim 500 - 1000$ Å) dislocated InAs clusters appears.

The reduction of the arsenic pressure affects the dots in a different way. At the arsenic pressure $1/6P^0_{As}$, dots completely disappear in favor of macroscopic two–dimensional InAs islands (~ 1000 Å).

The 3D–islands–to–2D–islands transition with the decrease of the arsenic pressure is reversible whereas the ordering–to–ripening transition with the increase of As pressure is accompanied by the onset of misfit dislocations and is irreversible.

All changes of the island morphology with the arsenic pressure can be explained on the basis of the surface energy. The stoichiometry of (001) surfaces of III–V semiconductors which are in equilibrium with the gas is known to depend on the pressure of the Vth group element in the gas phase. Thus, the change of As pressure leads to the change of the surface energy of GaAs(001) surface and even to the change of the surface reconstruction [79]. The wetting layer of InAs on a GaAs(001) surface may be considered as a complex surface of a semiconductor which is expected to exhibit the similar behavior with arsenic pressure. The decrease of the surface energy of the (001) surface results in the increase of the control parameter α in Eq.(26) and favors the transition from the ordering regime to the ripening regime. This is likely to be the case when the arsenic pressure increases from P^0_{As} to $3P^0_{As}$.

At low As pressure In is known to form segregate layers on the surface. This layer can be regarded as a quasi–liquid phase [80]. One may argue that this quasi–liquid phase has no strain–induced renormalization of the surface energy. Therefore, the increase of the surface area due to the island formation may lead to a large increase of the surface energy which makes the formation of 3D islands totally unfavorable.

3. Similarity in the island morphology between MBE–grown and MOCVD–grown islands

The lateral ordering of islands show a full agreement between MBE–grown and MOCVD–grown arrays in InGaAs/GaAs(001) and InAs/GaAs(001)–systems. The MBE–grown arrays of islands in InAs/GaAs(001) system exhibit the lateral ordering in a square lattice with primitive unit vectors in [100] and [010] directions [68,69]. Most recent result on MOCVD–grown InAs/GaAs(001)–system have revealed the local arrangement of islands in chains along $< 100 >$–type directions [83].

In the case of the growth of a solid solution, MBE–grown arrays of islands in InGaAs/GaAs(001) systems, show preferred arrangement in $< 110 >$–type directions [72,82]. Similar results have been obtained recently for MOCVD–grown arrays of InGaAs/GaAs islands which also show preferred arrangement in $< 110 >$–type directions [81].

Summing up these data on lateral arrangement of InAs/GaAs and InGaAs/GaAs islands, we note the following. The lateral arrangement is at least similar between MBE and MOCVD–grown samples whereas there is a difference between InAs/GaAs arrays and InGaAs/GaAs arrays. This indicates that the energetics which is the same for MBE–grown and MOCVD–grown arrays of islands and depends on the lattice mismatch seems to play a more important role in these processes than kinetics which is completely different between MOCVD and MBE.

Since these are both the lattice mismatch and the surface stress–induced forces at the edges which create the strain field, both these forces contribute to the interaction between islands via the substrate. The first, mismatch–induced interaction is anisotropic because of the cubic anisotropy of the bulk elastic moduli, and its orientational dependence exhibits the four–fold symmetry. This interaction favors the arrangement along the directions of the lowest stiffness $< 100 >$. The second, surface stress–induced interaction exhibits the symmetry of the tensor τ_{ij} which principle axes are $[\bar{1}10]$ and $[110]$. Therefore it might favor arrangement along one of the directions $< 110 >$. Since the mismatch–induced interaction is proportional to the square of the lattice mismatch, it decreases rapidly in the case of solid solution systems $In_xGa_{1-x}As/GaAs$ with the decrease of In composition x. The anisotropy of the surface stress–induced interaction is less sensitive to In composition since this anisotropy exists for pure semiconductors, too. Thus, the role of the surface stress in lateral arrangement increases with the decrease of In composition and with the corresponding decrease of the lattice mismatch. When both the mismatch–induced and surface stress–induced interaction of islands are important, the resulting arrangement can be, in principle, asymmetric.

The transition from $< 110 >$–oriented corrugation of the surface to $< 100 >$–oriented corrugation of the surface has been observed for InAs/GaAs system in the interval of coverages 1.0 — 1.5 monolayers before the onset of three–dimensional islands [87,88]. This phenomenon also results from the interplay of two contributions to the strain field.

This interplay of two interactions discussed may be a reason for a hexagonal arrangement of islands reported in Ref. [66]. Another possibility for such an arrangement is due to the fact that the hexagonal and the square arrays of islands are very close in total energies as can be seen from Fig. 14. Therefore, a hexagonal arrangement of islands can be one of possible metastable states of the system.

4. Equilibrium ordering versus kinetic–controlled ordering of islands

Discussing the ordering phenomena in the system of coherently strained islands on a lattice–mismatched substrate, one faces inevitably the question whether the ordering is an equilibrium phenomenon, or a kinetic–controlled one. A kinetic model has been proposed in Refs. [89,90] which accounts the atomistic structure of islands and proposes the self–limiting kinetics of the island growth. This self–limiting kinetics is due to the kinetic barrier for nucleation of a new facet when facets of the island are atomically smooth.

However, experimental results discussed in previous subsections are in favor of equilibrium ordering at least in the particular system of InAs/GaAs or InGaAs/GaAs strained islands. The evolution of the dot size up to the equilibrium value upon growth interruption, the arsenic–pressure driven phase transitions, the similarity in lateral arrangement between MBE–grown and MOCVD–grown arrays of dots indicate very likely that the ordering effects for the system in question are of thermodynamic nature.

From the theoretical point of view, the possibility of the equilibrium ordering of islands in size depends dramatically on effects of strain–induced renormalization of surface energies of the wetting layer and of islands facets. We note here that $(\Delta\Gamma)$ from Eq.(22) is very sensitive to the orientation of island facets. First, the bare surface energy is expected to be lower for low–index facets (e. g., (101)), and, second, the renormalization terms which are governed by the volume elastic relaxation, strongly depend on the tilt angle of island facets. For example, calculations of Ref. [84] based on the Keating model were performed for islands in the InAs/GaAs(001) system which have (104)–type side facets. These calculations did not reveal any minimum in the total energy versus the size of the island L. The analysis of the present paper where renormalization effects are *explicitly* taken into account in Eq.(27), shows that the total energy of the array of islands with low–energy (101)–type side facets *may have minimum as a function of the island size* L in general case for reasonable values of material parameters.

In this paper we have focused on the calculation of the total energy and did not consider the entropy term in the free energy. This entropy term, first, will lead to a finite concentration of the residual gas of adatoms on the surface and to a reduction of the total amount of material assembled in islands. Second, it is known for stepped vicinal surfaces (see, e. g., [85]) that the entropy term prevents the "collision" of steps and results in the effective repulsion between steps. Similarly, one may expect that for an array of islands, the entropy term will also lead to an effective repulsion between islands. A detailed study of entropy effects as well as of the distribution of islands in size will be presented elsewhere.

We emphasize that our conclusions do not depend on the approximations made while deriving Eq.(22). In a more detailed treatment, if the surface energy $\gamma(\hat{\mathbf{m}})$ does not have cusped local minima, then the tilt angle of facets φ will vary with the size of the island. The dependence of the island shape on the island volume has been studied by Pehlke *et al.* [86]. The authors calculated *ab initio* surface energies of different InAs facets, elastic relaxation energies for different shapes of islands, and revealed the equilibrium shape for InAs/GaAs(001) to be strongly volume–dependent. However, the strain–induced renormalization of surface energies was not taken into account in Ref. [86] which might be the reason why the calculated equilibrium shape of InAs/GaAs(001) islands does not coincide with the observed one [72]. If the shape of the islands is the function of their volume, then the scaling analysis of the total energy of island arrays is no longer possible. But, if there exists such a shape of islands, that the formation of islands leads to the *decrease of the renormalized surface energy*, then the ordering of islands occurs.

To conclude, we have shown that coherently strained *essentially three–dimensional* islands on lattice–mismatched substrate form, under certain conditions, *stable* periodic arrays of equal–sized islands. For islands on the (001)–surface of a cubic crystal, the arrangement in a 2D periodic square lattice with primitive lattice vectors oriented along the lowest–stiffness directions [100] and [010] is energetically preferred.

V. VERTICALLY CORRELATED GROWTH OF NANOSTRUCTURES

A. Correlated growth of nanostructures in the modulated strain field

The array of coherently strained islands (e. g., InAs) being covered with the substrate material (GaAs) creates the strain field in the surrounding matrix. If the thickness of the deposited material exceeds the height of islands, the islands are completely "buried". If one grows then the next layer of InAs, the growth proceeds in a completely new growth mode, which is the growth in a modulated strain field on the surface.

The modulated strain field on the surface leads to the modulation of the chemical potentials of surface adatoms [59]. The modulation of the chemical potential has a large impact on the growth mode. The migration of adatoms on the surface consist of diffusion and drift in the strain field (sometimes the term "directional migration" is used for the drift of adatoms). This drift is a driving force of the kinetic self–ordering in this complex growth mode. Such kinetic mechanism is responsible for the instability of the epitaxial growth of the homogeneous alloy and may result in the formation of composition–modulated structures [91,92].

The strain field which buried island generates on the surface may be calculated by using the continuum elasticity theory, following Maradudin and Wallis [93]. An island buried at a depth z_0 and lateral position $x = y = 0$ leads to a surface strain which hydrostatic part equals

$$\varepsilon(x, y, 0) = \frac{C}{(x^2 + y^2 + z_0^2)^{3/2}} \left[1 - \frac{3z_0^2}{x^2 + y^2 + z_0^2} \right] . \tag{30}$$

For In adatoms migrating on the GaAs surface, the minimum of the chemical potential corresponds to the maximum of the tensile strain which is achieved, according to Eq.(30), directly above the buried island, at $x = y = 0$. Therefore In adatoms are attracted to buried islands of InAs which results in vertical stacking of InAs islands in multi–sheet structures, the theoretical consideration has been carried out by Xie *et al.* [94]. The theory by Tersoff *et al.* [95] considers the preferential nucleation of islands at the surface position of maximum strain thus explaining vertical stacking in Ge/Si(001)–system. Experimental data and theoretical modeling [95] show that the lateral ordering in subsequent sheets of islands is better pronounced than in the first sheet. Self–ordering effects which are likely to be purely equilibrium ones for the first sheet of islands, become more pronounced for the growth of multi–sheet structures. In the latter case self–ordering effects are controlled both by energetics and by growth kinetics.

Since vertically stacked islands (quantum dots) just considered are essentially separated, both electron and hole wave functions are effectively localized inside each quantum dot. Therefore the regularity of the above arrangement does not result in a modification of the basic electronic properties of the structures, such as radiative lifetime, energy spectrum, carrier capture and relaxation mechanisms or material gain. This fact initiated attempts to fabricate electronically coupled quantum dots [7,83,96].

B. Shape Transformation Effects

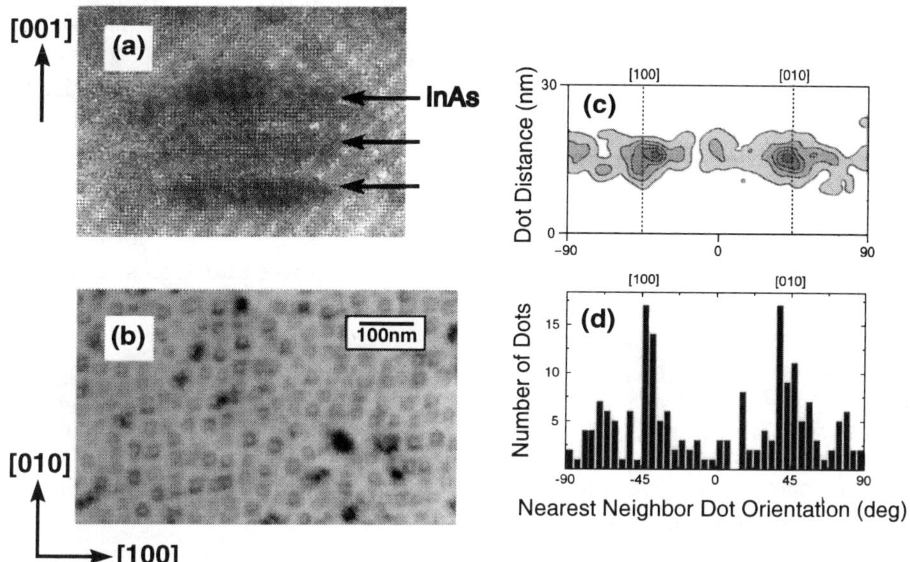

FIG. 16. Vertically coupled InAs quantum dots (VECODs) in a GaAs matrix: a) high resolution electron microscopy [010] cross–section image formed by 9 beams, defocusing is 60 nm; note the different spot density in InAs and GaAs regions. b) bright field plan–view transmission electron microscopy (TEM) micrograph under [001] zone axis illumination. c) 2D histogram of next two neighbors center to center distance and direction for the VECODs from Fig. 16b. d) Projection of Fig. 16c onto the angular axis. Maxima in [100] and [010] directions prove VECOD arrangement in a 2D square lattice.

In Fig. 16 we show cross–section (a) and plan-view (b) TEM images of InAs vertically coupled quantum dots (VECODs) formed by three–cycle InAs — GaAs deposition. The average thickness of the InAs deposited in each cycle equals to 5.5 Å. Each GaAs deposition cycle corresponds to an average thickness of 15 Å. As can be seen in Fig. 16b, the average lateral size of the islands at the top is close to 170 ± 10 Å. The islands have a square base with main axes along [100]– and [010]–directions. The histogram of nearest–neighbor dot orientation (Fig. 16c and 16d) proves that the dots are arranged in a 2D square lattice with primitive lattice vectors along the same directions. This ordering is clearly observed in all parts of the TEM image. Although the nature of lateral ordering can be much more complex than in the case of a single sheet of dots, we believe, that it still has the same main reason [74], i. e. repulsive interaction of strained islands via the elastically anisotropic substrate. Each VECOD is composed of three vertically aligned parts separated by 2–4 ML–thick GaAs regions (see Fig. 16a). The top parts have a larger lateral size ($\approx$ 170 Å) than the lower part ($\approx$ 110 Å). The resulting arrangement of InAs insertions can be considered as an artificial three–dimensional semiconductor crystal. Vertically–correlated stacking of well–separated InAs islands in a GaAs matrix is reported in literature since 1985 [54,94]. Recently stacking of spaced islands has been also reported [83,96]. However, even in that case, the thickness of the GaAs separating layer was larger than the pyramid height. In our case the height of the single pyramid–like island is (≈ 50 Å) [7,72]. The average thickness of only (≈ 15 Å) of GaAs deposited is not sufficient to cover the InAs pyramid. Thus, the deposition mode described here cannot be considered as simple vertically–correlated stacking [54,94,96] and another explanation for the final arrangement is required.

C. Energetics of Stranski-Krastanow growth for InAs and GaAs deposition cycles

Single cycle deposition of InAs on the GaAs (100) surface above the critical thickness (≈ 4.5 Å) leads to the formation of pyramid–shaped InAs islands on the InAs wetting layer. Further growth of GaAs on a surface with wetting layer and locally formed islands is affected by the inhomogeneous strain field due to the pyramids [59,97]. The strain field modulates the surface chemical potential for Ga adatoms as follows (see, e. g. [59]):

$$\mu^{Ga}(\mathbf{r}) = \left[\mu_0^{Ga} + \frac{Y\varepsilon_t^2(\mathbf{r})\Omega}{2}\right] + \gamma\Omega\kappa(\mathbf{r})\,. \tag{31}$$

Here μ_0^{Ga} is the chemical potential of Ga adatoms on the reference flat surface with the lattice parameter equal to that of bulk GaAs. The second term in the square parentheses is the elastic energy correction to μ_0^{Ga} where $\varepsilon_t(\mathbf{r})$ is the tangential component of the local strain defined with respect to unstrained GaAs, Y is Young's modulus, and Ω is the atomic volume. The third term is the surface energy contribution to the chemical potential, where γ is the surface energy and $\kappa(\mathbf{r})$ is the local curvature of the surface. Due to the elastic energy correction to μ_0^{Ga}, the incorporation of GaAs on the facets of elastically relaxed InAs pyramids is energetically unfavorable. The gradient of the surface chemical potential leads to a locally directional migration of Ga adatoms away from the InAs islands (Fig. 17a–17b). The latter results in a reduction of the growth rate and in a curved growth front in the vicinity of the islands [97]. When the InAs islands are partly covered by GaAs, the effect of the strain inhomogeneity on the profile of the growing surface decreases, and the planar growth front is completely re–established after the deposition of ≈ 60 Å GaAs [7,68,72].

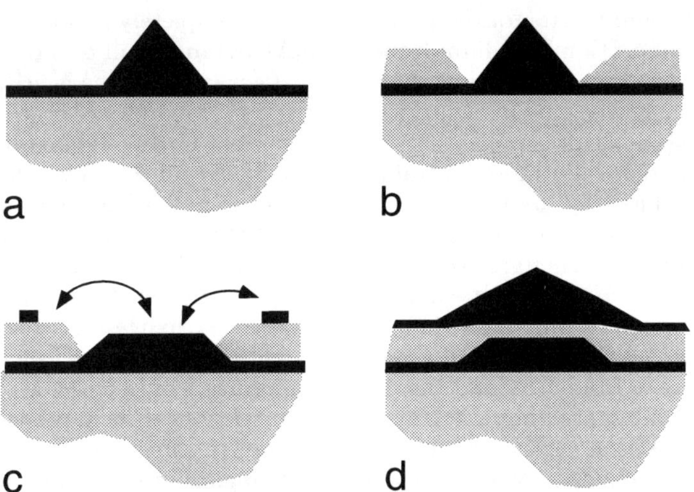

FIG. 17. a) — d) — Schematic diagram of progressive stages of the formation of the coupled quantum dot structure. Efficient exchange reactions on the surface due to the energetics of Stranski-Krastanow growth occur during growth interruption on stage (c).

The situation changes drastically if the GaAs growth is interrupted well before the dots are completely covered with GaAs, and another InAs deposition cycle is introduced (or just the growth interruption time is chosen to be long enough to let the system come to the equilibrium). According to the SK InAs–GaAs growth mode, it is energetically favorable for InAs to evaporate from the InAs islands and to cover the free surface of GaAs forming a second wetting layer. The surface chemical potential of In atoms is equal in this case to:

$$\mu^{In}(\mathbf{r}) = \left[\mu_0^{In} + \frac{Y\left(\varepsilon_t(\mathbf{r}) - \varepsilon_0\right)^2 \Omega}{2}\right] + \gamma\Omega\kappa(\mathbf{r}) - \frac{\zeta\Omega\vartheta(\mathbf{r})}{a}. \tag{32}$$

Here μ_0^{In} is the chemical potential of In adatoms on the unstressed (completely relaxed) surface, where the lattice parameter is equal to that of InAs. The elastic energy correction is minimum for a completely relaxed surface, where $\varepsilon_t(\mathbf{r}) = \varepsilon_0$ (ε_0 being the lattice mismatch between InAs and GaAs). The third term in Eq.(32) is the same as in Eq.(31). The last term represents the effect of wetting, where ζ is the energy benefit per unit area due to the formation of the wetting layer, a is the lattice parameter, and $\vartheta(\mathbf{r}) = 1$ on the GaAs surface and $\vartheta(\mathbf{r}) = 0$ on the InAs surface. The last term in Eq.(32) plays the dominant role in the directed migration of indium atoms before the second wetting layer is formed. According to this term there exists a thermodynamically favored tendency for indium atoms to be detached from the InAs island and to cover the free GaAs surface. As the InAs pyramids are only partly covered with GaAs, InAs of the upper part of the InAs island is available, and the top of the pyramid can be completely dissolved at this stage. Further evaporation of indium from the InAs pyramid will occur from the part laterally confined by GaAs unless the enhanced curvature of the nearby GaAs region, and, consequently, enhanced surface energy (last term in Eq.(31)) will make the planarization (partial or complete) of the GaAs surface by directional migration of Ga adatoms energetically more favorable. Then the rest of the InAs island will be completely confined by GaAs. Directional migration of Al and Ga adatoms was also found to result in spontaneous formation of the upper AlGaAs cladding on flat InGaAs islands during growth interruptions [98], indicating that in the 2D case (flat islands) this arrangement can be also energetically beneficial. Formation of split islands is possible only if the kinetics of surface exchange reactions is fast enough. After GaAs is deposited, and a new InAs deposition cycle has just started, most of the surface is InAs–free, and the directed migration of In and Ga adatoms continues. At our substrate temperature (480^0 C) and for the growth mode chosen (0.3 ML InAs deposition cycles with 10 s growth interruptions after each cycle), the kinetics is fast enough to produce severe morphological modifications even on a much larger geometrical scale [7]. Simultaneously with the shape transformation of the InAs islands, the deposition of extra InAs occurs, finally resulting in the formation of a complete InAs wetting layer ($\approx$ 1.5 ML). Then, after the second wetting layer is formed, the second term in Eq.(32) provides the tendency for excess In atoms to attach to the region of locally modulated lattice parameter due to existing pyramids. According to this energy term, the formation of InAs islands at new positions is energetically unfavorable. As a result, a vertical arrangement of two InAs islands separated by a several monolayer thick GaAs layer (split pyramid or VECOD) can be formed. Then the process can be continued. Schematically, the formation of the VECOD structure is illustrated in Figs. 17a — 17d.

D. Strain energy of VECOD structures

The possibility to fabricate the VECOD structure depends also on the energetics of the split pyramid in respect to other possible arrangements. We compare the energetics of several possible final states. Fig. 18a depicts the case where no splitting occurs. Fig. 18b — 18d refer to the situation where the "buried" part of the pyramid floats up partially (Fig. 18b, 18c) or completely (Fig. 18d) and is replaced by GaAs. We estimate here the gain in the elastic energy for structures of Figs. 18b — 18d with respect to that of Fig. 18a. This gain is due to the fact that a certain volume of InAs is transferred from the buried region, where it is not relaxed, to the uncovered pyramid, where it is partially relaxed. The elastic energy for each of the structures of Fig. 18 is determined by

$$E_{elastic} = \frac{1}{2}\lambda^{(2)}_{ijlm} \int\limits_{(2)} (\varepsilon_{ij}(\mathbf{r}) - \varepsilon_0\delta_{ij})\,(\varepsilon_{lm}(\mathbf{r}) - \varepsilon_0\delta_{lm})\,dV + \frac{1}{2}\lambda^{(1)}_{ijlm} \int\limits_{(1)} \varepsilon_{ij}(\mathbf{r})\,\varepsilon_{lm}(\mathbf{r})dV\,. \quad (33)$$

Here indices (1) and (2) denote GaAs and InAs, respectively; λ_{ijlm} is the tensor of elastic moduli; ε_0 is the lattice mismatch; $\varepsilon_{ij}(\mathbf{r})$ is the strain tensor defined via the displacement vector $u_i(\mathbf{r})$ as follows: $\varepsilon_{ij}(\mathbf{r}) = (1/2)\left[\partial u_i(\mathbf{r})/\partial r_j + \partial u_j(\mathbf{r})/\partial r_i\right]$. The displacement vector $u_i(\mathbf{r})$ in the heterophase system obeys equilibrium equations of the elasticity theory in each material,

$$\lambda_{ijlm}^{(1,2)} \frac{\partial^2 u_m(\mathbf{r})}{\partial r_j \partial r_l} = 0 \,, \tag{34}$$

obeys boundary conditions at the interface and at the free surface, and vanishes in the depth of the substrate. The boundary conditions at the InAs–GaAs interface read [2]:

$$\lambda_{ijlm}^{(2)} n_j(\mathbf{r}) \left[\frac{\partial u_m(\mathbf{r})}{\partial r_l} - \varepsilon_0 \delta_{lm} \right] \Bigg|^{(2)} = \lambda_{ijlm}^{(1)} n_j(\mathbf{r}) \left[\frac{\partial u_m(\mathbf{r})}{\partial r_l} \right] \Bigg|^{(1)} \,, \tag{35}$$

where $n_j(\mathbf{r})$ is the outer normal to the InAs–region at the interface, and the boundary conditions at the free surface of InAs are as follows [5]:

$$\lambda_{ijlm}^{(2)} m_j(\mathbf{r}) \left[\frac{\partial u_m(\mathbf{r})}{\partial r_l} - \varepsilon_0 \delta_{lm} \right] \Bigg|^{(2)} = 0 \,, \tag{36}$$

where $m_j(\mathbf{r})$ is the outer normal to the free surface. We solve equation (34) subject to boundary conditions (35,36) by the finite element method, calculate the strain tensor $\varepsilon_{ij}(\mathbf{r})$, evaluate the elastic energy of the two wetting layers and thus obtained the net elastic energy of each of the islands displayed in Fig. 18.

If we denote the net energy of the non–split pyramid of Fig. 18a as E_0, then the net energies of split pyramids are as follows: $0.89E_0$ for Fig. 18b, $0.76E_0$ for Fig. 18c, and $0.61E_0$ for Fig. 18d. To analyze other possible shape transformations of the pyramid, we have calculated $E_{elastic}$ for several split structures, where InAs from the buried part of the pyramid is transferred not to the uncovered part of the pyramid, but is redistributed between two planar wetting layers. Such a splitting does not lead to a reduction of the elastic energy, but, to the contrary, results in an increase of the elastic energy up to $1.20E_0$.

If the process described is repeated in a multi–cycle deposition mode, the volume of each island progressively increases with successive deposition cycles of InAs due to the transfer of InAs from the buried part of the structure to the uncovered part. The performed theoretical analysis demonstrates that the elastic energy for structures like those of Figs. 18b — 18d with respect to those of Fig. 18a is gained when at the final stage a certain volume of InAs is transferred from the buried region where it is not relaxed to the uncovered pyramid where it is partially elastically relaxed. This general conclusion is not affected by possible deviations of the real shape of islands from the simplified shapes of Fig. 18. Although the maximum energy gain would correspond to the complete transfer of the buried InAs to the uncovered part of the pyramid, the actual resulting shape of the island strongly depends on the kinetics of the growth process involved. Thus the strain energy calculations also support the total energy reduction due to splitting of the pyramids, in addition to the energetics of the Stranski–Krastanow growth itself. At the same time, for high deposition rates and low substrate temperatures, the splitting can be suppressed because of the slower kinetics of surface exchange reactions.

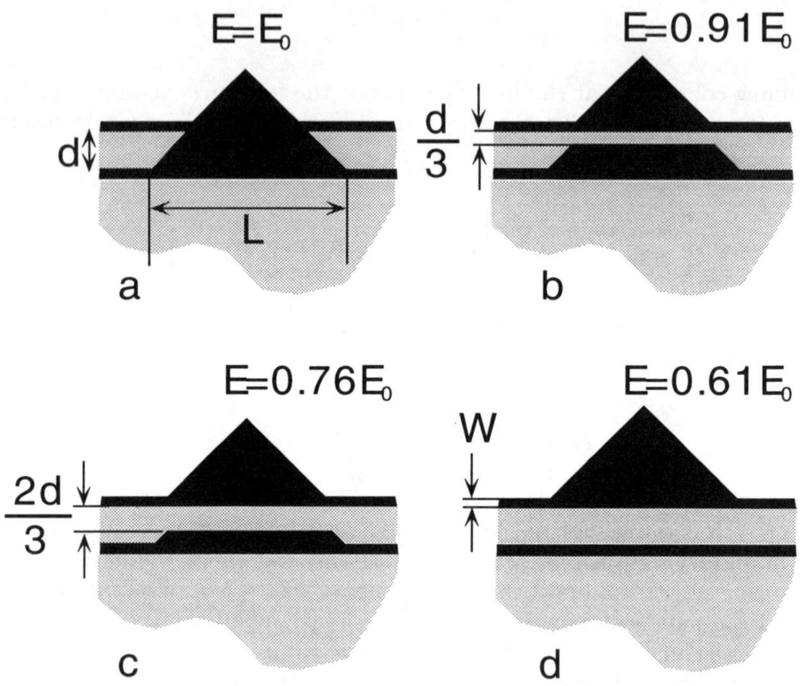

FIG. 18. a) — d) — Schematic representation of several possible final states for multilayer InAs (black) — GaAs (grey) deposition. Relative dimensions used in calculations of the elastic energy are as follows: $d = 3W$, $L = 24W$. The tilt angle of the pyramid facets is 45^0. We show that there occurs a reduction the elastic energy for structures of (b — d) with respect to that of (a) caused by InAs transfer from buried part to the uncovered part of the pyramid. No total energy reduction is induced by simple splitting without InAs transfer.

VI. CONCLUSIONS

To conclude, a review is given of the theory of spontaneous formation of periodic nanometer–scale structures on semiconductor surfaces. A general approach is used where several different classes of self–ordered nanostructures are considered as equilibrium structures of elastic domains. The long–range ordering in all cases is the result of long–range elastic interactions. A particular focus is given on a new class of self–ordered nanostructures, namely on ordered arrays of 3D coherent strained islands. A new important property of this class of nanostructures is the existence of a phase transition between ordering regime and ripening regime. For multi–sheet arrays of islands in the semiconductor matrix, the formation of vertically ordered array involves *both strain–driven equilibrium ordering* and *strain–driven growth kinetics.*

Self–ordering phenomena considered for all nanostructures in question are the basis for a novel technology of direct fabrication of quantum–wire and quantum–dot superlattices where a possibility exists to control geometrical and electronic parameters of nanostructures.

ACKNOWLEDGMENTS

The work was supported by Russian Foundation for Basic Research, Grant No. 96–02–17943a, by the Joint Grant of Russian Foundation for Basic Research (96–07–001686) and of Deutsche Forschungsgemeinschaft (Sfb 296), and by the Volkswagen Stiftung. One of us (N.N.L.) thanks the Alexander von Humboldt Foundation.

REFERENCES

* E-mail: shchukin@w422zrz.physik.TU-Berlin.DE and shchukin@ton.ioffe.rssi.ru

[1] Landau, L.D. and Lifshits, E.M. (1960) *Electrodynamics of Continuous Media*, Pergamon, New York.

[2] Khachaturyan, A.G. (1974) *Theory of Phase Transformations and the Structure of Solid Solution*, in Russian, Moscow, Nauka; and Khachaturyan, A.G. (1983) *Theory of Structural Transformations in Solids*, Wiley, New York.

[3] Roitburd, A.L. (1976) Equilibrium structure of epitaxial layers, *Phys. Status Solidi (a)* **37**, 329–339.

[4] Bruinsma, R. and Zangwill, A. (1986) Structural transitions in epitaxial overlayers, *J. Phys. Paris* **47**, 2055–2073.

[5] Ipatova, I.P., Malyshkin, V.G., and Shchukin V.A. (1993) On spinodal decomposition in elastically anisotropic epitaxial films of III–V semiconductor alloys, *J. Appl. Phys.* **74**, 7198–7210.

[6] Ipatova, I.P., Malyshkin, V.G., and Shchukin, V.A. (1994) Composition–modulated structures in epitaxial films of phase–separating semiconductor alloys, *Phil. Mag. B* **70**, 557–566.

[7] Ledentsov, N.N., Grundmann, M., Kirstaedter, N., Schmidt, O., Heitz, R., Böhrer, J., Bimberg, D., Ustinov, V.M., Shchukin, V.A., Kop'ev, P.S., Alferov, Zh.I., Ruvimov, S.S. Kosogov, A.O., Werner, P., Richter, U., Gösele, U., and Heydenreich, J. (1996) Ordered arrays of quantum dots: formation, electronic spectra, relaxation phenomena, lasing, in *Proc. 7th Int. Conf. on Modulated Semicond. Structures (MSS–7)*, Madrid, Spain, July 1995, — *Solid-State Electron.* **40**, 785–798.

[8] Bimberg, D., Ledentsov, N.N., Grundmann, M., Kirstaedter, N., Schmidt, O.G., Mao, M.H., Ustinov, V.M., Egorov, A.Yu., Zhukov, A.E., Kop'ev, P.S., Alferov, Zh.I., Ruvimov, S.S., Gösele, U., and Heydenreich, J. (1996) InAs–GaAs quantum dots: From growth to lasers, *Phys. Stat. Sol. (b)* **194**, 159–173.

[9] Wulff, G. (1901) *Z. Kristallogr. Mineral.* **34**, 449.

[10] Herring, C. (1951) Some theorems on free energies of crystal surfaces, *Phys. Rev.* **82**, 87–93.

[11] Mullins, W.W. (1963) Solid surface morphologies governed by capillarity, in *Metal Surfaces: Structure, Energetics and Kinetics*, American Society for Metals, Metals Park, P. 17.

[12] Chernov, A.A. (1961) The spiral growth of crystals, *Uspekhi Fiz. Nauk* **73**, 277–391, — *Sov. Phys. Uspekhi* **4**, 116–148.

[13] Rottman, C., and Wortis, M. (1984) Statistical mechanics of equilibrium crystal shapes: Interfacial phase diagrams and phase transitions, *Phys. Reports* **103**, 59–79.

[14] Eaglesham, D.J., White, A.E., Feldman, L.C., Moriya, N., and Jacobson, D.C. (1993) Equilibrium shape of Si, *Phys. Rev. Lett.* **70**, 1643–1646.

[15] Williams, E.D., Phaneuf, R.J., Wei, J., Bartelt, N.C., and Einstein, T.L. (1993) Thermodynamics and statistical mechanics of the faceting of stepped Si(111), *Surf. Sci.* **294**, 219–242.

[16] Mukherjee, S., Pehlke, E., and Tersoff, J. (1994) Calculation of temperature effects on the equilibrium crystal shape of Si near (100), *Phys. Rev. B* **49**, 1919–1927.

[17] Gibbs, J.W. (1928) *Collected Works, Vol. 1, Thermodynamics*, Longmans, London.

[18] Marchenko, V.I. and Parshin, A.Ya. (1980) Elastic properties of crystal surfaces, *Zh. Eksp. Teor. Fiz.* **79**, 257–260 — *Sov. Phys. JETP* **52**, 129–131.

[19] Needs, R.J. (1987) Calculations of the surface stress tensor at aluminium (111) and (110) surfaces, *Phys. Rev. Lett.* **58**, 53–56.

[20] Herring, C. (1951) in: W.E. Kingston (ed.) *The Physics of the Powder Metallurgy*, McGraw–Hill, New York.

[21] Shuttleworth, R. (1950) *Proc. Phys. Soc. London, Sect. A* **63**, 444.

[22] Fiorentini, V., Methfessel, M., and Scheffler, M. (1993) Reconstruction mechanism of fcc transition metal (001) surfaces, *Phys. Rev. Lett.* **71**, 1051–1054.

[23] Garcia, A. and Northrup, J.E. (1993) Stress relief from alternately buckled dimers in Si(100), *Phys. Rev. B* **48**, 17350–17353.

[24] Moll, N., Kley, A., Pehlke, E., and Scheffler, M. (1996) GaAs equilibrium crystal shape from first principles, *Phys. Rev. B* **54**, 8844–8855; Moll, N., Kley, A., Pehlke, E., and Scheffler, M. to be published.

[25] Andreev, A.F. and Kosevich, Yu.A. (1981) Capillarity phenomena in the theory of elasticity, *Zh. Eksp. Teor. Fiz.* **81**, 1435–1443, — *Sov. Phys. JETP* **54**, 761–765.

[26] Nozières, P. and Wolf, D.E. (1988) *Z. Phys. B* **70**, 399–407; *Z. Phys. B* **70**, 507–513.

[27] Landau, L.D. and Lifshits, E.M. (1959) *Theory of Elasticity*, Pergamon, New York.

[28] Andreev, A.F. (1980) Strictive superstructures in two–dimensional phase transitions, *Pis'ma Zh. Eksp. Teor. Fiz.* **32**, 654–656, — *JETP Letters* **32**, 640–642.

[29] Andreev, A.F. (1981), Faceting phase transitions of crystals, *Zh. Eksp. Teor. Fiz.* **80**, 2042–2052 — *Sov. Phys. JETP* **53**, 1063–1069.

[30] Marchenko, V.I. (1981) Theory of the equilibrium shape of crystals, *Zh. Eksp. Teor. Fiz.* **81**, 1141–1144, — *Sov. Phys. JETP* **54**, 605–607.

[31] Hibino, H., Fukuda, T., Suzuki, M., Hommo, Y., Sato, T., Iwatsuki, M., Miki, K., and Tokumoto, H. (1993) High–temperature scanning–tunneling–microscopy observation of phase transitions and reconstruction on a vicinal Si(111) surface, *Phys. Rev. B*

47, 13027–13030.

[32] Kasu, M., and Kobayashi, N. (1993) Equilibrium multiatomic step structure of GaAs(001) vicinal surfaces grown by metalorganic chemical vapor deposition. *Appl. Phys. Lett.* **62**, 1262–1264.

[33] Golubok, A.O., Gur'yanov, G.M., Petrov, V.N., Samsonenko, Yu.B., Tipisev, S.Ya., Tsyrlin, G.E., and Ledentsov, N.N. (1994) Formation of arrays of facets on vicinal surfaces of GaAs(100) during molecular–beam epitaxy, *Fiz. Tekh. Poluprovodn.* **28**, 516–519, — *Semiconductors* **28**, 317–319.

[34] Ledentsov, N.N., Gurianov, G.M., Tsyrlin, G.E., Petrov, V.N., Samsonenko, Yu.B., Golubok, A.O., and Tipisev, S.Ya. (1994) Effect of heat–treatment conditions on the surface morphology of gallium arsenide grown on vicinal GaAs(100) substrates by molecular beam epitaxy, *Fiz. Tekh. Poluprovodn.* **28**, 903–906, — *Semiconductors* **28**, 526–527.

[35] Watson, G.M., Gibbs, D., Zehner, D.M., Yoon, M., and Mochrie, S.G.J. (1993) Faceting transformation of the stepped Pt(001) surface, *Phys. Rev. Lett.* **71**, 3166–3169.

[36] Nötzel, R., Ledentsov, N.N., Däweritz, L., Hohenstein, M., and Ploog, K. (1991) Direct synthesis of corrugated superlattices on non–(100)–oriented surfaces, *Phys. Rev. Lett.* **67**, 3812–3815.

[37] Baski, A.A., and Whitman, L.J. (1995) Quasiperiodic nanoscale faceting of high–index surfaces, *Phys. Rev. Lett.* **74**, 956–959.

[38] Zuo, J.–K., Warmack, R.J., Zehner, D.M., and Wendelken, J.F. (1993) Periodic faceting on TaC(110): Observations using high–resolution low–energy electron diffraction and scanning tunneling microscopy, *Phys. Rev. B* **47**, 10743–10747.

[39] Koch, R., Borbonus, M., Haase, O., and Rieder, K.H. (1991) New aspects on the Ir(110) reconstruction: surface stabilization on mesoscopic scale via (331) facets, *Phys. Rev. Lett.* **67**, 3416–3419.

[40] Chernov, A.A. (1984) *Modern Crystallography III*, Springer Verlag, Berlin.

[41] Alferov, Zh.I., Egorov, A.Yu., Zhukov, A.E., Ivanov, S.V., Kop'ev, P.S., Ledentsov, N.N., Mel'tser, B.Ya., and Ustinov, V.M. (1992) Growth of GaAs–AlAs quantum clusters on non–(100)–oriented faceted GaAs surfaces by molecular beam epitaxy, *Fiz. Tekh. Poluprovodn.* **26**, 1715–1722, — *Sov. Phys. Semicond.* **26**, 959–963.

[42] Shchukin, V.A., Borovkov, A.I., Ledentsov, N.N., and Kop'ev, P.S. (1995) Theory of quantum wire formation on corrugated surfaces, *Phys. Rev. B* **51**, 17767–17779.

[43] Shchukin, V.A., Borovkov, A.I., Ledentsov, N.N., and Bimberg, D., Tuning and breakdown of faceting under externally applied stress, *Phys. Rev. B* **51**, 10104–10118.

[44] Marchenko, V.I. (1981) Possible structures and phase transitions on the surface of crystals, *Pis'ma Zh. Eksp. Teor. Fiz.* **33**, 397–399 — *JETP. Lett.* **33**, 381–383.

[45] Men, F.K., Packard, W.E., and Webb, M.B. (1988) Si(100) surface under externally applied stress, *Phys. Rev. Lett.* **61**, 2469–2972.

[46] Alerhand, O.L., Vanderbilt, D., Meade, R.D., and Joannopoulos, J.D. (1988) Spontaneous formation of stress domains on crystal surfaces, *Phys. Rev. Lett.* **61**, 1973–1976.

[47] Dabrowski, J., Pehlke, E., and Scheffler, M. (1994) Calculation of the surface stress anisotropy for the buckled Si(001)(1×2) and $p(2 \times 2)$ surfaces, *Phys. Rev. B* **49**, 4790–4793.

[48] Kern, K., Niehus, H., Schatz, A., Zeppenfeld, P., George, J., and Comsa, G. (1991)

Long–range spatial self–organization in the adsorbate–induced restructuring of surfaces: Cu{110}-(2×1)O, *Phys. Rev. Lett.* **67**, 855–858.

[49] Bressler–Hill, V., Lorke, A., Varma, S., Pond, K., Petroff, P.M., and Weinberg, W.H. (1994) Initial stages of InAs epitaxy on vicinal GaAs(001)–(2 × 4), *Phys. Rev. B* **50**, 8479–8487.

[50] Kwok–On Ng and Vanderbilt, D. (1995) Stability of periodic domain structures in a two–dimensional dipolar model, *Phys. Rev. B* **52**, 2177–2183.

[51] Vanderbilt, D. (1992) Phase segregation and work–function variations on metal surfaces: spontaneous formation of periodic domain structures, Surf. Sci. **268**, L300–L304.

[52] Zeppenfeld, P., Krzyzowski, M., Romainczuk, C., Comsa, G., and Lagally, M.G. (1994) Size relation for surface systems with long–range interaction, *Phys. Rev. Lett.* **72**, 2737–2740.

[53] Tersoff, J. and Tromp, R.M. (1993) Shape transition in growth of strained islands: Spontaneous formation of quantum wires, *Phys. Rev. Lett.* **70**, 2782–2785.

[54] Goldstein, L., Glas, F., Marzin, J.Y., Charasse, M.N., and Le Roux, G. (1985) Growth by molecular beam epitaxy and characterization of InAs/GaAs strained–layer superlattices, *Appl. Phys. Lett.* **47**, 1099–1101.

[55] Eaglesham, D.J. and Cerullo, M. (1990) Dislocation–free Stranski–Krastanow growth of Ge on Si(100), *Phys. Rev. Lett.* **64**, 1943–1946.

[56] Mo, Y.–W., Savage, D.E., Swartzentruber, B.S., and Lagally, M.G. (1990) Kinetic pathway in Stranski–Krastanov growth of Ge on Si(001), *Phys. Rev. Lett.* **65**, 1020–1023.

[57] Asaro, R.J. and Tiller, W.A. (1972) Interface morphology development during stress corrosion cracking: Part I. Via surface diffusion, *Metall. Trans.* **3**, 1789–1796.

[58] Grinfield, M.A. (1986) Instability of the separation boundary between a non–hydrostatically stressed elastic body and a melt. *Dokl. Acad. Nauk SSSR* **290**, 1358, — *Sov. Phys. Dokl.* **31**, 831.

[59] Srolovitz, D. (1989) On the stability of surfaces of stressed solids, *Acta Metall.* **37**, 621–625.

[60] Spencer, B.J., Voorhees, P.W., and Davis, S.H. (1991) Morphological instability in epitaxially strained dislocation–free solid films, *Phys. Rev. Lett.* **67**, 3696–3699.

[61] Vanderbilt, D. and Wickham, L.K. (1991) Elastic energies of coherent germanium islands on silicon, in C.V. Thompson, J.Y. Tsao, and D.J. Srolovitz (eds.), *Evolution of Thin-Film and Surface Microstructure*, MRS Proceedings Vol. 202, MRS, Pittsburgh, p. 555.

[62] Jesson, D.E., Pennycook, S.J., Baribeau, J.–M., and Houghton, D.C. (1993) Direct imaging of surface cusp evolution during strained–layer epitaxy and implications for strain relaxation, *Phys. Rev. Lett.* **71**, 1744–1747.

[63] Yang, W.H. and Srolovitz, D.J. (1993) Cracklike surface instabilities in stressed solids, *Phys. Rev. Lett.* **71**, 1593–1596.

[64] Ratsch, C. and Zangwill, A. (1993) Equilibrium theory of the Stranski–Krastanow epitaxial morphology, *Surf. Sci.* **293**, 123–131.

[65] Leonard, D., Krishnamurthy, M., Reaves, C.M., Denbaars, S.P., and Petroff, P.M. (1993) Direct formation of quantum–sized dots from uniform coherent islands of InGaAs on GaAs surfaces, *Appl. Phys. Lett.* **63**, 3203–3205.

[66] Moison, J.M., Houzay, F., Barthe, F., Leprince, L., André, E., and Vatel, O. (1994) Self–organized growth of regular nanometer–scale InAs dots on GaAs, *Appl. Phys. Lett.* **64**, 196–198.

[67] Ledentsov, N.N., Grundmann, M., Kirstaedter, N., Christen, J., Heitz, R., Böhrer, J., Heinrichsdorf, F., Bimberg, D., Ruvimov, S.S., Werner, P., Richter, U., Gösele, U., Heydenreich, J., Ustinov, V.M., Egorov, A.Yu., Maximov, M.V., Kop'ev, P.S., and Alferov, Zh.I. (1995), in D.J. Lockwood (ed.) *Proc. 22nd Int. Conf. Phys. Semicond, Vancouver, Canada, August 1994*, World Scientific, Singapore, Vol. 3, p. 1855.

[68] Grundmann, M., Christen, J., Ledentsov, N.N., Böhrer, J. Bimberg, D., Ruvimov, S.S., Werner, P., Richter, U., Gösele, U., Heydenreich, J., Ustinov, V.M., Egorov, A.Yu., Zhukov, A.E., Kop'ev, P.S., and Alferov, Zh.I. (1995) Ultranarrow luminescence lines from single quantum dots, *Phys. Rev. Lett.* **74**, 4043–4046.

[69] Grundmann, M., Ledentsov, N.N., Heitz, R., Eckey, L., Christen, J., Böhrer, J., Bimberg, D., Ruvimov, S.S., Werner, P., Richter, U., Gösele, U., Heydenreich, J., Ustinov, V.M., Egorov, A.Yu., Zhukov, A.E., Kop'ev, P.S., and Alferov, Zh.I. (1995) InAs/GaAs quantum dots radiative recombination from zero–dimensional states, *Phys. Stat. Sol. (b)*, **188**, 249–258.

[70] Cirlin, G., Guryanov, G.M., Golubok, A.O., Tipissev, S.Ya., Ledentsov, N.N., Kop'ev, P.S., Grundmann, M., and Bimberg, D. (1995) Ordering phenomena in InAs strained layer morphological transformation on GaAs (100) surface, *Appl. Phys. Lett.* **67**, 97–99.

[71] Bimberg, D., Grundmann, M., Ledentsov, N.N., Ruvimov, S.S., Werner, P., Richter, U., Heydenreich, J., Ustinov, V.M., Kop'ev, P.S., and Alferov, Zh.I. (1995) Self–organization processes in MBE–grown quantum dot structures, *Thin Solid Films* **267**, 32–36.

[72] Ruvimov, S.S., Werner, P., Scheerschmidt, K., Richter, U., Gösele, U., Heydenreich, J., Ledentsov, N.N., Grundmann, M., Bimberg, D., Ustinov, V.M., Egorov, A.Yu., Kop'ev, P.S., Zh.I. Alferov, Zh.I. (1995) Structural characterization of (In,Ga)As quantum dots in a GaAs matrix, *Phys. Rev. B* **51**, 14766–14769.

[73] Shchukin, V.A., Ledentsov, N.N., Grundmann, M., Kop'ev, P.S., and Bimberg, D. (1996) Strain–induced formation and tuning of ordered nanostructures on crystal surfaces, *Surf. Sci.* **352–354**, 117–122.

[74] Shchukin, V.A., Ledentsov, N.N., Kop'ev, P.S., and Bimberg, D. (1995) Spontaneous ordering of arrays of coherent strained islands, *Phys. Rev. Lett.* **75**, 2968–2971.

[75] Mansfield, M. and Needs, R.J. (1990) Application of the Frenkel–Kontorova model to the surface reconstruction, *J. Phys. Condens. Matter* **2**, 2361–2374.

[76] Narasimhan, S. and Vanderbilt, D. (1992) Elastic stress domains and the herringbone reconstruction on Au(111), *Phys. Rev. Lett.* **69**, 1564–1567.

[77] Wolf, D. (1993) Should all surfaces be reconstructed? *Phys. Rev. Lett.* **70**, 627–630.

[78] Moll, N., Kley, A., Pehlke, E., and Scheffler, M., to be published.

[79] Qian, G.–X., Martin, R.M., and Chadi, D.J. (1988) First–principle study of the atomic reconstructions and energies of Ga– and As–stabilized GaAs(100) surfaces, *Phys. Rev. B* **38**, 7649–7663.

[80] Ivanov, S.V., Kop'ev, P.S., and Ledentsov, N.N. (1990) Thermodynamic analysis of segregation effects in molecular beam epitaxy, *Crystal Growth* **104**, 345–354.

[81] Heinrichsdorff, F., Krost, A., Grundmann, M., Bimberg, D., Kosogov, A., and Werner, P. (1996) Self–organization process of InGaAs/GaAs quantum dots grown by metalorganic chamical vapor deposition, *Appl. Phys. Lett.* **68**, 3284–3286.

[82] Gurianov, G.M., Tsyrlin, G.E., Petrov, V.N., Samsonenko, Yu.B., Gubanov, V.B., Polyakov, N.K., Golubok, A.O., Tipisev, S.Ya., Musikhina, E.P., and Ledentsov, N.N. (1995) Self–organization of strained quantum–size $In_xGa_{1-x}As$ structures grown on misoriented (100) surfaces of GaAs during submonolayer

molecular–beam epitaxy, *Fiz. Tekh. Poluprovodn.* **29**, 1642–1648, — *Semiconductors* **29**, 854–857.

[83] Heinrichsdorff, F. *et al.*, unpublished.

[84] Priester, C. and Lannoo, M. (1995) Origin of self–assembled quantum dots in highly–mismatched heteroepitaxy, *Phys. Rev. Lett.* **75**, 93–96.

[85] Joós, B., Einstein, T.L., and Bartelt, N.C. (1991) Distribution of terrace widths on a vicinal surface within the one–dimensional free–fermion model, *Phys. Rev. B* **43**, 8153–8162.

[86] Pehlke, E., Moll, N., and Scheffler, M. (1996) The equilibrium shape of InAs quantum dots grown on a GaAs(001) substrate, in *Proc. 23nd Int. Conf. Phys. Semicond.* Berlin, Germany, July 1996, in print.

[87] Guryanov, G.M., Cirlin, G.E., Golubok, A.O., Tipisev, S.Ya., Ledentsov, N.N., Shchukin, V.A., Grundmann, M., Bimberg, D., and Alferov, Zh.I. (1996) An intermediate (1.0–1.5) monolayers stage of heteroepitaxial growth of InAs on GaAs(100) during submonolayer molecular beam epitaxy, *Surf. Sci.* **352–354**, 646–650.

[88] Guryanov, G.M., Cirlin, G.E., Petrov, V.N., Polyakov, N.K., Golubok, A.O., Tipisev, S.Ya., Gubanov, V.B., Samsonenko, Yu.B., Ledentsov, N.N., Shchukin, V.A., Grundmann, M., Bimberg, D., and Alferov, Zh.I. (1996) STM and RHEED study of InAs/GaAs quantum dots obtained by submonolayer epitaxial techniques, *Surf. Sci.* **352–354**, 651–655.

[89] Chen, K.M., Jesson, D.E., Pennycook, S.J., Thundat, T., and Warmack R.J. (1995) Self–limiting growth kinetics of 3D coherent islands, *MRS Proceedings*, **399**, 271–281.

[90] Jesson, D.E., Chen, K.M., Pennycook, S.J., Thundat, T., and Warmack R.J. (1996) Morphological evolution of strained films by cooperative nucleation, *Phys. Rev. Lett.* **77**, 1330–1333.

[91] Malyshkin, V.G. and Shchukin, V.A. (1993) Development of composition inhomogeneities in layer–by–layer growth of an epitaxial film of a solid solution of III–V semiconductors, *Fiz. Tekh. Poluprovodn.* **27**, 1932, *Semiconductors* **27**, 1062–1068.

[92] Ipatova, I.P., Malyshkin, V.G., Maradudin, A.A., Shchukin, V.A., and Wallis, R.F. (1996) Kinetic phase transitions in the epitaxial growth of compound semiconductors, in: *Proc. 23rd Int. Symposium on Compound Semiconductors*, St. Petersburg, September 1996.

[93] Maradudin, A.A., and Wallis, R.F. (1980) Elastic interaction of point defects in a semi–infinite medium, *Surf. Sci.* **91**, 423–439.

[94] Xie, Q., Madhukar, A., Chen, P., and Kobayashi, N. (1995) Vertically self–organized InAs quantum box islands on GaAs(100), *Phys. Rev. Lett.* **75**, 2542–2545.

[95] Tersoff, J., Teichert, C., and Lagally, M.G. (1996) Self–organization in growth of quantum dot superlattices, *Phys. Rev. Lett.* **76**, 1675–1678.

[96] Solomon, G.S., Trezza, J.A., Marshall, A.F., and Harris, J.S., Jr. (1996) Vertically aligned and electronically coupled growth induced InAs islands in GaAs, *Phys. Rev. Lett.* **76**, 952–955.

[97] Xie, Q., Chen, P., and Madhukar, A. (1994) InAs island–induced–strain driven adatom migration during GaAs overlayer growth, *Appl. Phys. Lett.* **65**, 2051–2053.

[98] Nötzel, R., Temmyo, J., and Tamamura, T. (1994) Nature (London) **369**, 131.

[99] Ledentsov, N.N, Shchukin, V.A., Grundmann, M., Kirstaedter, N., Böhrer, J., Schmidt, O., Bimberg, D., Ustinov, V.M., Egorov, A.Yu., Zhukov, A.E., Kop'ev, P.S., Zaitsev, S.V., Gordeev, N.Yu., Alferov, Zh.I., Borovkov, A.I., Kosogov, A.O., Ruvimov, S.S., Werner, P., Gösele, U., and Heydenreich, J. (1996) Direct formation of vertically coupled dots in Stranski–Krastanow growth, *Phys. Rev. B* **54**, 8743–8752.

FABRICATION OF QUANTUM DOTS FOR SEMICONDUCTOR LASERS WITH CONFINED ELECTRONS AND PHOTONS

Yasuhiko Arakawa
University of Tokyo
7-22-1 Roppongi, Minato-ku, Tokyo 153, Japan
arakawa@iis.u-tokyo.ac.jp

1. Introduction

A reduction of dimensionality of the electron motions in quantum nano-structures brings new phenomena in semiconductor physics. Moreover, it allows new device concepts to be considered and permits improvements in performance of the transistors and lasers. In 1982, the quantum wire laser and the quantum dot laser were proposed, predicting significant improvement of lasing characteristics[1]. When the concept was proposed, it was considered that the quantum wires and quantum dots would be realized in the 21 century. However, the progress has been made much faster than expected.

The fabrication technologies so far demonstrated include selective growth technique, self-assembling growth technique, and etching technique. In particular, recent progress of the self-assembling technique with the SK growth mode for fabricating the quantum dots has received great attention[2-4]. In fact, quantum dots with a size of 15-20nm have been already available using the SK growth mode, although the size fluctuation can not be avoided[5,6]. It should be noted that the selective growth technique is still important as a seeded self-assembling growth technique for controlling position and size.

The use of microcavity structures is expected to achieve substantial reduction of the threshold current due to the controlled spontaneous emission mode[7,8]. Also, strong interactions between excitons (or zero-dimensional electrons) and photons in the microcavity bring important phenomena including the vacuum Rabi oscillation[9,10]. Thus the control of both photon modes and electron modes in confined structures is important for the ultimate light sources[11].

In this chapter, we discuss our recent progress on fabrication of the quantum dots and their optical properties for semiconductor lasers with confined electrons and photons. The discussion will include investigation of formation process of the quantum dots with the SK growth mode using an AFM together with a self-alignment of the quantum dots. In addition, quantum dots with a lateral size of 10nm in two-dimensional V-groove

G. Abstreiter et al. (eds.), Optical Spectroscopy of Low Dimensional Semiconductors, 303–313.

(2DVG) structures are also demonstrated using the selective growth technique. Using these results, we succeeded in fabricating a vertical microcavity quantum dot laser as the first step toward an ultimate semiconductor laser. Finally, the phonon bottleneck issue in the quantum dots is discussed.

2. Formation process of self-assembling quantum dots grown by MOCVD

The SK growth mode has been intensively investigated for the fabrication of self-assembled quantum dots by MBE and MOCVD. In this growth mode, the three-dimensional (3D) islands are naturally formed by the strain effect after the initial 2D growth of a few monolayers (MLs). However, detailed studies of the growth mode including the formation process of the 3D islands have not been clarified. Here, we investigated the formation process using the AFM which is useful for observation of grown surfaces with an atomic scale.

The sample was grown by a low pressure (76 Torr) MOCVD. First, a GaAs layer was grown on GaAs (001) substrate at 700 C. Then InGaAs was grown on the GaAs layer at 500 C. The InGaAs surface grown on the GaAs layer was investigated by the AFM with the contact mode in the air at room temperature. The growth rate of InGaAs was estimated to be 0.9 s / ML by the AFM observation of the surface.

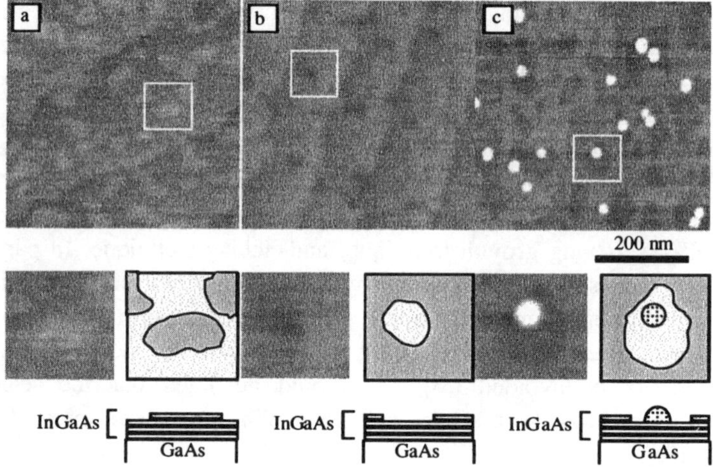

Fig. 1: The AFM images and schematic illustrations of InGaAs surface. The deposition times are (a) 3.0 s (3.3 ML), (b) 3.5 s (3.9 ML) and (c) 3.7 s (4.1 ML).

Figure 1 shows AFM images and illustrations of InGaAs surface grown on GaAs for 3.0 s, 3.5 s, and 3.7 s, respectively. As shown in Fig. 1 (a), there exist 2D islands on terraces in the sample grown for 3.0 s. The formation of the 2D islands is due to the wide terraces and the strain effects. Note that no such 2D island was observed when only GaAs was grown: the step-flow growth always occurs at the step edge of the substrate in GaAs. The 2D islands were at the level of the 4th layer of InGaAs. When the growth time is 3.5 s, many holes appeared on the surface, because the 4th layer occupied most of the surface.

When the growth time is 3.7 s, the 3D islands (i.e., quantum dots) appear as show in Fig. 1 (c). Figures (b) and (c) suggest that the critical thickness is 4 MLs under this growth condition. The diameter and the height of the quantum dots were 20 nm and 4.5 nm, respectively. Fig. 1 (c) also shows that the 3rd layer was exposed on the surface around the quantum dots in some areas. These results indicate that the quantum dots are formed by desorbing some atoms nearby at the 4th layer. This formation mechanism is different from that previously discussed. Photoluminescence spectra for these samples exhibit consistent results with the AFM observation.

3. Spontaneous alignment of quantum dots

One of the inherent problems for the S-K mode quantum dots is that random distribution of the quantum dots can not be avoided. In this section, we show a successful demonstration of the spontaneous alignment of the quantum dots.

In order to align the quantum dots, we grew the quantum dots on the step edges of multi-atomic steps, which results from the step bunching through the generation of facets with two different orientations during annealing. By carefully choosing the growth condition, we found that this multi-atomic step edge can be formed on almost straight lines. In addition, we have already found that the dot density on GaAs substrate misorientated is higher than that on GaAs (001), which indicates that the quantum dots are easily formed on the step edge compared to the terraces. Therefore, by artificially forming the multi-atomic steps during the growth, the position of the quantum dots can be controlled.

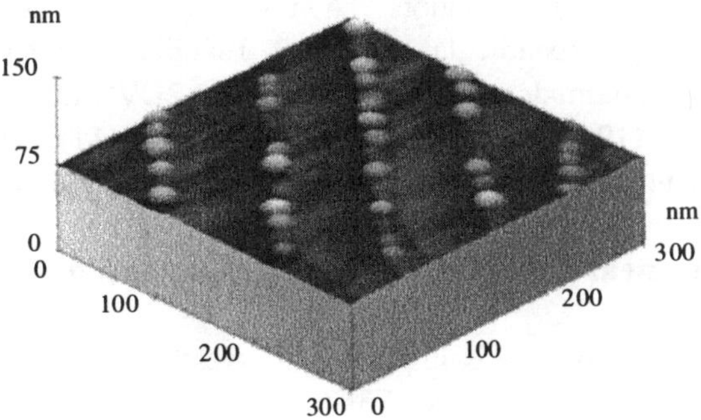

Fig. 2: AFM images of the self-aligned quantum dots grown on the [010]-misorientated surface.

In fact, by choosing growth conditions, the multi-atomic step edge was obtained on a straight line. Figure 2 shows AFM images of the quantum dots grown on the [010]-misorientated surface. The growth time for the InGaAs dots was 3.5 sec. As indicated in the figure, the InGaAs quantum dots are well aligned at the multi-atomic step edges. The diameter of the quantum dots is less than 20nm

4. Fabrication of seeded self-assembling quantum dots

In order to overcome the difficulty to control the position as well as the size fluctuation, the selective growth technique is still promising. By forming an appropriate pattern which corresponds to seeds prior to the growth, the position and growth kinetics can be controlled. Previously we achieved the GaAs quantum dots on AlGaAs plinths which were grown on SiO2 patterned substrate with a lateral width of 25nm[12]. The masks consist of 100nmx100nm windows with a period of 140nm. First, $Al_{0.4}$ $Ga_{0.6}$ As plinths are formed on the SiO_2 masks. Then, the GaAs is grown on the top of the AlGaAs plinths, followed by the growth of $Al_{0.4}$ $Ga_{0.6}$ As so that the GaAs quantum dots are embedded by $Al_{0.4}$ $Ga_{0.6}$ As.

On the other hand, the growth of quantum wires in V-groove structures has been successful[13,14]. Therefore, as a natural extension of this technique, the growth of quantum dots in two-dimensional V-grooves (2DVG) should be investigated. However, such two-dimensional V-grooves are difficult to be formed on (100) substrate using chemical etching technique. We succeeded in growing quantum dots of a size of 10nm in the V-grooves which are formed by epitaxial growth on patterned substrates. The principle of the formation of the 2DVG is quite similar to that of the one-dimensional V-groove developed by our group[14]. The only difference is that the SiO_2 pattern is a two-dimensional arrays instead of one-dimensional lines. Figure 3 shows a SEM micrograph of cross section of a GaAs quantum dots grown in the 2DVGs. The cross sectional plane is perpendicular to (0-11) orientation. A similar structure was also observed when the cross sectional plane is perpendicular to (011). Just like the quantum wires in the V-groove structures, the quantum dots can be formed in the 2DVG utilizing the difference of the growth rate along (100)orientation from that along (111)A,B orientations. The minimal size of the quantum dots with well controlled position around 10nm.

5. Low temperature near-field spectroscopy of quantum dots

Nanometer-scale probing technique is important to investigate dynamics of electrons and photons in the quantum nanostructures. For this purpose, the near-field spectroscopy technique is one of promising tools. In this section, we discuss our recent progress on nanoprobing technique for the quantum dots using a low-temperature near-field spectroscopy.

The sample used for the measurement is GaAs quantum dots which are fabricated

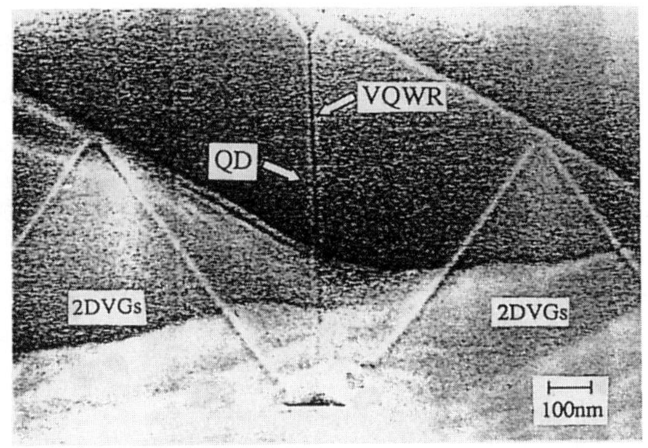

Fig.3: Quantum dots grown on a two-dimensional V-groove.

by the selective epitaxial growth on SiO2-patterned GaAs (100) substrates. The difference of the growth rate of the crystal plane gives rise to three-dimensional confinement of the carriers. This growth technique has the advantage of its damage-free formation and a high regularity of the dots position arrangement. Near-field optical microscope is placed inside a He-flow cryostat. Chemical etched nanometric fiber tips are used 3. Since the quality of the near-field spectroscopic image depends not only on the spatial resolution but on the efficiency of illuminating the sample, it is not useful for the present study to fabricate the smallest aperture size. This trade off was overcome by applying selective resin coating method 4 which is one of the advantageous methods to optimize the size of the aperture. In this experiment, the tips with the aperture sizes from 100nm to 200nm were chosen by considering the size of dot.

The photoluminescence (PL) spectrum was shown in Fig. 4 (a). The image was obtained by positioning the tip 200nm above the top of a quantum dot. This provided us with the PL spectrum from the carriers excited in the whole QD structure. Figures 6 (b), (c) and (d) show the spectrally-resolved optical images corresponding to each energy region indicated by arrows labeled A, B and C in Fig. 4 (a). In order to map the luminescent region, simultaneously monitored shear force image was taken (Fig.4 (e)). These images indicate clearly that energy regions A, B and C in Fig. 4 (a) are originated from the GaAs bulk, the GaAs quantum dot and the GaAs quantum well (QW), respectively. It should be noted that no luminescent sites are appeared in the QD region in Fig.6 (b). From this result, we can say that the carriers diffuse and are captured effectively in the quantum dot and the QW regions. Monitoring the carrier diffusion and obtaining the effective excited region of the quantum dot can be carried out by varying intensity of the excitation light.

We also performed the NSOM measurement for the self-assembling InGaAs quantum dots. Figure 5 shows one example of the measure spectra. Many sharp spectral peaks come from individual quantum dots.

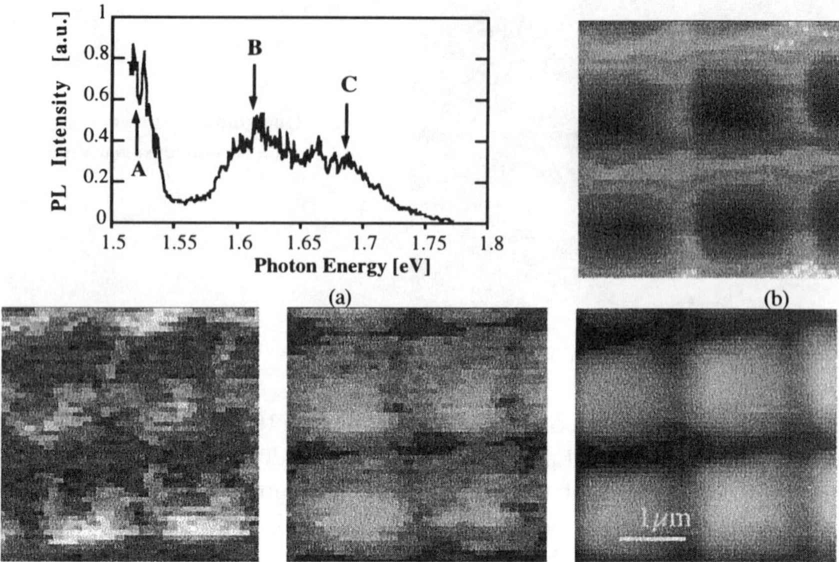

Fig. 4. (a)Spatially-resolved PL spectrum maintaining the tip with 200nm above the QD, and spatial profiles of spectrally-resolved PL images at 18K with the energy region (b) labeled A, (c) B and (d) C in (a). (e) Simultaneously observed topographic image.

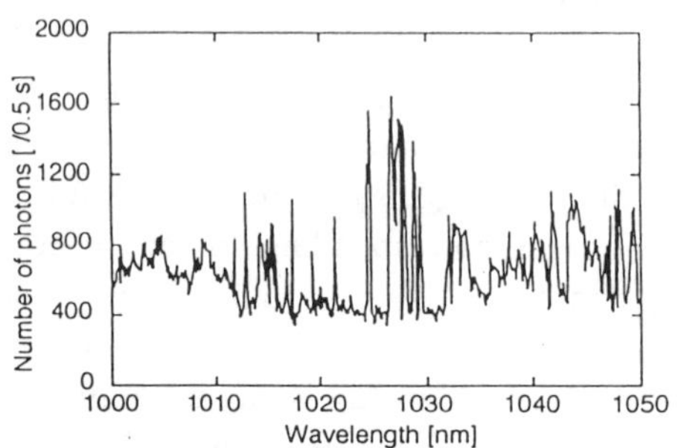

Fig.5: PL spectrum from self-assembling InGaAs quantum dots measured by NSOM

6. Fabrication of Vertical Microcavity Lasers with InGaAs/GaAs Quantum Dots Formed by Spinodal Phase Separation

The self-assembling growth using the SK mode has been already applied to a trial of fabrication of the quantum dot lasers. In addition, fabrication of a microcavity (λ-cavity) quantum do laser in which photons are also confined was also demonstrated utilizing the

SK mode by our group. In this section, we demonstrate a novel self-assembling growth mode for the quantum dot formation utilizing the temperature gradient (TG) method through the spinodal phase separation (SPS). Moreover, we succeeded in fabricating a vertical microcavity quantum dot laser with lasing oscillation at 20K by an optical pumping.

In the SPS growth mode, the quantum dots are grown in InGaAs layers with a nominal In-content as low as x=0.03. For the growth of the quantum dots the temperature was lowered from the initial temperature of 550° C by 100° C during the deposition of InGaAs, with a rate of temperature decrease of 0.28° C/sec. This TG method leads to phase separation into island-like structures at the growth front.

Using the SPS growth mode, we succeeded in fabricating a vertical microcavity laser. A diagram of the sample structure is shown in Fig. 6. The microcavity consists of an InGaAs quantum dot layer, located between two AlAs/Al0.2Ga0.8As distributed Bragg- reflector mirrors that are separated by a 4λ spacer (λ=950 nm). The number of layer pairs is 20 in the upper mirror and 23.5 in the lower one. The quantum dots are located at an antinode of the resonant mode in order to enhance the coupling of the spontaneous emission into the lasing mode. AFM pictures yield clear evidence for the formation of quantum dots. Fig. 7 shows an AFM image of a sample in which the growth was interrupted after 10 nm. The etched surface displays quantum dots with an average diameter of ~24 nm and an area density of ~2.2 X 10^{10}/cm^2. The full width with half maximum of photoluminescence spectrum was 25meV, which is much narrower than that by the SK growth mode.

In order to investigate the lasing activity of the structures, we optically pumped used a PL setup with a Ti:sapphire laser at 20K. Figure 8 shows light input-output characteristics. The result clearly demonstrate a lasing oscillation in the microcavity quantum dot lasers.

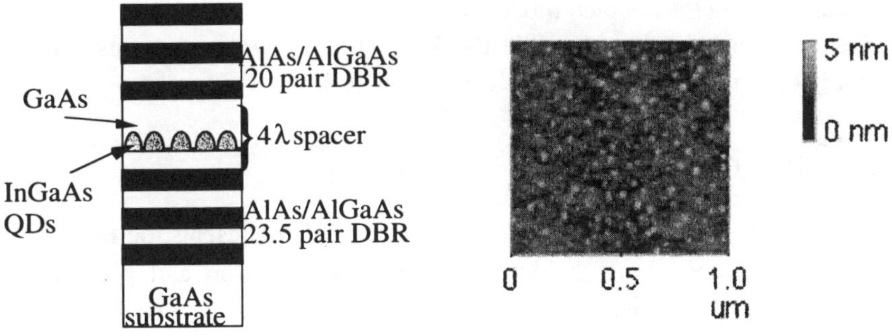

Fig. 6: Sample structure

Fig.7: AFM image of quantum dots

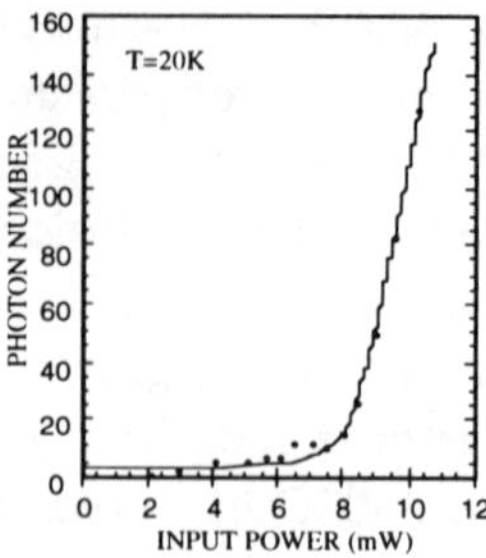

Fig. 8: Input-Output characteristics

7. Phonon bottleneck problem: Is it a serious problem ?

Many efforts have been devoted to realize quantum dot structures because significant improvement of lasing characteristics is expected[1]. However, the phonon bottleneck problem was proposed in the quantum dot structures although there has been no experiment which supports this prediction. In this paper, we theoretically discuss the interaction of electrons with LO phonons using the coupled mode equations in a system in which many LO phonon modes and quantized electronic states are coupled. The results indicate that electrons can interact with LO phonon in a wide energy range (~50meV), which implies that the bottleneck problem can be ignored in the quantum dot lasers. On the basis of these results, threshold current and modulation dynamics are discussed.

In the discussion of the phonon bottleneck problem, Fermi's golden rule is usually used for the relaxation rate. However, we found that a simple use of Fermi's golden rule is not correct because the interaction between electrons and LO phonon is strong. In fact, the uncertainty principle is not satisfied in the relation between the relaxation rate and the possible energy range of the transition. Here, we directly solve the time dependent Shroedinger equation using the coupled mode equations.

The electron transition rate through the interaction between electrons and LO phonons is usually described by Fermi' s Golden rule as follows,

$$\tau^{-1} = \frac{2\pi}{\hbar} |\langle i|H'|f\rangle|^2 \rho(E)\delta(E_i - E \mp \hbar\omega)dE \tag{1}$$

where $\rho(E)$ is the density of states function at final state and H' represents the Froehlich's perturbation hamiltonian describing the interaction between electron and polar optical phonon. The initial state 'i' and the final state 'f' are assumed as (1, 1, 1) and (1, 1, 2), respectively, in a cubic quantum dot with infinite potential barrier. The transition rate calculated as a function of level spacing ΔE of the electronic states using eq.(1) is shown in Fig. 9. The rate has a peak where the level spacing ΔE is 36.43meV and decreases to zero immediately in both higher and lower energy sides. The value of the highest rate is 10^{16} 1/sec which is much larger than that of carrier recombination. This result shows that phonon bottleneck is unavoidable unless the level spacing is adjusted to this energy, because the full width at half maximum (FWHM) of the spectrum is only 0.01meV. However, this situation is not satisfying the uncertainty relation. Since Δt ~ 10^{-16} and ΔE ~ 1.9×10^{-21}, the product is much smaller than h. This contradiction results from the use of the Fermi's Golden rule under the condition where electrons and LO-phonons are strongly coupled.

Here, we started the analysis using the time dependent Schrodinger equation. The |a> and |b> are the initial and final states of electron, respectively. The |{n}> represents the state where the n-th confined LO phonon mode is excited, while |{0}> represents there is no excited LO phonon. The quantum states is totally described as a linear combination of these states using a time dependent coefficient function C(t),

$$\psi = C_{a,\{0\}}(t)|a>|\{0\}> + \sum_n C_{b,\{1_n\}}(t)|b>|\{1_n\}> \qquad (2).$$

By substituting the wave function of eq.(2) into the Schrodinger equation we obtain coupling equations between coefficient functions.

$$\begin{aligned} &\frac{\partial}{\partial t}C_{b,\{1_n\}}(t) = -ig_n^* \exp(i(\omega_n - \Delta\omega)t)C_{a,\{0\}}(t) \\ &\frac{\partial}{\partial t}C_{a,\{0\}}(t) = -\sum_n ig_n \exp(-i(\omega_n - \Delta\omega)t)C_{b,\{1_n\}}(t) \\ &\hbar g_n = \langle i|H'|f\rangle \end{aligned} \qquad (3)$$

where g is a coupling constant and perturbation hamiltonian H' is assumed as Frolich's perturbation hamiltonian described for confined phonon mode. $h\omega_n$ is the energy of the n-th confined phonon mode which is assumed to be given by the results for the bulk mode[8]. The hΔω is equal to the level spacing ΔE in electron states.

The transition rate spectrum calculated with and without considering the dissipation is shown in Fig. 9, where the conditions are the same as that used in Fig. 10. The solid curve is the spectrum of the intrinsic rate tint by defining the tint as the mean time of the transition rate calculated from first half cycle of Rabi oscillation in resonance. The dashed line is the rate calculated assuming LO phonon dissociation rate is 4×10^{11} which is the value at 300K.

The intrinsic rate spectrum has a peak of 6x1012 1/s when level spacing DE is 15meV. The rate spectrum assuming LO phonon dissociation has a peak of 2×10^{11} 1/s when ΔE = 36meV and the FWHM is 15meV. In both case the relation between maximum and FWHM satisfies the uncertainty relation. Moreover, the spectrum keeps larger rate than carrier recombination rate in wide range (even the case with phonon dissociation, it is about ~100meV). Thus, the interaction between electron and LO phonon can be regarded as one of the effective carrier path from barrier to ground state in smaller quantum dots.

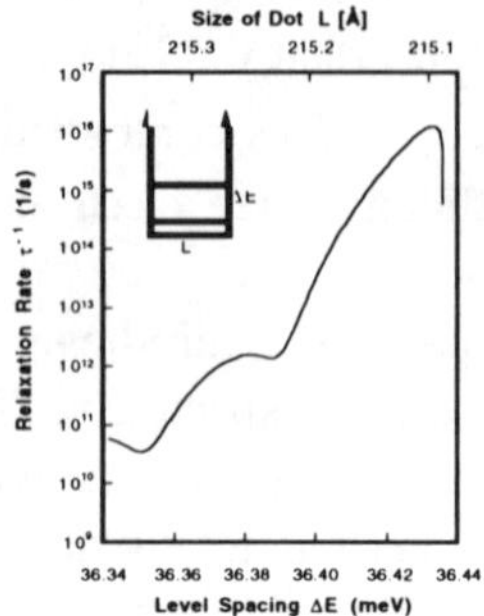

Fig.9: Carrier relaxation rate calculated using Fermi's golden rule as a function of level spacing ΔE in electron levels..

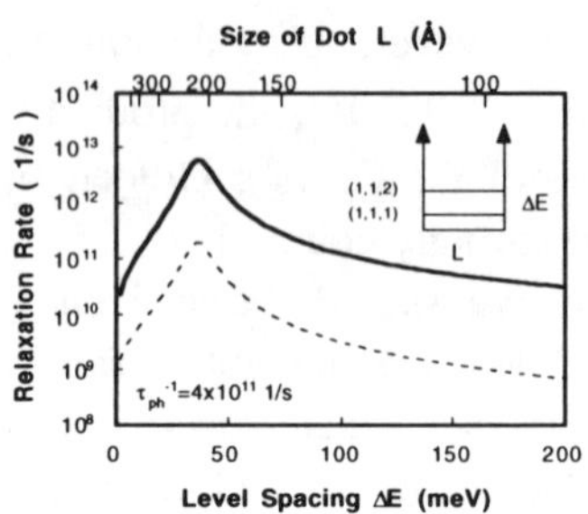

Fig.10: Electron relaxation rate calculated by Schrodinger equation as a function of the level spacing between quantized energy levels

8'. Conclusion

We discussed fabrication of the self-assembling quantum dot structures using the SK growth mode. Although it is easy to fabricate the structures under an appropriate growth condition, the randomness of position and the size fluctuation are serious problems. Here we demonstrated a trial to realize self-alignment of the quantum dots without any lithographic pre-processing. On the basis of these results, as the first step toward the ultimate semiconductor lasers in which both electrons and photons are fully quantized, a vertical microcavity quantum dot lasers was demonstrated. Finally we discussed the interaction between electrons and LO phonons theoretically by solving the time dependent Shroeinger equation through the coupled equation. The results indicate that the phonon bottleneck problem can be ignored because a strong interaction is realized in a wide energy range.

The author would like to thank M. Nishioka, S. Ishida, H. Nakayama, M. Kitamura, F. Sogawa, R. Schur, Y. Nagamune for their intensive collaborations on the research described in this paper.

References

[1] Y. Arakawa and H. Sakaki, Appl. Phys. Lett. 40, 939 (1982).
[2] I. N. Stranski and Von L. Krastanow, Akad. Wiss. Lit. Mainz Math.-Natur. Kl. IIb 146, 797 (1938).

[3] C. W. Snyder, B. G. Orr, D. Kessler and L. M. Sander, Phys. Rev. Lett. 66,3032 (1991).
[4] S. Guha, A. Madhukar and K. C. Rajkumar, Appl. Phys. Lett. 57, 2110 (1990).
[5] D. Leonard, M. Krishnamurthy, C. M. Reaves, S. P. Denbaars and P. M. Petroff, Appl. Phys. Lett. 63, 3203 (1993).
[6] J. Oshinowo, M. Nishioka, S. Ishida and Y. Arakawa: Appl. Phys. Lett 65 1421.(1994)
[7] E. Yablonovitch, T. J. G. Gmitter, and R. Bath, Phys. Rev. Lett. 61,2546 (1988).
[8] G. Bjoerk and Y. Yamamoto, IEEE J. Quantum Electron. 27, 2386 (1991).(1995) 3663.
[9] C. Weisbuch M. Nishioka, A.Ishikawa, and Y. Arakawa, Phys. Rev. Lett. 69, 3314 (1992).
[10]T. Norris, J-K. Rhee, C-Y. Sung, Y. Arakawa, M. Nishioka, and C. Weisbuch: Phys. Rev. B, 50, 14663 (1994)
[11] Y. Arakawa, Extended Abstracts of the 1990 International Conferenceon Solid State Devices and Materials (The Japan Society of Applied Physics,Sendai 1990), p. 745
[12] Y. Nagamune, S. Tsukamoto, M. Nishioka, and Y. Arakawa, Appl. Phys. Lett. 64, 2495 (1994).
[13] S. Tsukamoto, Y. Nagamune, M. Nishioka, and Y. Arakawa, Appl. Phys. Lett., 63, 310 (1993)
[14] H. Benisty, C. M. Sotomayor-Torres and C. Weisbuch, Phys. Rev., B44, 10945 (1991)

InGaAs/GaAs QUANTUM DOT LASERS

D. BIMBERG, N. KIRSTAEDTER, N.N. LEDENTSOV
TU-BERLIN
HARDENBERGSTR. 36, 10623 BERLIN
GERMANY
ZH.I. ALFEROV, P.S. KOP'EV, V.M. USTINOV, S.V. ZAITSEV, M.V. MAXIMOV
IOFFE PHYSICAL-TECHNICAL INSTITUTE
POLITEHNICHESKAYA 26, 194021, ST.PETERSBURG
RUSSIA

1. Introduction

Quantum dot (QD) lasers are expected to have superior lasing characteristics compared to quantum well (QW) and quantum well wire lasers [1], [2]. Due to the three dimensional confinement of carriers in a quantum dot with a size below or equal to the exciton Bohr radius the density of states becomes a delta function [3]. Consequently no thermal or excitation dependent broadening of the gain function is possible. QD lasers with such an atomic like density of states are expected to show ultra low threshold current densities, ultrahigh temperature stability of threshold current, ultrahigh differential gain and chirpfree operation under direct current modulation. Potential device applications therefore range from high power semiconductor lasers to high speed light sources for fiber based data transmission.
In the last decade two different approaches have been explored to fabricate quantum dots. The first one originally developed for quantum wire lasers uses different variations of patterning techniques either prior to growth for selective growth or after growth (e.g. chemical etching with following overgrowth). Using patterning techniques InGaAs/InGaAsP QD lasers were successfully realized, but they still show high threshold current density of 7.6 $kAcm^{-2}$ at 77K [4]. The second approach simply uses strain induced self-organisation of InGaAs/GaAs quantum dots [5], [6] with threshold current densities as low as ~60Acm^{-2} at room temperature [7]. Self-organisation is also very successfully applied to the growth of InP on GaInP and GaSb or InSb on (100)-GaAs resulting in the formation of 3D-islands. Despite the fact that the growth on (311)B-GaAs [6] also shows the formation of large quantum dots with a disk like shape their diameter of 60nm is too large to show clear QD effects. Since the first presentation of a QD injection laser [5] based on self-organised dots as gain medium large efforts were undertaken to decrease the threshold current density to its ultimate limit determined by the transparency condition of the dot states. For this purpose one has to increase the internal efficiency and the carrier capture rate into the dots which are

G. Abstreiter et al. (eds.), Optical Spectroscopy of Low Dimensional Semiconductors, 315–330.

considered to be the most serious limitations for QD lasers [8]. Since 1994 the threshold density has indeed decreased by 1½ orders of magnitude as visualized in Table (1).

TABLE 1. History of major steps in the decrease of threshold current density of QD lasers. The second column lists the dot and substrate material (QD/Sub), the number of quantum dot layers (dot layers) grown on top of each other, the type of dot formation (formation), the size of the dot baselength (dot size), the dot growth temperature (T_{growth}), the area dot density (dot density) and the growth reactor type (reactor). The third column lists the threshold at various temperatures and the fourth column the corresponding lasing wavelength and the type of the lasing transition. WL indicates lasing transition via wetting layer states, QD* via excited quantum dot states and QD via quantum dot ground state.

Year	Material: QD/Sub	Dot Size	Dot Layers	T_{growth}	Formation	Dot Density	Reactor	Threshold	Temperature	Wavelength	Transition	Reference
1994	InAs/GaAs	7nm	1 layer	(460-490)°C	self-organised	$4x10^{10}cm^{-2}$	MBE	$1.0kAcm^{-2}$	300K	0.95μm	WL	[5]
								$0.1kAcm^{-2}$	77K	0.95μm	QD*	
1994	InGaAs/InGaAsP	30nm	1 layer	…	etched	$2x10^{10}cm^{-2}$	MBE	$7.6kAcm^{-2}$	77K	1.26μm	QD	[4]
1995	$In_{0.5}Ga_{0.5}As$/GaAs	20nm	1 layer	510 °C	self-organised	…	MBE	$0.8kAcm^{-2}$	85K	0.92μm	WL	[9]
1996	InP/GaInP	25nm	1 layer	…	self-organised	…	MBE	$25kWcm^{-2}$ (optical excitation)	300K	0.7μm	…	[10]
1996	$In_{0.3}Ga_{0.7}As$/GaAs	…	1 layer	515°C	self-organised	$(2-3)x10^{10}cm^{-2}$	MBE	$0.5kAcm^{-2}$	300K	1.2μm	QD*	[11]
								$1.2kAcm^{-2}$	300K	1μm	QD*	
1996	$In_{0.4}Ga_{0.6}As$/GaAs	12nm	1 layer	(500-550)°C	self-organised	…	MBE	$0.65kAcm^{-2}$	300K	1μm	QD*	[12]
1996	$In_{0.5}Ga_{0.5}As$/GaAs	…	10 layers	485°C	self-organised	…	MBE	$0.06kAcm^{-2}$	300K	1μm	QD	[7]
1996	InAs	…	5 layers	475°C	self-organised	…	MOCVD	$1kAcm^{-2}$	300K	1.1μm	QD	[13]

The threshold is now approaching the physical limit ($\sim 2Acm^{-2}$) given by the actual dot size distribution (see section (3) on threshold characteristics) and there will be now a stronger focus on important fundamental properties of QD lasers like the temperature and excitation dependence of the gain spectra and in particular the modulation characteristics. The modulation bandwidth is to first order proportional to the square root of the differential gain. Since the differential gain is greatly enhanced in QD lasers [14] one expects a much higher modulation bandwidth. On the other hand the capture rate of carriers into the quantum dot states was suggested to be strongly reduced due to the phonon bottleneck effect [8]. By careful designing the energy difference between barrier and dot energy the relaxation rate might however exceed values of $\sim 10^{12}$ electrons per second permitting a modulation bandwidth up to 200GHz [15].

2. Realisation of QD Lasers

A schematic view of the band structure of a typical quantum dot laser is presented in Fig. (1). The ideal QD laser consists of a 3D-array of dots with equal size and shape surrounded by a higher band gap material which confines the injected carriers. The whole structure is embedded in an optical waveguide consisting of lower and upper cladding layers.

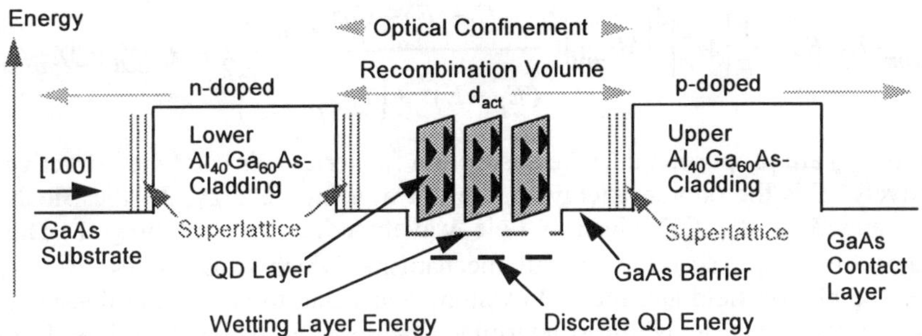

FIGURE 1. Schematic bandstructure of a quantum dot laser with self-organised dots under forward bias. A 3D-array of dots vertically aligned along the growth direction which is formed during the growth of multiple QD layers is illustrated schematically. Typically the dot area density in the (100)-plane is $4x10^{10}cm^{-2}$ and the dot size distribution is around 10%. The distance between the dot layers is 5nm and the real dot density in the recombination volume with a thickness of 200nm is $6x10^{15}cm^{-3}$ for three QD layers.

Spontaneous formation of QDs for currently operating QD lasers (see Table (1)) was successfully demonstrated at growth temperatures between 460°C and 550°C. The QDs are formed using a Stranski-Krastanov growth mode. The area dot densities range between $2x10^{10}cm^{-2}$ [11] and $1x10^{11}cm^{-2}$ [3] with a typical size distribution of ~10% [14]. The low growth temperature and the low dot density cause several problems concerning threshold and gain.

The cladding layer and the GaAs QD barrier grown at such low temperatures are a possible source for current leakage and nonradiative recombination. The QDs show intermixing with the surrounding barrier material if temperatures of 700°C are used to grow high quality cladding layers.

The limited dot area density sets an upper value for the QD modal gain. The modal gain can be increased by stacking several layers of QDs upon each other [16]. Using this approach ground state lasing [7] at room temperature has been demonstrated.

An ideal QD laser has only one electron level and one hole level. Several problems arise from fact that more than one hole state exists in typical $In_xGa_{1-x}As/GaAs$ QDs [17]. First the lasing wavelength shows a blueshift when increasing the injection level which is unwanted for light sources operated by direct current modulation. Second if excited hole states contribute to the gain the gain spectrum becomes asymmetric prohibiting chirpfree operation.

3. Gain of self-organised QD Lasers

Some of the most important properties which strongly influence the threshold current density and modulation bandwidth of a semiconductor laser are the material gain and the differential material gain. The material gain of a QD laser can be derived from the complex susceptibility χ of an ensemble of atoms interacting with a time-dependent electromagnetic field. In case the dots show only one electron and one hole level the susceptibility is as follows:

$$\chi_{Atom}(E_l,E) \propto \frac{1}{E}\left|M_b\right|^2\left|M_{env}\right|^2 \frac{(E_l - E) - j\hbar/\tau_{phase}}{(E_l - E)^2 + \left(\hbar/\tau_{phase}\right)^2} = \chi_{real} + j\chi_{imag} \quad (1)$$

χ_{imag} and χ_{real} are proportional to the absorption (material gain) and the refractive index, respectively. E is the photon energy, E_l the dot transition energy, M_b the Bloch matrix element and M_{env} the QD electron hole wavefunction overlap integral. The phase coherence time τ_{phase} accounts for all mechanisms which cause a loss of coherence between the photon field and the QD exciton. According to eq. (1) the lineshape of the gain is Lorentzian and its width is determined by the phase coherence time. In the case of self-organised dots the transition energies of different dots are not identical due to size or shape fluctuations. Thus the gain function is inhomogenously broadened and one can define a function $p(E)$ so that the probability of an dot having its emission energy between E_l and E_l+dE is $p(E_l)dE$. $p(E_l)$ is normalized such that the probability of finding a dot emission within an infinite energy range becomes unity. If $p(E_l)$ is multiplied with the ideal dot density of a 3D-dot array in the recombination volume one gets the density of states $D(E_l)$ for an inhomogenously broadened dot ensemble. The ideal dot density is the number of dots per volume where the number of dots is chosen as the total recombination volume divided by the dot volume (V_{dot}).

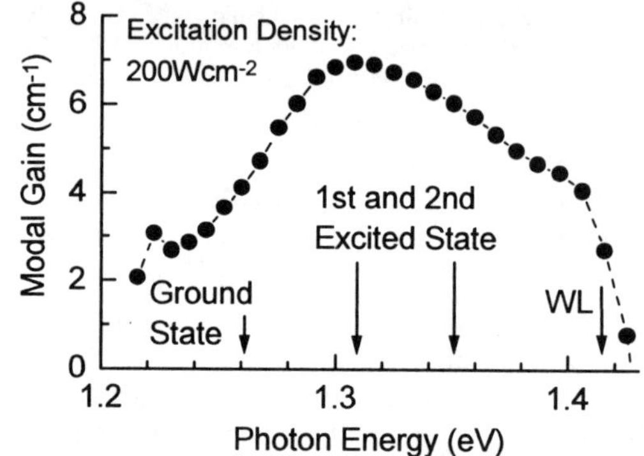

FIGURE 2. Modal gain spectrum of a single layer of quantum dots derived from single pass amplification. The transition energies of the QD ground state and excited states and the WL are marked.

$$D(E_l) = \frac{2}{V_{Dot}} \frac{1}{\sigma_{gain}\sqrt{2\pi}} \exp\left[-(E_l - E_{\max})^2 / \left(2\sigma^2_{gain}\right)\right] \quad (2)$$

The factor of 2 in eq. (2) accounts for the spin degeneracy. E_{max} is the maximum of the dot distribution. The width of the Gaussian is given by σ_{gain}. By combining eq. (2) and eq. (1) one gets the material gain g_{mat} for an inhomogenously broadened dot ensemble.

$$g_{mat}(E) = C_g \frac{1}{E} \int_{-\infty}^{\infty} \left|M_b\right|^2 \left|M_{env}\right|^2 D(E')\left[f(E', E_{fc}) - f(E', E_{fv})\right] L(E', E) dE' \quad (3a)$$

$$C_g = \frac{\pi e^2 \hbar}{m_0^2 \varepsilon_0 c_0 n_r} \quad (3b)$$

$$L(E_l, E) = \frac{1}{\pi} \frac{\hbar/\tau_{phase}}{(E_l - E)^2 + \left(\hbar/\tau_{phase}\right)^2} \quad (3c)$$

The Fermi functions f determine the occupation probability of electrons and holes in single dot states. If the minimum spectral distance between transitions from two different dots is much smaller than the single dot emission line width (determined by the phase coherence time) the gain spectrum of eq. (3) is determined by the density of states of the dot ensemble. Since the total number of dots in the recombination volume of a typical ridge-waveguide laser exceeds a value of $\sim 10^6$ the very narrow gain spectrum of a single dot transition (demonstrated by Grundmann *et al.* [3]) transforms to a broad gain spectrum as shown in Fig. (2).

Due to the lateral separation of the dots carriers in different dots are thermally coupled only via wetting layer (WL) and GaAs barrier states [16]. This fact is directly proved by the intensity ratio of QD and WL electroluminescence (EL) at different temperatures and constant injection level which follows an Arrhenius dependence (see Fig. (3b)). The intensity ratio fits to an activation energy for the QD exciton to the WL of 25meV. It was suggested however, by Moritz *et al.* [10] that the carriers in InP/InGaP QDs and the WL are not in thermal equilibrium with each other.

Based on the structural parameters of the laser and the transition energies of ground and excited states one can calculate the material gain spectra according to eq. (3). Since the Bloch matrix element and the overlap integral are not exactly known in the case of a QD laser one has to calibrate the gain spectra [14] to absolute values by using the well known equilibrium value of the absorption coefficient of bulk GaAs.

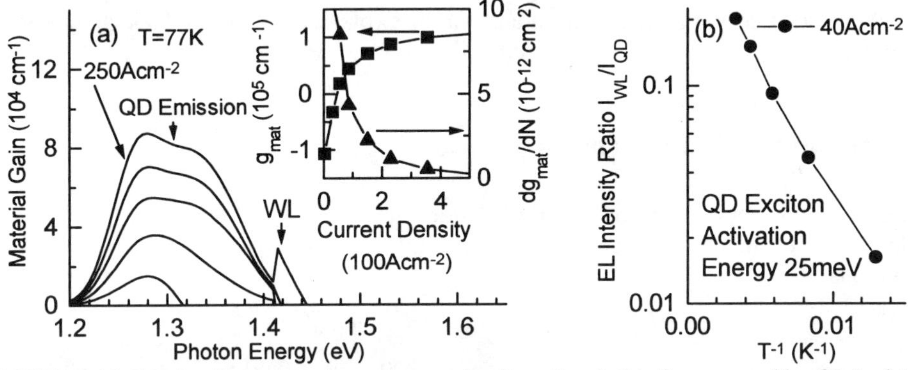

FIGURE 3. (a) Calculated material gain spectra as a function of excitation for an ensemble of InAs dots having a base length of 7nm and a size distribution of 13%. The WL is assumed to be a ~0.5nm thick single quantum well. The inset shows the peak material gain (g_{mat}) at ~1.28eV and the differential material gain (dg_{mat}/dN) as a funcion of current density. (b) Arrhenius plot of WL- and QD electroluminescence (EL) well below threshold current density proving thermal equilibrium of carriers between QD and WL states.

The material gain spectra in Fig. (3) are asymmetric due to the contribution of excited QD states. The overlapping gain functions of ground and excited states enhance the material gain to a value of $9\text{x}10^4\text{cm}^{-1}$ at $\sim 250\text{Acm}^{-2}$ in the single layer QD laser investigated here. This value is one order of magnitude larger than for a 8nm $In_{0.2}Ga_{0.8}As$/GaAs single QW laser [14] and in agreement with the prediction of Asada *et al.* [2]. The calculation of the material gain versus current density in Fig. (3a) is based on the experimentally determined decay time of carriers τ_{eff} in the recombination volume (see Fig. 1) in order to calculate the carrier density N in that volume from the injection current density J [18].

$$N = \frac{\tau_{eff} J}{ed_{act}} \sum_{i=0}^{n} C_i^{QD} f(E_i^{QD}, E_{fc,fv}) \int D_i^{QD}(E')dE' + \frac{\tau_{eff} J}{ed_{act}} C^{WL,GaAs} \int D^{WL,GaAs}(E') f(E', E_{fc,fv}) dE' \quad (4)$$

Here d_{act} is the thickness of the recombination volume (see Fig. (1)), C is a factor which accounts for the different volumes occupied by QD, WL and GaAs[14], D is the density of states, f is the Fermi function and e the electron charge.
The current density J_{sat} for the onset of gain saturation shown in Fig. (3a) is much larger than the value estimated from the laser rate equation [19] in the case that all carriers are confined to the QD states.

$$J_{sat} = 3\frac{2\rho e}{\tau_{QD}} \quad (5)$$

Here ρ is the dot area density and τ_{QD} is the exciton decay time of the InAs QDs. The factor of three accounts for one ground state and two excited states. From eq. (5) the gain saturation level (assuming contributions by one ground and two excited states) is $19 \mathrm{Acm}^{-2}$ for a dot area density of $4 \times 10^{10} \mathrm{cm}^{-2}$ and a QD decay time of 2ns [14]. This saturation level is exceeded in real QD laser structures by one order of magnitude. This effect is discussed in the next section and can be explained by current leakage due to the population of barrier states (WL, GaAs) and nonradiative recombination.

4. Threshold Characteristics

From the gain versus current density relation and the lasing condition one can easily calculate the threshold (transparency) current density J_{thr} (J_{tr}) by the following equations.

$$\Gamma g_{mat}(J_{thr}) = \alpha_{tot}, \qquad J_{thr} = J_{thr}^{QD} + J_{thr}^{leak} + J_{thr}^{barr} \quad (6a)$$

$$\Gamma g_{mat}(J_{tr}) = 0, \qquad J_{tr} = J_{tr}^{QD} + J_{tr}^{leak} + J_{tr}^{barr} \quad (6b)$$

Here J_{thr}^{QD} (J_{tr}^{QD}) is the fraction of the current density injected into QDs and contributing to radiative recombination, J_{thr}^{bar} (J_{tr}^{bar}) is the fraction of the current density from the barriers contributing to radiative recombination and J_{thr}^{leak} (J_{tr}^{leak}) the fraction of the current density contributing to the leakage current at threshold (transparency). α_{tot} is the sum of internal loss and mirror losses (see eq. (10)). Although the maximum material gain of the QD ensemble reaches a value of $10^5 \mathrm{cm}^{-1}$ according to eq. (3) the confinement factor Γ defined as

$$\Gamma = \int_{\text{exciton volume}} E_\nu^2(r)dr \Big/ \int_{\text{laser volume}} E_\nu^2(r)dr = \Omega \frac{V_{exc}}{V_{phot}} \quad (7)$$

is rather low for a single layer QD laser. Here Ω denotes the number of dots in the recombination volume (defined in Fig. (1)), V_{phot} is the integral over the laser volume of the electric field vector $E_v(r)$ of the optical mode which is normalised to unity at its maximum value. Using a typical dot lateral coverage of 2% in the (100)-plane (dot area density of $4x10^{10}cm^{-2}$, dot base length 7nm) and the QD exciton volume V_{exc} (~$80nm^3$) [14] the confinement factor for a single layer dot laser structure is estimated to be ~$1x10^{-4}$. Thus for a maximum material gain of ~10^5cm^{-1} a typical single layer quantum dot laser structure is expected to have a modal gain of ~$10cm^{-1}$ which is just sufficient for laser action in long cavity lasers. Short cavity lasers require a higher dot volume density. A way to overcome this limitation is the introduction of multiple layers of dots [16], [7].

The maximum material gain g_{mat}^{max} is calculated from the steady state rate equation with the radiative current density J^{QD} (assuming that all injected carriers are 100% confined in the dots and no leakage current exists) and the population inversion factor Λ.

$$g_{mat}^{max}(E,\Lambda)=\frac{\pi e^2\hbar}{m_0^2\varepsilon_0 c_0 n_r}\frac{1}{E}\frac{2}{V_{Dot}}\frac{1}{\sqrt{2\pi}\sigma_{gain}}|M_b|^2\Lambda \tag{8a}$$

$$\Lambda=\frac{4N_{exc}^{QD}-2N_{3D}^{QD}}{2N_{3D}^{QD}} \tag{8b}$$

$$N_{exc}^{QD}=\frac{\tau_{QD}J^{QD}}{ed_{act}} \tag{8c}$$

Here N_{3D}^{QD} is the dot density and N_{exc}^{QD} is the density of QD excitons calculated from the steady state rate equation using the QD carrier decay time τ_{QD} and the thickness d_{act} of the recombination volume (see Fig. (1)). Using eq. (8) and the laser condition (see eq. (6)) one can calculate the threshold current density as a function of the dot density (Fig. (4)) in the case the current is injected by 100% into the dots and only one dot energy level exists for electrons and holes.

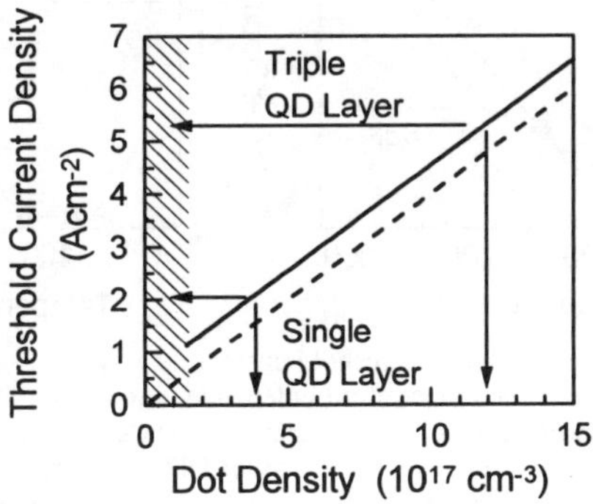

FIGURE 4. Threshold current density (solid line) as a function of the dot density. The calculation is done for a typical laser structure shown in Fig. (1) with an active layer thickness of 200nm and total loss of $8cm^{-1}$. The threshold current densities for single and triple QD layer lasers are marked. The transparency current density is also shown (dashed line). The dashed area marks the forbidden region due to gain saturation.

For a total loss of $8cm^{-1}$ (e.g. for a long laser cavity of 2mm and internal loss of $\sim3cm^{-1}$) the threshold current density for a single layer QD laser structure with a gain width of ~50meV and a QD carrier decay time of 2ns is as low as $2Acm^{-2}$. Thus even for a QD laser with typical dot size distribution of ~10% and dot area density of $4x10^{10}cm^{-2}$ the minimum threshold current density neglecting any current leakage is more than one order of magnitude lower than in high quality QW lasers which exhibit lowest threshold current densities of $50Acm^{-2}$ at 300K [20].
From eq. (8) we see that longer decay times (i.e. lower oscillator strength) which are expected for dots with size smaller or equal the exciton Bohr radius decreases the transparency current. But simultaneously the reduced oscillator strength reduces the maximum gain thus increasing the threshold current to maintain the threshold gain. Note that the transparency current is proportional to the dot volume density and also to the number of dot layers since the carriers in the QDs are not thermally coupled.

4.1. TEMPERATURE DEPENDENCE OF THRESHOLD AND LOSS MECHANISM

A nonequilibrium distribution of carriers between all dots in a dot ensemble should prohibit thermal broadening of the gain spectrum. If this condition is strictly fullfilled the peak gain should remain constant as a function of temperature (*T*) at constant injection level. Therefore the phenomenological description of the threshold temperature dependence according to the following equation results in an infinite characteristic temperature T_0.

$$J_{thr} \propto \exp(T/T_0) \tag{9}$$

The first observation of an enhancement of T_0 due to carrier confinement was reported by Arakawa *et al.* [1] for bulk lasers placed in a magnetic field. A dramatic increase of T_0 for a true single layer QD laser up to 425K is shown in Fig. (5).

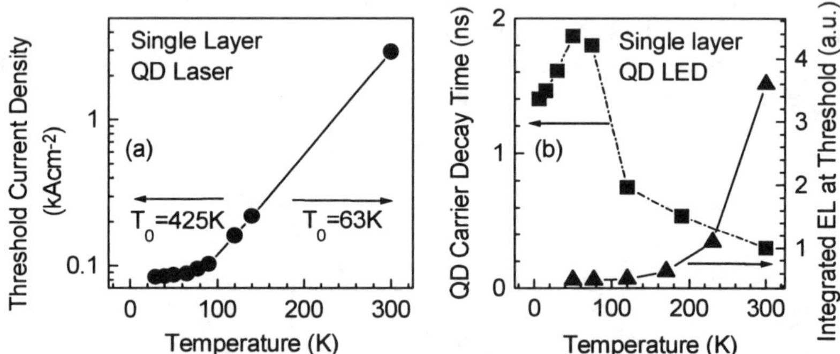

FIGURE 5. (a) Temperature dependence of threshold current density for a single layer QD laser. The large T_0 between 50K and 100K reveals the temperature independent peak gain of the QD ensemble. (b) Above 100K barrier population and nonradiative centers decrease the decay time of carriers in the QD ground state. The integrated electroluminescence (EL) intensity of QD, WL barrier and GaAs barrier at threshold unambiguously indicates the onset of current leakage above 100K resulting in the decrease of T_0.

Unfortunately in these structures the breakdown of the nonequilibrium carrier distribution and the temperature dependent current leakage reduces the value of T_0 at temperatures above 100K.

It was discussed in the last section that the carriers in the dots are thermally coupled to the WL and the GaAs barrier. Since the activation energy for a QD exciton (confined to a 7nm InAs/GaAs dot) to the WL is only ~25meV (see Fig. (3)) one can expect two loss mechanisms.

First of all with increasing temperature the injected carriers start to populate the barrier states thus increasing the injection current to maintain the threshold gain of the QD laser. Secondly nonradiative recombination in the barrier might increase current leakage. This leakage current is related to the quality of the expitaxial barrier layer (in particular the low temperature GaAs) and does not present a principal limitation of the threshold current.

It has been shown [21] that higher growth temperatures of the cladding layers and optimization of the dot size (leading to an increase of the QD exciton thermal activation energy) shifts the onset of current leakage to 220K with an ultrahigh T_0 of 530K between 80K and 220K. Another way to decrease the leakage current is to place the dot layers in a GaAs/AlGaAs QW thus decreasing the escape probability of carriers from the dots. With this approach the onset of current leakage has been increased up to room temperature while maintaining a high T_0 of 385K between 80K and 300K [22].

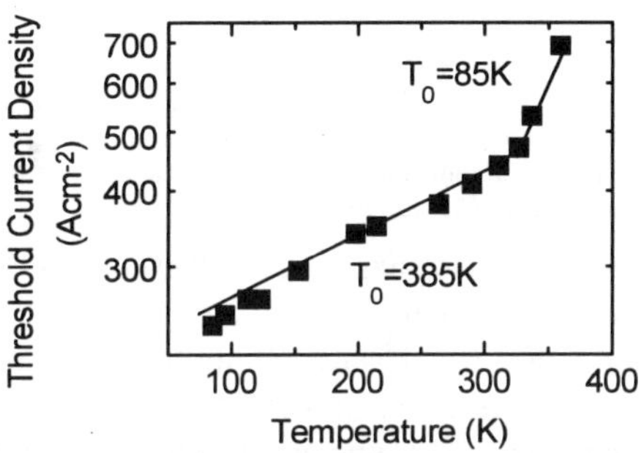

FIGURE 6. Threshold current density as a function of temperature for a single dot layer laser structure. The dots are placed in a 16nm wide GaAs/AlGaAs QW to reduce the escape probability of QD excitons into the barrier. This enhances T_0 up to 385K while maintaining a low threshold current density of ~400Acm^{-2} at 300K.

An important factor characterizing the laser quality is the differential internal efficiency η_i of the laser. This value can be derived from the dependence of the external differential quantum efficiency η_{ext} on mirror losses α_{mirr} according to the following equation.

$$\frac{1}{\eta_{ext}} = \frac{1}{\eta_{\text{int}}}\left[\frac{a_{\text{int}}}{a_{mirr}} + 1\right] \quad (10a)$$

$$a_{mirr} = \frac{1}{L}\ln\left(\frac{1}{R}\right) \quad (10b)$$

Here R is the laser facet reflectivity, L the cavity length and α_{int} the internal loss of the laser. The presently observed differential internal efficiency of a single layer QD laser is about 40%. This value was improved to 50% by introducing multi layer QD structures (see Fig. (7)). Values of 70% for a $In_{0.5}Ga_{0.5}As$/GaAs QD laser and of 81% for a $In_{0.3}Ga_{0.7}As$/GaAs QD laser were recently reported by Mirin *et al.* [11] and Ustinov *et al.* [23]. These values are still lower than the recently reported internal efficiency of QW lasers of 99% [24].

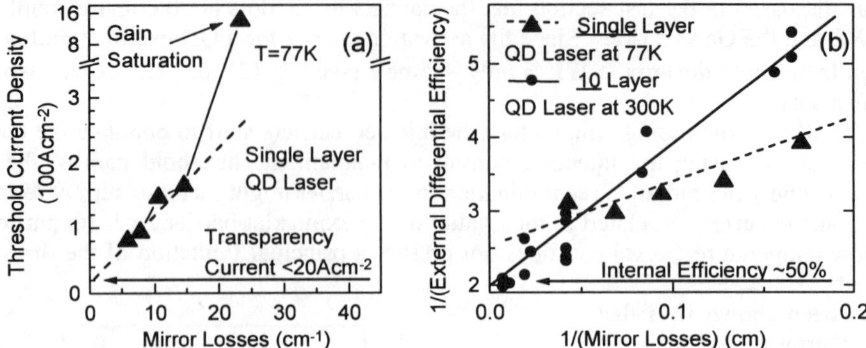

FIGURE 7. (a) The threshold density versus mirror losses yields a transparency current density below $20Acm^{-2}$ and demonstrates clearly gain saturation for small cavity length. (b) External differential efficiency for single and multiple layer QD laser as a function of mirror losses. The maximum internal efficiency obtained is about 50% at 300K.

The more typical internal efficiency around 50% of current QD lasers demonstrates that the difference between the calculated threshold current density of $\sim 2Acm^{-2}$ (see Fig. (4)) for an quasi ideal QD laser (with no leakage current and one electron and one hole state) and the measured value of $\sim 100Acm^{-2}$ for a single layer QD laser having both a total loss of $8cm^{-1}$ is still connected to current leakage. On the other hand the transparency current density for a single layer QD laser is below $20Acm^{-2}$ indicating that ultralow threshold current densities are possible.

5. Spectral Characteristics

The emission of QD lasers consisting of a dot ensemble with only one energy state for electrons and holes is expected to occur at the maximum of the PL emission and should not show any energy shift with injection current. In the case of real structures emission from both ground and excited states is often observed. Thus the line shape of the gain spectrum is no longer independent of the injection level causing a strong blueshift of the laser emission with injection current. This blueshift is demonstrated in Fig. (8). Obviously excited QD states provide higher modal gain. Upon complete QD gain saturation the laser emission jumps to the WL state which has a much higher modal gain due to the larger confinement factor. One of the most interesting features of a QD laser is its potential to produce longitudinal single mode operation when the gain spectrum is more narrow than the longitudinal mode spacing defined by the laser cavity length. None of the currently existing QD lasers exhibit a gain width below 1meV which is necessary to fullfill this condition for stripe geometry lasers with a typical laser cavity length of ~0.5mm. For vertical cavity surface emitting lasers (VCSEL) there is a much larger chance to obtain single mode lasing.

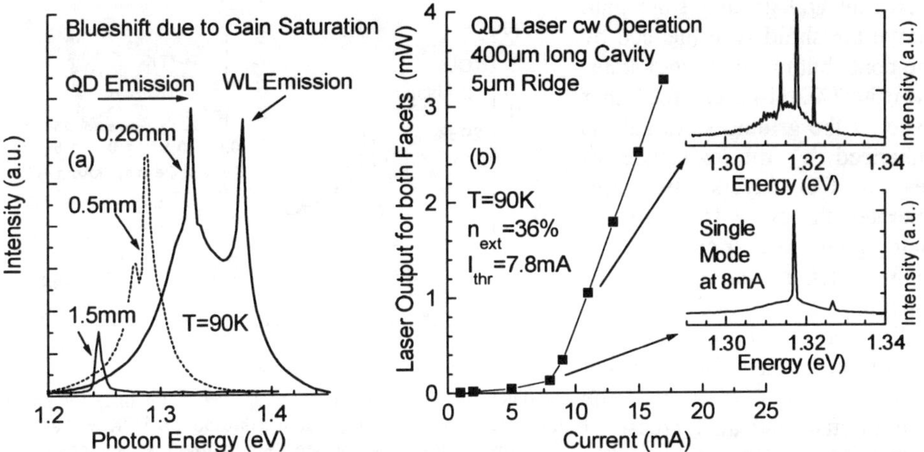

FIGURE 8. (a) Blueshift of the QD emission due to state filling. Each spectrum is marked with the corresponding laser cavity length. For short cavity lasers the QD gain is completely saturated and the emission jumps to WL states. (b) Light current characteristic of a 6 layer QD laser with a 400μm long and 5μm wide laser cavity showing single mode operation just above the threshold.

The full width half maximum of the QD PL ranges from 33meV [12] to 50meV [14] which is sufficiently narrow for VCSELs to show single mode operation. Nevertheless the experiment on Fabry-Perot lasers show an unexpected single mode operation just above the threshold (see Fig. (8)). For QW lasers it is well known that they can switch to one single mode at injection levels far above threshold despite their broad gain spectra. Two possible mechanism might be responsible for the single mode operation reported here. Due to inhomogenities of dot distribution the gain spectra may have local minima near the main gain maximum. This effect will effectively increase the threshold of other longitudinal modes near the main emission mode. Consequently the first suppressed mode will start lasing only at higher injection levels. In addition dots of different sizes might have different carrier capture times. Thus under stimulated emission conditions only dots which are rapidly refilled contribute.

The laser emission from QD states can be tuned by varying the $In_xGa_{1-x}As$ composition and by varying the size of the dots. With this approach the QD PL emission could be tuned between 0.95μm and 1.37μm at 300K [22]. Thus the first optical transmission window of glass fibres is already covered by emission from QDs grown on GaAs substrate. It was shown by several groups that $In_xGa_{1-x}As$ QD lasers in the composition range of $0.3 \leq x \leq 1$ can be grown by MBE [16], [7] but growth by MOCVD was thought to be more complicated because of the different growth kinetics as compared to MBE [25]. Recently we discovered a MOCVD growth window for InAs/GaAs QDs and demonstrated a five layer InAs/GaAs QD laser. The threshold at 77K is as low as $100 Acm^{-2}$ and lasing occurs via the QD ground state (see Fig. (9)). The threshold current density increases to $1.3 kAcm^{-2}$ at 293K and the laser emission is red shifted by ~45meV going from 77K to 293K which is only slightly lower than the InAs band gap change. In single sheet QD lasers state filling of excited QD and barrier (WL and GaAs) states leads to a carrier redistribution with increasing temperature and induces a strong blue shift of the laser emission [11].

Here the QD ground state gain above threshold is influenced by the contribution of excited states even at 77K [14]. On the other hand, if the ground state gain is enhanced by the existence of several QD layers the laser operates far away from ground state gain saturation and all higher lying energy levels, especially the excited states are well below the transparency condition. Therefore lasing occurs in this case via the ground state and an increase of temperature does not increase the occupation of the excited states above the transparency level. Then the QD laser emission should follow the band gap change of the dot material with temperature.

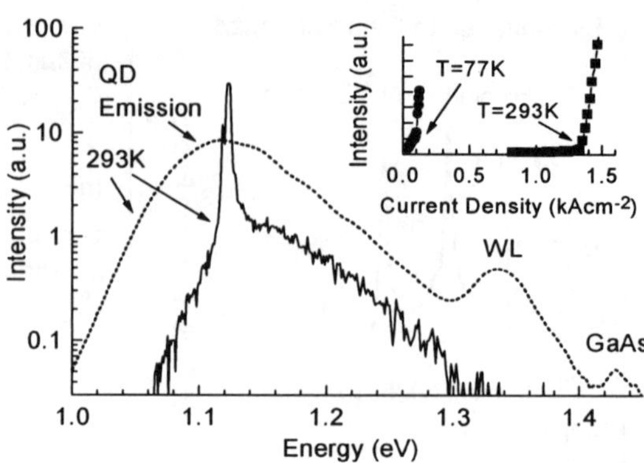

FIGURE 9. Electroluminescence spectra (solid line) for an InAs QD laser with 5 dot layers above threshold at 77K and at 293K. The inset shows the pulsed light current characteristic. The photoluminescence spectra (dotted line) at 293K indicates that laser emission is due to the QD ground state. WL and GaAs transitions are also marked.

6. Dynamic Properties and Modulation Bandwidth

The orders of magnitude larger differential gain of QD lasers should lead to an increased modulation bandwidth. Small signal analysis of standard rate equations yields an approximation of the modulation bandwidth defined as the 3dB value of the ratio of photon density S and current density J [26].

$$\frac{S(\omega)}{I(\omega)} \propto \left| \left(\frac{1}{1 + j2\pi\omega\tau_{cap}} \right) \frac{1}{4\pi^2\omega_R^2 - 4\pi^2\omega^2 + j2\pi\omega\gamma} \right| \tag{11a}$$

$$\omega_R^2 \cong \frac{v_g \Gamma_{act} g^1 S_0}{\tau_{phot}(1 + \varepsilon S_0)} \frac{1}{4\pi^2} \tag{11b}$$

$$\gamma = K\omega_R^2 + 1/\left[(1+\chi)\tau_{QD}\right] \tag{11c}$$

$$K = 4\pi^2\left[\tau_{phot} + (1+\chi)\varepsilon/\left(v_g g^1\right)\right] \tag{11d}$$

Here ω is the modulation frequency, τ_{cap} the QD capture time, τ_{QD} the QD carrier decay time, τ_{phot} the photon round trip time in the laser cavity, g^1 ($=dg_{mat}/dN$) the differential material gain, Γ the confinement factor, v_g the group velocity in the laser, ε the gain compression coefficient, S_0 the photon density in the laser cavity at the bias level, χ the ratio of QD capture and emission time, K the proportionality constant between the damping factor γ and the square of the relaxation oscillation frequency ω_R. By measuring the relaxation oscillation frequency as a function of the photon density (i.e. the power in the laser cavity) one can determine experimentally the differential gain g^1 and the gain compression coefficient ε which is equal to the inverse of the saturation photon density coefficient S_{sat} in the laser resonator. From these values the K-factor is obtained. If ω_R and γ remain linear functions of power the maximum modulation bandwidth ω_{max} is given by $f_{max} = \sqrt{2}(2\pi/K)$.

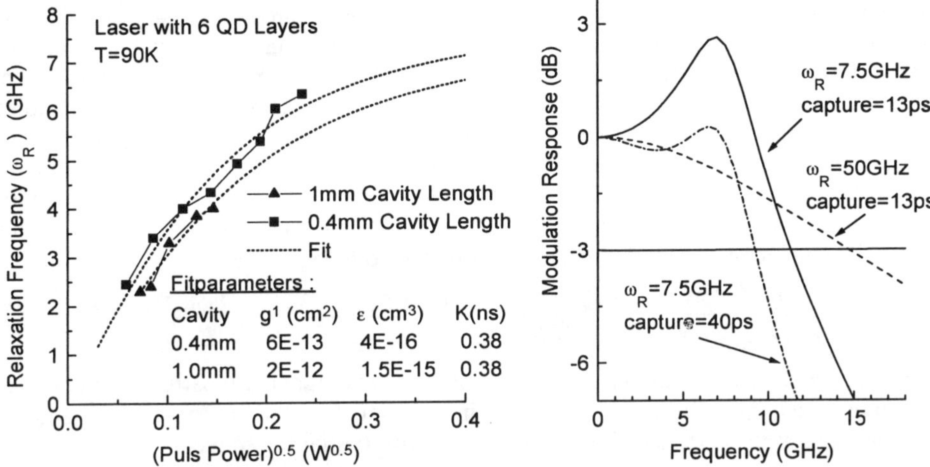

FIGURE 10. (a) Relaxation oscillation frequency ω_R of a QD laser as a function of the square root of pulse power in the laser cavity for two different cavity lengths. The excitation was done by a square pulse of 5ns length and rise time below 100ps. The ridge waveguide laser structure (ridge 5μm) consists of 6 QD layers. (b) Calculated modulation bandwidth for different relaxation oscillation frequencies and capture times according to eq. (11) using the parameters given in Figure (10a).

The carrier density N used for the calculation of the differential gain (g^1) in Fig. (10) is related to the recombination volume defined by the lower and upper cladding layer of the laser structure (see Fig. (1)). The experimental dependence of the relaxation oscillation frequency on pulse power shows a deviation from the expected proportionality (see eq. (11)) to the square root of optical pulse power indicating the onset of gain compression at a power level of ~50mW. The gain compression coefficient has a minimum of $4x10^{-16}cm^3$ at 90K equivalent to a power saturation coefficient of ~100mW in the laser cavity. The gain compression coefficient is more than one order of magnitude higher than typical values (~$10^{-17}cm^3$) in InGaAs/InGaAlAs QW lasers [27]. The differential gain of $2x10^{-12}cm^2$ obtained from the fit according to eq. (11) is close to the value derived in the previous section from the QD parameters.

Since the gain compression is closely correlated to the capture time of carriers into the dots the physical origin of the enhanced gain compression in QD laser structures may be connected to the slower relaxation rate into the QDs. This mechanism reduces the modulation bandwidth even when the differential gain is as high as shown in Fig. (3). The measured K-factor of 0.38ns (corresponding to a maximum modulation bandwidth of 23GHz) is a factor of 2-3 higher than best values reported for InGaAs QW lasers [26] at 100K indicating that the modulation bandwidth is lower in present QD lasers than in QW lasers. The PL rise time of the QD luminescence is reduced from 40ps to 13ps by increasing the coupling strength between QDs, grown on top of each other with very thin barriers [28]. This gives the possibility to suppress the limitation due to the carrier capture time into the QDs.

7. Chirp

Potential applications of a QD laser as directly modulated light source for data transmission via glass fibers requires minimum chirp. The chirp is determined from the shift of the wavelength a laser exhibits during current modulation. The physical origin of this shift is related to the coupling of the real and imaginary part of the complex susceptibility in the laser medium (see eq. (1)). A gain (imaginary part of the susceptibility) variation due to a change of the injection level changes the refractive index (real part of the susceptibility) which modifies the phase of the optical mode in the laser cavity. The coupling strength between real and imaginary parts is defined by the linewidth enhancement factor α which is defined as follows:

$$\alpha = \frac{\partial \chi_{real} / \partial N}{\partial \chi_{imag} / \partial N} = \frac{\partial n_{real} / \partial N}{\partial n_{imag} / \partial N} = -2 \frac{\hbar c_0}{E} \frac{\partial n_{real} / \partial N}{\partial g_{mat} / \partial N} = -2 \frac{\hbar c_0}{E} \frac{dn_r}{g^1} \tag{12}$$

Here N is the carrier density, n_{real} (n_{imag}) are the real and imaginary parts of the complex refractive index, c_0 is the velocity of light in vacuum, E is the photon energy, g_{mat} the material gain, g^1 the differential gain and dn_r the differential refractive index.

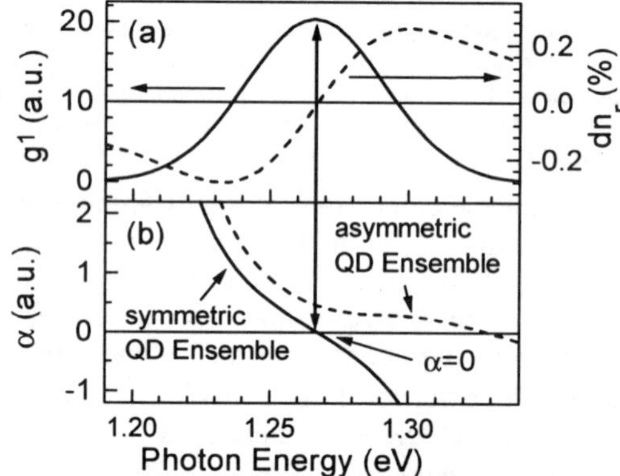

FIGURE 11. (a) Differential gain g^1 and differential refractive index dn_r for a quasi ideal QD ensemble with a symmetric gain curve and only one energy level for electrons and holes. (b) Calculated linewidth enhancement factor α for the quasi ideal QD ensemble and the experimentally reported single layer QD ensemble with an asymmetric gain spectrum shown in Fig. (3).

The linewidth enhancement factor can be calculated from the gain spectrum via the Kramers Kronig (KK) relation. In the case of a QD laser with a dot ensemble showing a perfect Gaussian energy distribution and only one energy level for electrons and for holes the gain spectrum is perfectly symmetric around the peak gain energy. In this case the differential gain is also symmetric to the peak gain energy position where lasing occurs. Thus the differential refractive index change computed via the KK relation is exactly zero at the lasing energy (i.e. peak gain position). Therefore the linewidth enhancement factor is zero at the lasing energy and QD lasers with a quasi ideally

distributed (symmetric) ensemble of QDs exhibit chirpfree operation. This property is shown in Fig. (11).
In the case of real QD lasers the contribution of excited states might cause a non symmetric gain curve (see Fig. (3)) which increases the linewidth enhancement factor depending on excitation level to about 0.5. This value is considerably lower than values of 1-2 [29] which are considered as being very good for QW laser structures.

8. Conclusion

In summary we have described lasing characteristics of (InGa)As QD lasers. Already the first preliminary results indicate that most of the laser key parameters are significantly improved as compared to QW lasers due to the introduction of QDs as an active medium. The single layer QD structures exhibit an ultrahigh material gain of $10^5 cm^{-1}$ only limited by the dot size distribution. A threshold density of ~$2 Acm^{-2}$ for a laser structure with a QD ensemble having a gain width of ~50meV and no current leakage is predicted. The threshold current density of currently operating lasers is as low as $40 Acm^{-2}$ at 77K and $62 Acm^{-2}$ at 300K while a T_0 of 385K is obtained up to room temperature.
The main current leakage is connected to barrier population and nonradiative loss in the barrier. Both effects can be suppressed in the future by growth optimization and careful band gap engineering of the laser structure. The QD emission wavelength can be tuned up to 1.3µm which is important for fibre optic communication. The linewidth enhancement factor of currently operating laser is reduced due to a nearly symmetric gain spectrum to a value of 0.5 which is a factor of 3 better than in current QW lasers.

9. References

1. Arakawa, Y., and Sakaki, H. (1982) Multidimensional Quantum Well Lasers and Temperature Dependence of its Threshold Current, *Appl. Phys. Lett.* 40, 939-941.
2. Asada, M., Miyamoto, Y., and Suematsu, Y. (1986) Gain and the Threshold of Three-Dimensional Quantum-Box Lasers, *IEEE J. Quantum Electron.* 22, 1915-1921.
3. Grundmann, M., Christen, J., Ledentsov, N.N., Böhrer, J., Bimberg, D., Ruvimov, S.S., Werner, P., Richter, U., Gösele, U., Heydenreich, J., Ustinov, V.M., Egorov, A.Yu., Zhukov, A.E., Kope'v, P.S., and Alferov, Zh.I. (1995) Ultranarrow Luminescence Lines from Single Quantum Dots, *Phys. Rev. Lett.* 74, 4043-4046.
4. Hirayama, H., Matsunaga, K. Asada, M., Suematsu, Y. (1994) Lasing Action of $Ga_{0.67}In_{0.33}As$/GaInAsP/InP tensile-strained Quantum-Box Lasers, El. Lett. 30, 142-143.
5. Kirstaedter, N., Ledentsov, N.N., Grundmann, M., Bimberg, D., Ustinov, V.M., Ruvimov, S.S., Maximov, M.V., Kop'ev, P.S., and Alferov, Zh.I. (1994) Low Threshold, large T_0 Injection Laser Emission from (InGa)As Quantum Dots, *Electron. Lett.* 30, 1416-1417.
6. Temmyo, J. Kuramochi, E., Sugo, M., Nishiya, T., Nötzel, R., and Tamamura, T. (1995) Quantum Disk Lasers with Self-Organised Dot-like Active Regions, Proc. of the Lasers and Electro-Optic Society 1995 Vol. 1, 77-78.
7. Ledentsov, N.N., Shchukin, V.A., Grundmann, M., Kirstaedter, N., Böhrer, J., Schmidt, O., Bimberg, D., Ustinov, V.M., Egorov, A.Yu., Zhukov, A.E., Kop'ev, P.S., Zaitsev, S.V., Gordeev, N.Yu., and Alferov, Zh.I. (in press) Direct Formation of vertically coupled Quatum Dots in Stranski-Krastanow growth, *Phys. Rev. B* 54.

8. Benisty, H., Sotomayor-Torres, C.M., and Weisbuch, C. (1991) Intrinsic Mechanism for the poor luminescence properties of quantum-box systems, *Phys. Rev B 44*, 10945-10948.
9. Shoji, H., Mukai, K., Ohtsuka, N., Sugawara, M., Uchida, T., and Ishikawa, H. (1996) Lasing at Three-dimensionally Quantum-Confined Sublevel of Self-Organised $In_{0.5}Ga_{0.5}As$ Quantum Dots by Current Injection, *Photon. Tech. Lett.* 7, 1385-1387.
10. Moritz, A., Wirth, R., Hangleiter, A., Kurtenbach, A., and Eberl, K. (1996) Optical Gain and Lasing in Self-Organised InP/GaInP Quantum Dots, *Appl. Phys. Lett. 69*, 212-214.
11. Mirin, R., Gossard, A., and Bowers, J. (in press) Room Temperature Lasing from InGaAs Quantum Dots, Proc. of the International Conference on Quantum Devices and Circuit (1996).
12. Kamath, K., Bhattacharya, P., Sosnowski, T., Norris, T., and Phillips, J. (1996) Room temperature Operation of $In_{0.4}Ga_{0.6}As$/GaAs Self-Organised Quantum Dot Lasers, *El. Lett. 32*, 1374-1375.
13. Heinrichsdorf, F. et al., to be published in *Appl. Phys. Lett.*
14. Kirstaedter, N., Schmidt, O.G., Ledentsov, N.N., Bimberg, D., Ustinov, V.M., Egorov, A.Yu., Zhukov, A.E., Maximov, M.V., Kop'ev, P.S., and Alferov, Zh.I. (1996) Gain and Differential Gain of Single Layer InAs/GaAs-Quantum Dot Injection Lasers, *Appl. Phys. Lett.* 69, 1226-1228.
15. Nakayama, H., and Arakawa, Y. (in press) Calcuation of Lasing Characteristics in Quantum Dot Lasers Considering Interaction of Electrons with LO Phonons, Proc. of the 15th IEEE International Semiconductor Laser Conference (1996) in Haifa (Israel).
16. Schmidt, O.G., Kirstaedter, N., Ledentsov, N.N., Mao, M.-H., Bimberg, D., Ustinov, V.M., Egorov, A.Y., Zhukov, A.E., Maximov, M.V., Kop'ev, P.S., and Alferov, Zh.I. (1996) Prevention of Gain Saturation by Multi-Layer Quantum Dot Lasers, *Electron. Lett.* 32, 1302-1303.
17. Grundmann, M., Stier, O., and Bimberg, D. (1995) InAs/GaAs pyramidal Quantum Dots: Strain distribution, Optical Phonons, and Electronic Structure, *Phys. Rev.* B 52, 11969-11980.
18. Böttcher, E.H., Kirstadter, N., Grundmann, M., and Bimberg, D. (1992) Nonspectroscopic approach to the determination of the chemical potential and band-gap renormalization in quantum wells, *Phys. Rev.* B 45, 8535-8541.
19. Lau, Y.L. (1993) in *Quantum Well Lasers*, edited by Zory, P.S., Academic Press, San Diego, p. 219.
20. Chand, N., Becker, E.E., van der Ziel, J.P., Chu, S.N.G., and Dutta, N.K. (1991) Excellent Uniformity and very low ($<50Acm^{-2}$) threshold current density strained quantum well diode lasers on GaAs substrate. *Appl. Phys. Lett.* 58, 1704-1706.
21. Alferov, Zh.I., Gordeev, N.Yu. Zaitsev, S.V. Kop'ev, P.S., Kochnev, I.V., Komin, V.V., Krestnikov, I.L., Ledentsov, N.N., Lunev, A.V., Maximov, M.V., Ruvimov, S.S., Sakharov, A.V., Tsapulnikov, A.F., and Shernyakov, Yu.M. (1996) A low Threshold Injection Heterojunction Laser based on Quantum Dots, produced by Gase-Phase Epitaxy from Organometallic Compounds, *Semiconductors* 30, 197-200.
22. Ledentsov, N.N. (in press) Ordered Arrays of Quantum Dots, Proc. of the International Conference of Semiconductors (ICPS) 1996.
23. Ustinov, V.M., Egorov, A.Yu., Kovsh, A.R., Zhukov, A.E., Maximov, M.V., Tsatsulnikov, A.F., Gordeev, N.Yu., Zaitsev, S.V., Shernyakov, Yu.M., Bert, N.A., Kope'ev, P.S., Alferov, Zh.I., Ledentsov, N.N., Böhrer, J., Bimberg, D., Kosogov, A.O., Werner, P., and Gösele, U. (in press) Low-Threshold Injection Lasers Based on Vertically Coupled Quantum Dots, Proc. of the 9th International Conference on MBE in California (USA) 1996.
24. Zhang, G. (1994) High Power and High Efficiency Operation of Al-free InGaAs/GaInAsP/GaInP GRINSCH SQW Lasers (λ=0.98μm), *Electron. Lett.* 30, 1230-1232.
25. Heinrichsdorf, F., Krost, A., Grundmann, M., Bimberg, D., Kosogov, A., and Werner, P. (1996) Self-Organisation Process of InGaAs/GaAs Quantum Dots grown by Metalorganic Chemical Vapor Deposition, *Appl. Phys. Lett.* 68, 3284-3286.
26. Nagarajan, R., Ishikawa, I., Fukushima, T., Geels, R., Bowers, J.E. (1992) High Speed Quantum Well-Lasers and Carrier Transport Effects, *IEEE J. Quantum Electron.* 28, 1990-2008.
27. Grabmeier, A. Hangleiter, A., Fuchs, G., Whiteaway, J.E.A., and Glew, R.W. (1991) Low nonlinear gain in InGaAs/InGaAlAs seperate confinement multiquantum well lasers, *Appl. Phys. Lett.* 59, 3024-3026.
28. Kirstaedter, N. (1996) Statische und Dynamische Eigenschaften von (InGa)As/GaAs Quantenpunktlasern, PhD Thesis at the Technical University of Berlin, D 83, p 45.
29. Raghuraman, R., Yu, N., Engelmann, R., Lee, H., and Shieh, C.L. (1993) Spectral Dependence of the Differential Gain, Mode Shift, and Linewidth Enhancement Factor in a InGaAs-GaAs Strained-Layer Single-Quantum-Well Laser Operated Under High-Injection Conditions, *IEEE J. Quantum Electron.* 29, 69-75.

RAMAN SCATTERING AS A DIAGNOSTIC TOOL OF SEMICONDUCTOR NANOFABRICATION

C.M. Sotomayor Torres
Institute of Materials Science
University of Wuppertal
42097 Wuppertal
Germany

1. Introduction

Nanostructuring of semiconductor wafers to realise quantum wires and quantum dots has been an active area of research due to its scientific impact and industrial relevance. The former has been driven partly by a wealth of new physical phenomena involving, for example, confined charged carriers and excitons, whereas the latter has its main impetus in the pursue of the zero threshold laser and single electron electronics to name but two nm-scale applications under investigation.

There are several techniques to fabricate semiconductors with physical size in the nm-range. Among pattern writing techniques holography, focused ion beam, X-rays and electron beams are the better known ones. To transfer the pattern, wet etching, quantum well intermixing and dry etching are commonly used. The continuous progress in the field on nanofabrication can be followed in the proceedings of the annual conference series Micro- and Nano-Electronics published in the Journal of Vacuum Science and Technology. The desired size is determined by the application envisaged. For example, for ballistic transport studies feature sizes of $\leq$ 1 μm are needed, whereas for single electron metallic devices a few nm suffice. Likewise, optical structures making use of confined photons require features of about 1 μm whereas to confine excitons 10s of nm are needed. For phonon confinement the challenge is even greater requiring feature sizes in the 1-5 nm regime.

In general one has to distinguish between the physical (feature) size and the "effective" size, where by effective size is meant the length, width or height in which the quasi-particle is confined. In the case of excitons the difference between the physical and the effective size is given by the thickness of an optical "dead" layer where excitons do not exist due to the high density of non-radiative states resulting from nanofabrication surface damage.

In what follows I will concentrate in the use of Raman scattering to assess electron beam lithography (EBL) and reactive ion etching (RIE) for the fabrication of structures with 10-100s nm feature sizes. First I will describe our earlier work on the assessment of dry etched planar and patterned surfaces of GaAs and GaAs-GaAlAs. Here the main finger prints are the emergence of otherwise forbidden phonons to label damaging surface etching of nominally undoped material and the use of coupled LO phonons-plasmons modes to determine the depletion layer thickness for various RIE gas

G. Abstreiter et al. (eds.), Optical Spectroscopy of Low Dimensional Semiconductors, 331–354.

mixtures and process parameters in doped GaAs. The use of complementary techniques gives us an insight into the change of the strain in the etched material and into the possible role of defects in the as-grown material.

We use Raman scattering and photoluminescence excitation to study optical process in deep dry etched GaAs-GaAlAs quantum dots and found that multiphonon processes were very efficient. I will give an overview of this work and attempt to put it in the context of localised *and* confined excitons.

Having used doped semiconductors for the study of RIE processing we examined doped quantum wires and dots of GaAs-GaAlAs using Resonant Raman scattering to reveal electronic excitations. I summarise our work on electronic excitations in 1D and 0D which has lead to the observation of energy levels akin to the shell structure of the atom in GaAs-GaAlAs quantum dots.

2. Raman scattering from dry etched planar surfaces

2.1. SOME BASIC ISSUES CONCERNING NANOFABRICATION

Nanostructures fabricated using electron beam lihtography and dry etching are subjected to a variety of process each of which potentially affecting the properties of the material. For example, the operating voltage (in keV) and the beam current (in pA), are perhaps the most important parameters in terms of potential material degradation due to electron momentum transfer to the lattice, containing information on the electron radiation received by the sample. Likewise, the energy of the ions in reactive ion etching, which is given in terms of the bias voltage or the RF power during dry etching, is the main parameter to compare dry etching processes. These aspects have been reviewed by Wilkinson and Beaumont [1]. Meaningful comparisons of damage resulting from different nanofabrication processes are virtually impossible unless all relevant parameters are specified. To name but a few: the mask type and thickness, gas mixture, gas volume flow rate, gas pressure, etching rate, substrate temperature, duration of the etching, post etching treatment, etc. In practice, it is only possible to make a reasonable comparison of processing techniques if a single set of equipment is used in very well characterised material following a well controlled procedure. In general, one aims for nanostructures with minimum electrical, optical and structural damage. However, experiments have shown time and again that a process that is promising for optically active nanostructures can still be detrimental for electrical studies. Moreover, some structures that look almost perfect in a scanning electron micrograph (SEM micrograph), are optically dead, sometimes not necessarily due to nanofabrication damage but as a result of contamination while in the SEM chamber.

An indication of the residual surface damage is provided by measurements of the ideality factor of a Schottky diode fabricated on a freshly etched surface. Interpretation of these measurements are usually complicated by the additional effect of surface oxides in, for example, Al containing materials, however, as a rough indication of surface damage they provide a guideline. Measurements of the photoluminescence intensity of quantum wells of different thickness grown close to the surface to be etched have also been used to determine the depth of surface damage for a given fabrication process. This method provides a guideline only as there are too many unknowns in the luminescence process which involves photoexcitation, carrier relaxation, carrier diffusion and subsequent recombination. In particular, the use of this technique is hampered by the possibility of exciton localisation at interface states which varies from point to point in a given sample and between samples. Surface compositional changes have been monitored using XPS to obtain information on, for example, changes of the relative oxide composition between as-grown and etched surfaces. A comprehensive study of ion etching induced damage in GaAs and GaAlAs flat surfaces and sidewalls gathering the information obtained by various complementary techniques, including Raman scattering, has been carried out by Cheung *et al.* [2]

2.2 EMERGENCE OF FORBIDDEN PHONONS

In phonon Raman scattering the finger prints of nanofabrication damage are revealed by the emergence of forbidden phonons for a given crystal surface and scattering geometry, the relative intensities of allowed and forbidden phonons and the linewidth of these phonons. Previous work on nominally undoped GaAs was carried out by, for example, Semura *et al.* [3] who reported Raman scattering data from GaAs etched in CCl_4/H_2 and CCl_4/O_2. Watt *et al.* [4, 5] showed that the TO phonon (269 cm^{-1}) appeared in the near-backscattering Raman spectrum from flat etched [001] GaAs surfaces processed with $SiCl_4$ RIE, with increasing intensity for increasing RIE RF power, compared to its absence in an unetched control sample as seen in Figure 1. Moreover, the LO phonon (292 cm^{-1}) exhibited asymmetric broadening.

Considering the extreme case of an amourphous layer of GaAs on the surface, the long range order (wavevector q=0) would no longer hold and therefore the selection rules would break down. This in turn would mean that all phonons with a finite density of states would participate in the Raman process, which was not the case observed in these experiments. Probing the surfaces with different wavelengths (different penetration depths) it was found that the intensity of the TO phonon was stronger closer to the surface. The asymmetric linewidth of the LO phonon was interpreted in terms of a spatial correlation length following the formalism of Tiong *et al.* [6] from which an average length of undamaged regions is obtained. In our tests, the shifts of the LO phonon to lower energies were less than 2.4 cm^{-1} and the linewidth asymmetry was less than 1.17 which lead us to conclude that the minimum undamaged correlation length was about 25 nm depending on dry etching conditions. The isotropic nature of the damage was confirmed by rotating the sample about the axis parallel to the incident light direction. The intensity of the LO phonon was observed to follow the cosine function whereas the TO phonon intensity remained unchanged as shown in Figure 2.

2.3. DEPLETION LAYER THICKNESS DETERMINATION

Dry etching processing of doped GaAs lends itself to Raman scattering studies to determine the depletion length and carrier concentration by means of the coupled LO phonon-plasmon mode as demonstrated by, among others, Chung *et al.* [7], Kirillov *et al.* [8] and Roughani *et al.* [9]. The idea behind the technique is the following: At the surface of GaAs there is a depleted space charge region with a surface barrier height of typically 0.7 eV which, depending on the donor concentration, can vary from 20 to 100 nm into the material. Since the penetration depth of typical Argon ion laser lines in GaAs varies from some 200 to 30 nm one can choose a laser line with a penetration depth longer than the depletion depth. The light will then be scattered by both,

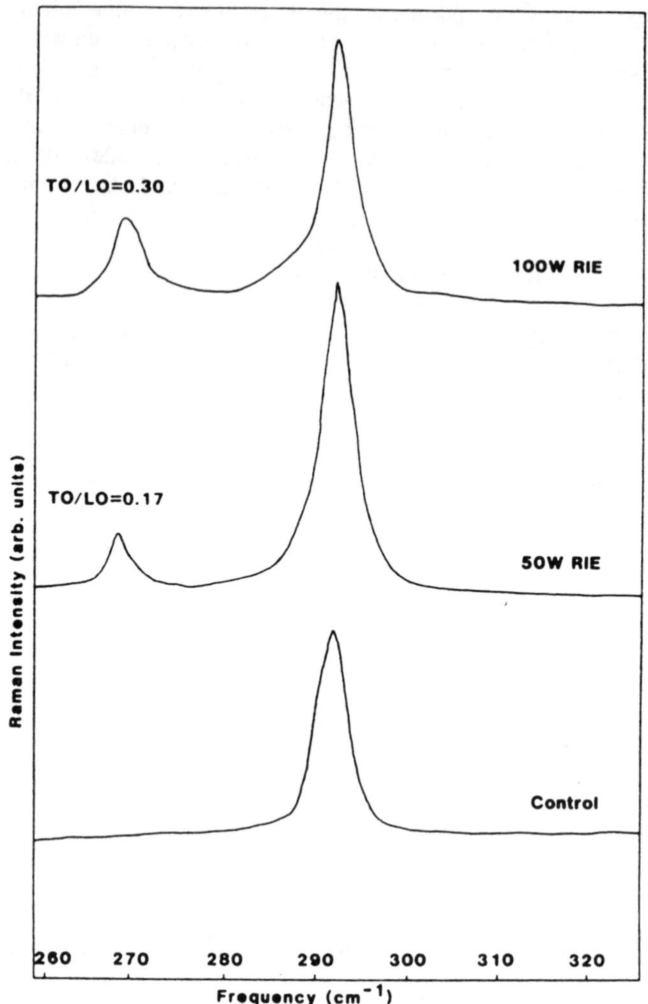

Figure 1. Raman scattering spectra from reactive-ion etched GaAs for two etch powers. In both etched samples the symmetry forbidden TO phonon is observed and the LO phonon is seen to broaden asymmetrically to lower energies. The resolution is 2.4 cm^{-1} and the laser line is 482.5 nm.[4]

unscreened LO phonons in the surface depletion region and by coupled LO phonon-plasmon modes in the bulk region, where the plasmon is representative of the free carrier concentration in the bulk. From a comparison of their relative intensities, information can be extracted on the surface band bending and the bulk carrier concentration following the formalism discussed by Abstreiter *et al.* [10] and references therein. I will try to summarize the arguments. The coupling between plasmons and LO phonons splits the optical vibrational modes into L_1 (low frequency) and L_2 (high frequency). Solving the system dielectric function $\varepsilon(\omega) = 0$ for LO modes one finds:

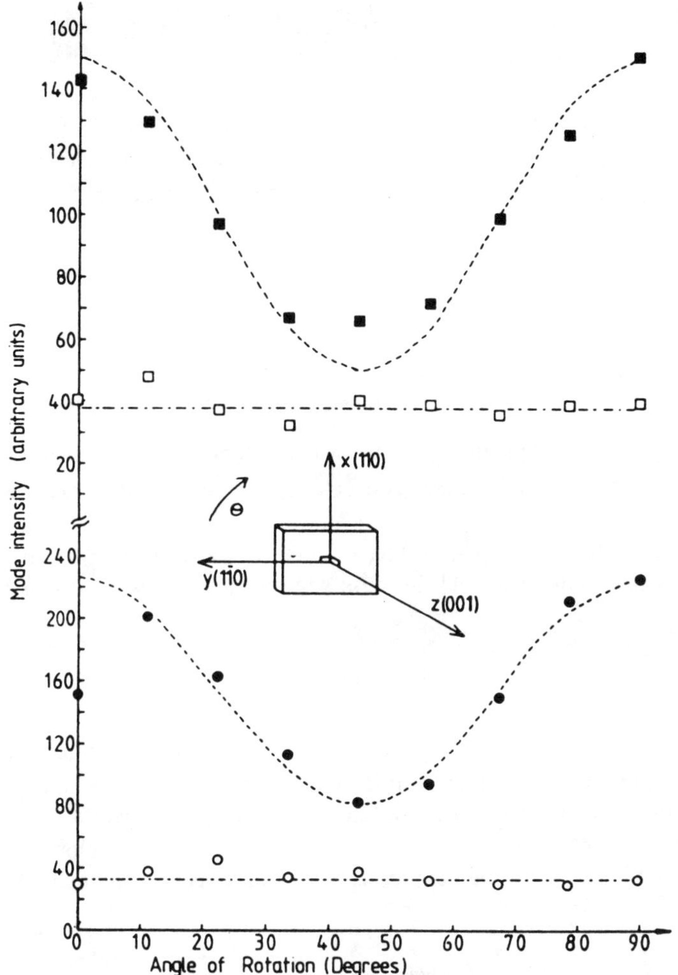

Figure 2. LO (full squares and circles) and TO (open squares and circles) phonon intensities as a function of rotation angle (see inset). The upper (lower) figure corresponds to 100W (50W) RIE. The dashed lines are a guide to the eyes. [4]

$$\varepsilon(\omega) = \varepsilon_\infty \left(\frac{\omega^2 - \omega_L^2}{\omega^2 - \omega_T^2} - \frac{\omega_p^2(q)}{\omega^2} \right)$$

where $\omega_L = \hbar\omega_{LO}$, $\omega_T = \hbar\omega_{TO}$, ε_∞ is the high frequency dielectric constant and $\omega_p(q)$ the wavevector dependent plasma frequency given by:

$$\omega_p^2(q) = \frac{N_e e^2}{\varepsilon_0 \varepsilon_r m^*} + \frac{3}{5} q^2 v_f^2$$

in backscattering configuration $q = 2\nu\eta$, where ν is the wavevector of the exciting light, η the refractive index at the excitation frequency, N_e the carrier concentration, m^* the electron effective mass of the material and v_f the Fermi velocity from:

$$v_f = \frac{\hbar}{m^*}\left(3\pi^2 N_e\right)^{\frac{1}{3}}$$

for parabolic bands.

The term $\frac{3}{5} q^2 v_f^2$ is a second order one arising from non-local effects due to the non-compressibility of the system. For low wavevectors q, ω_p is independent of q in 3D. For GaAs ω_p is between 100 and 700 cm^{-1} for carrier densities between 10^{17} and 10^{19} cm^{-3}.

If one considers the depletion layer thickness to be d, with a typical interface between 4 and 5 nm between it and the bulk region of the crystal, d can be obtained from:

$$I(LO) = I_o(LO)\left(1 - e^{-2\alpha d}\right)$$

where I(LO) is the integrated intensity of LO phonons, $I_o(LO)$ is the integrated intensity of the LO phonon in the undoped or semi-insulating region (< $1\text{x}10^{14}$ cm^{-3}) and α the absorption coefficient of GaAs at the excitation wavelength. Thus, as the doping level increases I(LO) decreases. The assumption made here is that $I_o(LO)$- I(LO) is proportional to the intensity of the coupled LO phonon-plasmon intensity $I(L_1)$, and in this case:

$$\frac{I(LO)}{I(L_1)} = \beta\left(e^{-2\alpha d} - 1\right)$$

where β is a scattering constant from:

$$I_o(LO) - I(LO) = \beta(I_1)$$

In a simple picture the unscreened LO phonons will participate in light scattering in the depletion region, whereas the L_1 coupled modes will participate in the scattering in the bulk region. Therefore a relative change in I(LO) and $I(L_1)$ gives information on the change in the depletion layer thickness d, which in turn helps to assess RIE-induced damage. Figure 3 shows unpolarised Raman spectra of two GaAs samples with different doping showing the L_1, LO and L_2 modes.

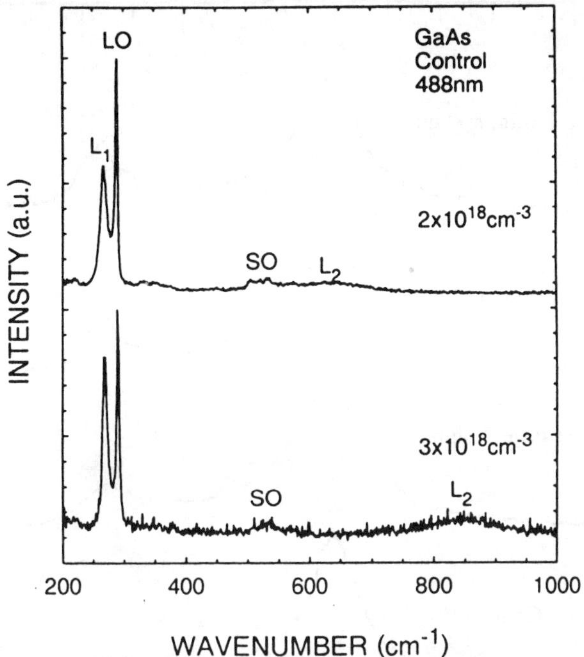

Figure 3. Unpolarised Raman scattering spectra at room temperature of n-type GaAs. The upper (lower) spectrum corresponds to a sample with doping 2 x 10^{18} cm^{-3} (3 x 10^{18} cm^{-3}). The coupled LO phonon-plasmon modes L_1 and L_2 and the unscreened LO phonon are seen as well as second order (SO) scattering. [12]

There are some important symmetry considerations which have been reviewed in a number of Raman scattering handbooks, most notably the series edited by M. Cardona and G. Güntherodt [11]. For example, in near-backscattering configuration, i.e., $z(x,y)\bar{z}$ and $z(x',y')\bar{z}$, where x, y, z, x' and y' are associated with the crystallographic directions $[0\bar{1}1]$, [011], [100], [010] and [001], respectively. Cubic crystals like GaAs posses a group symmetry called T_d, and the selection rules [11] for light scattering from a (100) surface allow only the LO phonon, whereas the TO phonon is forbiddden.

Wang *et al.* [12] and Murad *et al.* [13] have studied the impact of various dry etching processes of GaAs by Raman scattering using the couple LO phonon-plasmon mode, assuming that the selection rules for LO phonons also apply to coupled modes. Figure 4 shows the Raman spectra of GaAs etched in $SiCl_4$ as a function of etched depth and time. The depletion depths obtained from these and other measurements are plotted in Figure 5.

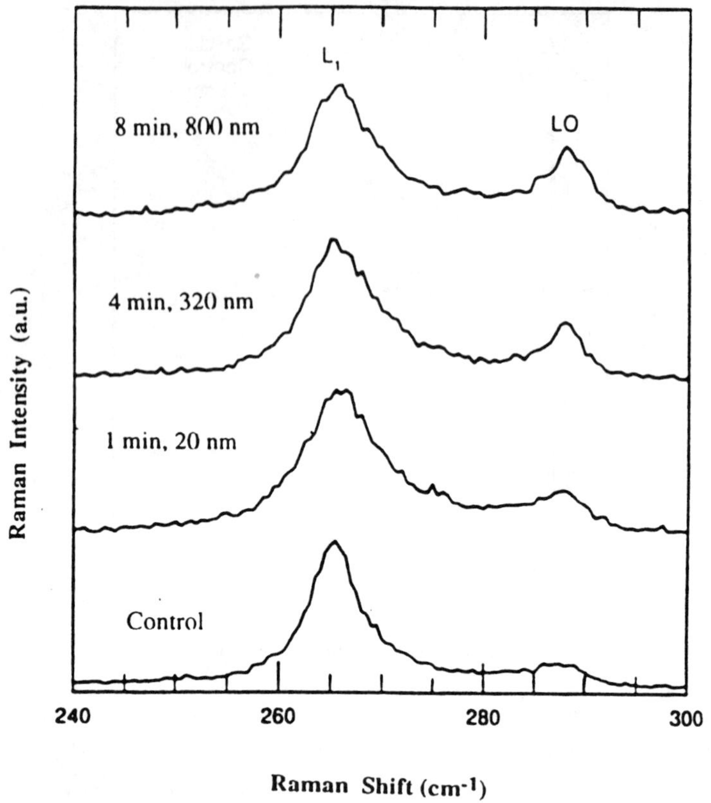

Figure 4. Raman spectra of GaAs at room temperature using the 488 nm laser line as a function of etched depth and etching time. The gas mixture was $SiCl_4$ (15W, 6 sccm, 8-9 mTorr, 50-70V). Damage saturation is observed at an etch depth of 320 m with an etch time of 4 minutes. [13]

Data for reactive ion etching, ion beam etching and electron cyclotron resonance radio frequency reactive ion etching, using $SiCl_4$, Ne, Ar, CCl_2F_2/H_2 and CH_4/H_2 and wet etching is reported in reference [12].

3. Raman scattering from nanostructured surfaces

3.1. GENERAL COMMENTS

In what follows we will consider nanostructured surfaces which means that in addition to "flat surfaces" there will also be "sidewalls". So far the sidewall damage is less understood than flat surface damage as direct measurements are more challenging.

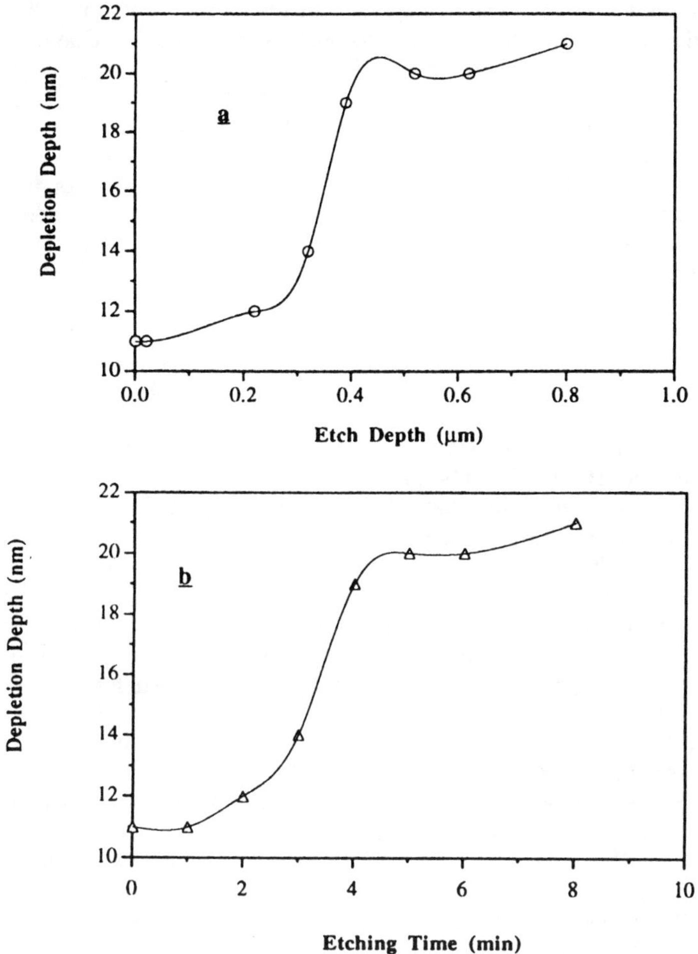

Figure 5. Depletion depth in GaAs as a function of (a) etched depth and (b) etching time, partly from the data shown in Figure 4. [13]

Several new terms come on board such as the aspect ratio (ratio of the lateral size to the height of the nanostructure (eg. 1:1, 1:8) and the overall shape if not a well defined geometrical one (eg. undercut, overcut, keyhole, mushroom), which bears a direct graphic analogy to what is observed in a SEM micrograph. In general one expects to have shape uniformity of the nanostructures and, more importantly, uniform structural, electrical and optical properties in a nanostructure array over certain area (from μm^2 to cm^2). The packing density, pitch or area coverage ranges typically from 2 to 20% and provides information on how much of the original material is left on the patterned surface. In general sidewall damage is typically two to four times the depth of the flat surface damage from Schottky diode techniques as shown by Cheung et al. [2] and from conductivity cut-off measurements [13].

3.2. SURFACE PHONONS IN GaAs nm-SIZE CYLINDERS

In addition to the a weak TO phonon, the intensity of which depends on dry etching conditions, the Raman spectrum of an array of GaAs submicron sized cylinders exhibits a new feature between the TO and LO phonon frequencies. Light scattering away from the array area yields spectra similar to that shown in Figure 1. However, the spectrum from the nanostructured surface shows the TO phonon intensity comparable to that of the LO phonon as light scattering under near back scattering configuration takes places from the (100) surface and the sidewalls. This breakdown of the selection rules means that a polarisation analysis yields little information and thus spectra reported in this section are unanalysed.

Watt *et al.* [14] studied dry etched GaAs cylinders and observed surface phonons appearing between the TO and LO phonon frequencies in samples with vertical wall, diameter $\leq$ 80 nm and height greater than about 200 nm. Figure 6 shows the scattering geometry in the inset, a spectrum of the patterned area and of the cylinder arrays at three angles of incidence θ. The larger the angle formed by the incident laser beam and the cylinder axis, the stronger the surface phonon. Surface phonons arise from the geometrical perturbation to the electromagnetic field on the surface, in our case by the presence of highly regular cylindrical structures, which invalidates the infinite crystal approximation in the length scale of the material volume probed by light scattering. Although there have been previous observations of surface phonons in GaP nm size spheres produced by gas-evaporation techniques [15], these observations were the first ones in deep dry etched nanostructures.

The analysis followed the model of Rupin and Englman [16] based on the dielectric continuum with the phonon frequencies being derived from first principles. The cylindrical geometry is introduced by selecting suitable boundary conditions. In the full model, polaritons are included, but in the case of sample features being small (here 80 nm diameter) compared to the infrared radiation wavelength (here approximately 30 µm), the coupling between electromagnetic waves and lattice vibrations (retardation effects) can be neglected resulting in a simplified model. The main equation relating the surface phonon frequency to the cylinder radius R and the dielectric constants ε_o, and ε_∞ is:

$$\frac{\omega_{nh}^2}{\omega_{TO}^2} = \frac{\varepsilon_o - \varepsilon_m \rho_{nh}}{\varepsilon_\infty - \varepsilon_m \rho_{nh}}$$

where:

$$\rho_{nh} = \frac{K_n'(hR) I_n(hR)}{K_n(hR) I_n'(hR)}$$

and where ω_{nh} is the surface phonon frequency of order n for a propagation constant h,

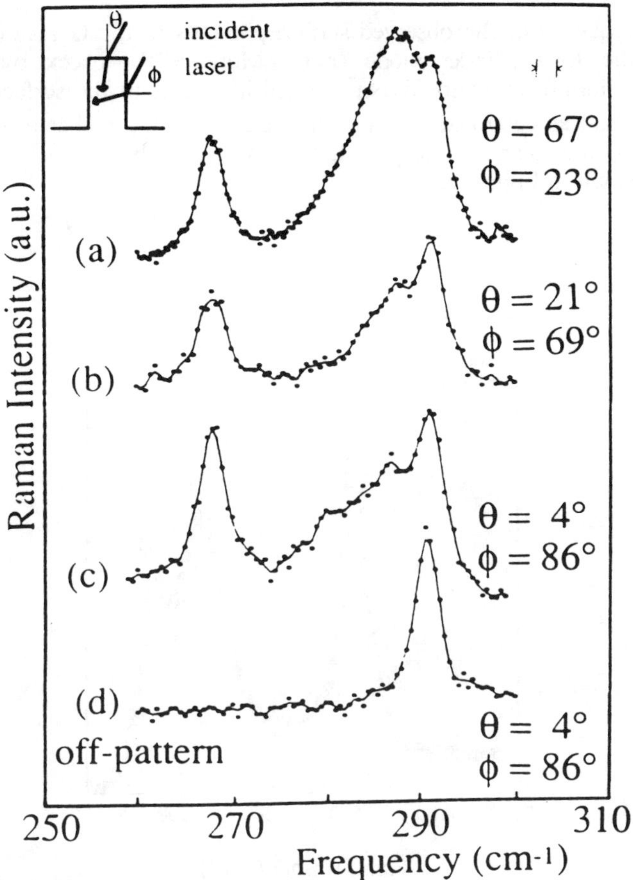

Figure 6. Raman scattering from an array of GaAs cylinders width 80 nm diameter and 310 nm height fabricated by electron beam lithography and dry etching in SiCl4. The laser line was 488 nm. The bottom spectrum is recorded off the patterned area and shows the LO phonon (292 cm^{-1}). The three upper spectra are from the cylinder array taken with the laser at different angles of incidence θ. The TO phonon is seen at 269 cm^{-1} and the surface phonons at around 285 cm^{-1}.

ε_m is the dielectric constant of the surrounding medium, I_n and K_n are modified Bessel functions of order n, $I_n^{'}$ and $K_n^{'}$ are their first derivatives, and

$$h = \frac{2\pi\eta\cos\theta}{\lambda}$$

is the propagation constant along the cylinder.

The frequency of the observed surface phonons in the GaAs cylinders is well described by the lowest order mode (n=0). Moreover, a check by changing the surrounding medium from air to Si_3N_4 resulted in a shift of the surface phonon mean frequency of 3 cm^{-1} as described in detail in reference [14]. Another example of surface phonons in 100 nm diameter deep etched GaAs dots obtained with 633 nm laser excitation is shown in Figure 7.

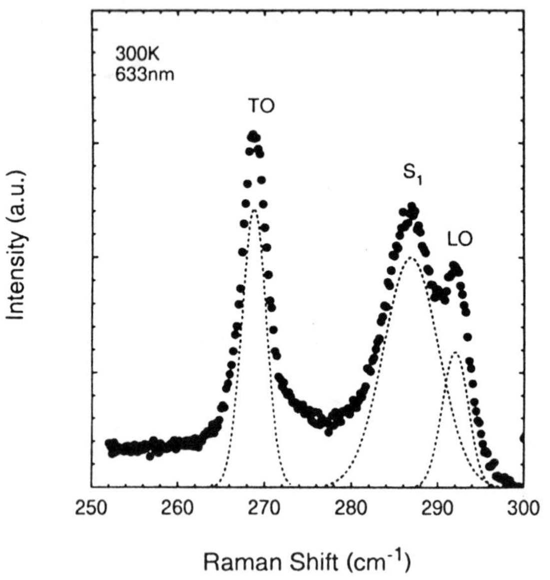

Figure 7. Raman spectrum of 100 nm diameter GaAs dots. S_1 denotes the GaAs surface phonon peak. Dotted lines show the deconvolution of the LO and surface phonon.[14]

Surface phonons in 500 nm dots fabricated by deep etching in GaAs-$Al_{0.3}Ga_{0.7}As$ quantum wells have also been observed (see Figure 8) in addition to the GaAs LO and TO phonons, the "GaAs"- and "AlAs"-like phonons of the AlGaAs barriers at 291, 267, 280 and 377 cm^{-1}, respectively. The surface phonons are labelled S1, S2 and S3 for the "GaAs"-like, GaAs and "AlAs"-like surface phonons, respectively.

3.3. RAMAN MICROSCOPY OF nm SIZED DOTS

The technique of micro-Raman has already been used to investigate a number of physical systems. We used it to produce images of phonon scattering in dot arrays of GaAs [17]. The 1.36 µm dots were produced by electron beam lithography and dry etching in $SiCl_4$.

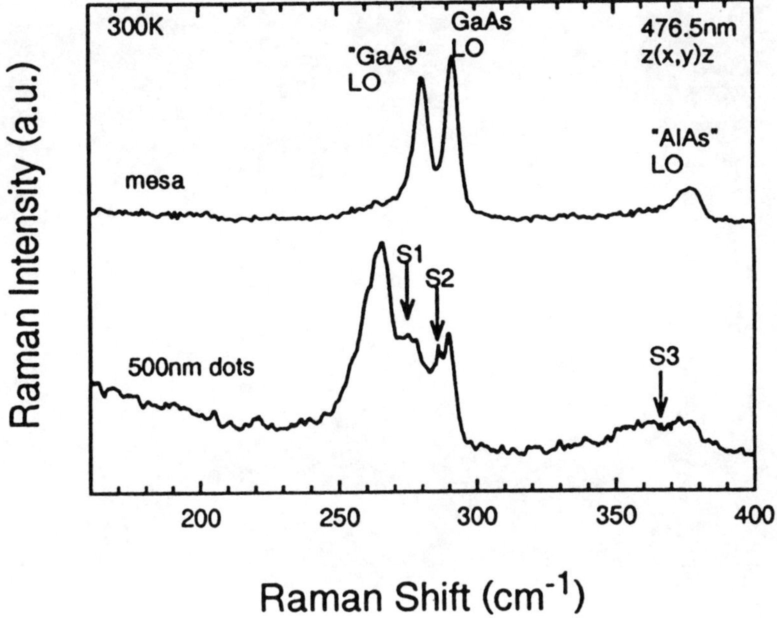

Figure 8. Raman scattering of GaAs-$Al_{0.3}Ga_{0.7}As$ quantum wells (4, 6 and 8 nm) in a mesa structure (top spectrum) and in an array of 500 nm diameter dots (bottom spectrum). S1, S2 and S3 are surface phonons discussed in the text.

The intensity contrast of the TO phonon peak between the dot and its surrounding area is used for imaging. In order to produce the image, the background image signal recorded at 200 cm^{-1} is substracted from the image recorded at 262 cm^{-1}. In Figure 9 a surface intensity plot is shown where the brightest peaks correspond to the location of maximum scattering of TO phonons by the dots in the array. The image mimics a SEM micrograph.

3.4. STRAIN IN ETCHED nm-SIZED CYLINDERS

Nanostructures patterned in strained quantum wells or superlattices experience a change in their strain compared to the as-grown strain, mainly due to the removal of material around each dot, wire or wall. A recent review of the topic can be found in reference [18]. The change in strain is best illustrated in the case of Si-SiGe as has been shown by, for example, Dietriech et al [19] using Raman scattering.

Si-SiGe quantum dots of 50 nm diameter and different Ge content were examined by micro-Raman scattering. The layers were grown on a Si substrate which means that the SiGe layers are biaxially compressed before patterning. After patterning the Ge containing layer expands along with the thin Si layers in the stack. This in turn leads to an elastic deformation of the structure in the spatial scale of the layer thickness.

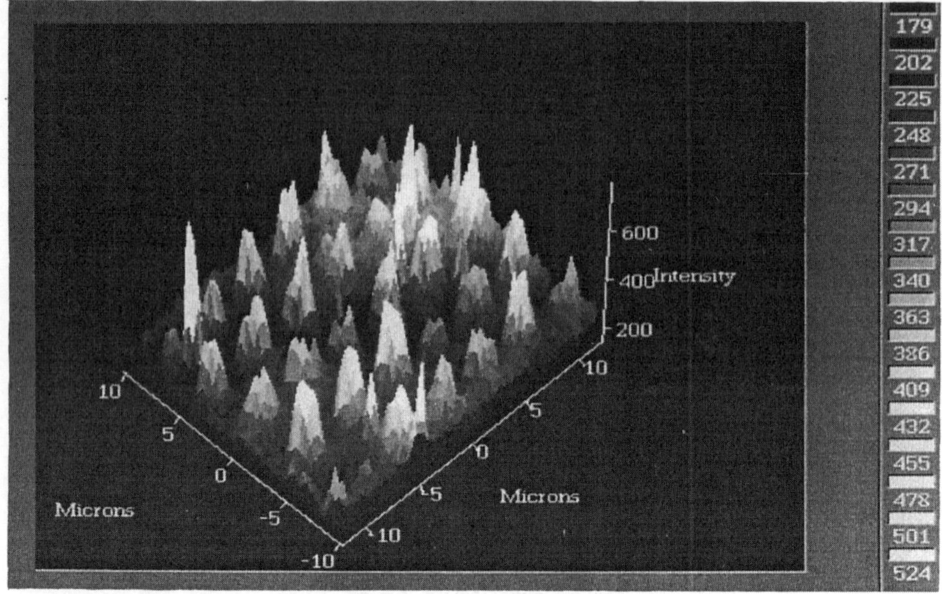

Figure 9. Raman microscopy imaging of an array of 1.36 μm diameter GaAs dots using the scattered light from TO phonons. [17]

Such deformation results in a modified bandstructure and associated electronic transitions. Using micro-Raman scattering spectra were obtained from groups of about 10 quantum dots to reveal that after patterning the quantum dots have a strain value halfway between the as-grown pseudomorphic one and complete relaxation (see Figure 10).

Using X-ray reciprocal space mapping, Ni *et al.* [20] showed that in Si-SiGe dots the in-plane lattice constant is reduced to the mid-point between pseudomorphic and relaxed SiGe values whereas the lattice constant of the Si layers increased by the same amount. Moreover the symmetry axis of the dots was found to be 1.5° off the [110] direction compared to [110] for the as-grown material.

The relevance of understanding and controlling the strain in semiconductor nanostructures is illustrated in Figure 11, where an array of Si-SiGe quantum dots was coated by SiN_x with different levels of stress. The impact of varying the strain of the surrounding SiN_x can be seen in the changes in the room temperature integrated luminescence. The maximum luminescence intensity is obtained for a coating with the minimum stress, i.e., with conditions as close as possible to the free standing, as-etched quantum dots [21].

3.5. COMPLEMENTARY TECHNIQUES

The study of strain in semiconductor nanostructures has benefitted from the use of alternative and complementary techniques. Among these photoreflectance, X-ray reciprocal space mapping and luminescence spectroscopy are perhaps the better known.

The reader is directed to references on quantum dots examined by photoreflectance of CdTe-CdMnTe [22] and GaAs-GaAlAs [22, 23]. Si-SiGe quantum

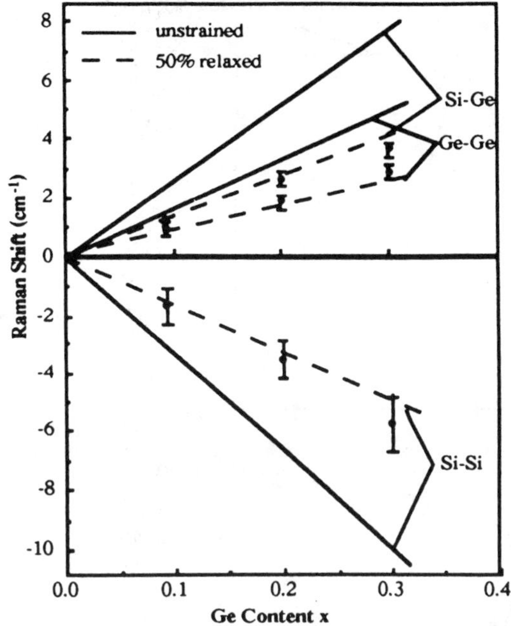

Figure 10. Raman shifts of the Si-Si, Ge-Ge and Si-Ge modes in Si-SiGe 50 nm diameter deep etched quantum dots as a fuction of Ge content. [18]

wires have also been examined by photoreflectance [24] showing clearly the interplay between carrier confinement and strain. In addition, the comprehensive work of Holy *et al* [25] on X-ray reciprocal space mapping corroborated the findings in GaAs-based quantum dots, namely that there is a resulting random strain in the core of the dots after patterning but more significantly, that there is a layer of about 15±5 nm of non-diffracting material on the sidewalls of the dots. The width of this layer was obtained from simulations of the x-ray data.

The changes observed in the strain after nanopatterning are only partly accounted for by the relaxation of the built-in strain. In reference [18] possible causes of fabrication-induced strain are discussed. Among them, although still not quantified, are: (i) The formation of vacancies, for example, V_{As} are suspected to occur on dry etched GaAs surfaces and V_{Cd} in CdTe, CdMnTe surfaces and deeper into the material; these in turn could offer sites for impurity incorporation, such as oxygen and carbon. (ii) Ion channelling to a depth of about a micron is likely to cause point defects and deep levels, which could in principle be probed by Deep Level transient spectroscopy but offer a substancial challenge in data analysis [26] and; (iii) Contamination by polymer and etched material redeposition.

It is almost an indisputable fact that there is a residual strain present in freshly etched lattice matched and lattice mismatched samples and that this strain is smaller for smaller lateral sizes. Moreover, the residual strain can be modified by post-etching treatments such as thermal annealing and/or overgrowth. From optical spectroscopic data in the references given in this section, there is strong evidence of a quasi-parabolic confining potential in quantum dots, both in doped and undoped samples. However, the

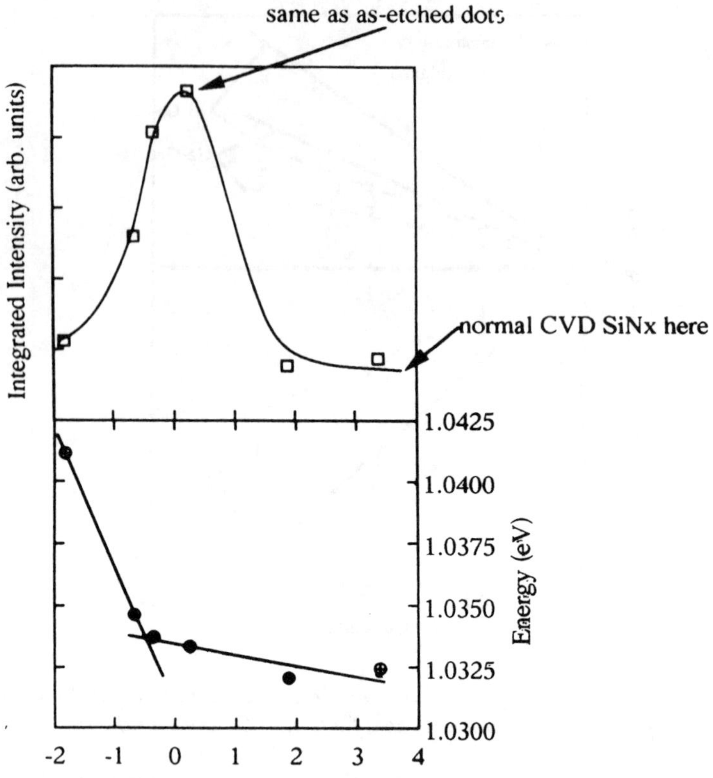

Figure 11. Integrated emission intensity at room temperature from 50 nm diameter Si-SiGe quantum dots coated with SiN_x with different stress. [21].

manner in which combined strain, surface damage and surface depletion contribute to the resulting potential shape remains under investigation.

3.6. MULTIPHONON PROCESSES IN GaAs-GaAlAs QUANTUM DOTS

Quantum dots were fabricated in a 100 period 8 nm GaAs-12 nm $Al_{0.3}Ga_{0.7}As$ multiple quantum well sample grown by MBE. The smallest diameter was 200 nm and the etch depth was 2.1 μm. Wang et al [27] reported the details of sample preparation and light scattering studies. Figure 12 shows the photoluminescence excitation (PLE) of arrays of 250 nm diameter quantum dots detecting the signal at the heavy-hole exciton energy. The heavy-hole exciton was shifted from its 2D transition energy (1.573 eV) by 1.47, 5.81 and 9.04 meV for arrays of dots with 500, 300 and 250 nm diameter, respectively. Since 200nm dots did not exhibit any luminescence it was assumed that the sidewall damage of this deep (and long) etching was closer to 100-110 nm. This means that the effective size of the 250 nm dot was less than 50 nm diameter.

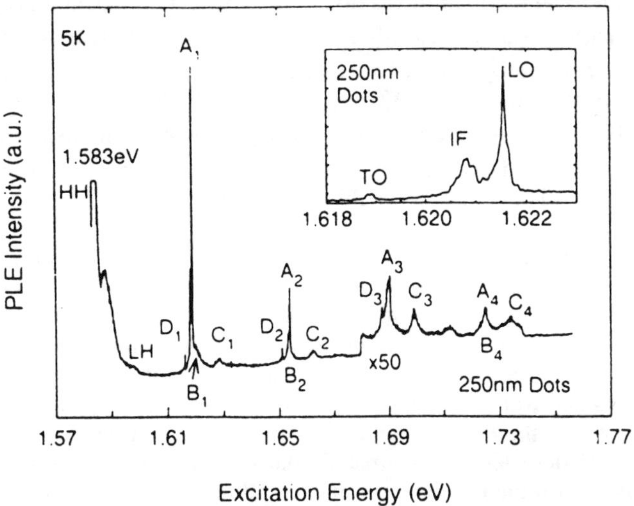

Figure 12. PLE spectrum of 250 nm quantum dots of GaAs-$Al_{0.3}Ga_{0.7}As$. The labels are discussed in the text. The inset shows the fine structure of the A_n peaks associated with interface phonons (IF).

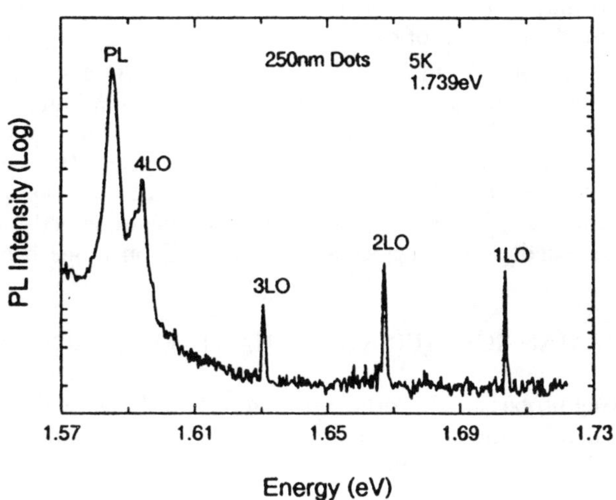

Figure 13. Photoluminescence at 5K of an array of 250 nm GaAs-GaAlAs quantum dots excited with a photon energy just above the sum of the heavy hole exciton and four LO phonons. Up to fourth-order multiphonon processes are distinguised.

The sharp features in the PLE of the dots form four series labelled A_n, B_n, C_n and D_n which were interpreted as follows: The A_n series corresponds to scattering via LO phonons due to its periodicity and has an associated IF phonon to the low energy side of the A_n peaks separated by 0.75 meV. The D_n series is related to the TO phonon. B_n and C_n were ascribed to hot exciton luminescence since their intensity and width remains invariant with order and because they exhibit a high energy tail typical of a Maxwellian distribution. They are probably associated with acoustic phonon thermalisation.

The origin of the series was discussed in terms of the phonon bottleneck [28] and from in-going and out-going Raman resonances together with the temperature dependence of their intensity, the observations were found to be consistent with the presence of localised excitons in the dots, whereas the hot luminescence path was probably enhanced by the poor relaxation in the 0D excited states. Figure 13 shows the photoluminescence (resonant Raman scattering) of the 250 nm dot array under excitation with an energy just greater than the ground state plus 4 LO phonons. Up to four multiphonon processes are detected in a manner similar to II-VI wide gap semiconductor alloys. A recent discussion on the phonon bottleneck and its relation to confinement and localisation in quantum dots is forthcoming [29].

4. Electronic Raman scattering from doped quantum wires and dots

4.1. PROGRESS IN LOW DAMAGE REACTIVE ION ETCHING

Developments in low damage RIE processes based on a magnetically confined plasma and the use of $SiCl_4$ with the addition of oxygen [30, 31], led to very high aspect ratios and vertical sidewall, which in turn allowed the observation of 1D confinement in the PLE spectrum of deep etched GaAs-AlAs quantum wires as shown in Figure 14.

We fabricated doped and undoped multiple quantum wells into quantum wires down to 60 nm width and quantum dots down to 100 nm diameter. The samples were characterised by Shubnikov-de Haas (doped structure only), PLE, magneto-optics and Raman scattering and exhibited the usual Fermi shake-up transitions and Landau level transitions [32]. Three subbands were occupied and above 15 T only one band remained occupied.

4.2. ELECTRONIC RAMAN SCATTERING IN QUANTUM STRUCTURES

Electronic light scattering probes single particle excitations (SPE) as well as collective excitations. Some manifestations of many-body effects include band gap renormalisation, plasmons, the electron-hole liquid, the electron-hole plasma, the Fermi-edge singularity, biexcitons, trions, spin density excitations (SDE), charge density excitations (CDE), etc. In order to study these excitation in reduced dimensions, quatnum wires and dots need to have electrons, requiring nanofabrication of doped structures. The following aspects become therefore important: lateral confinement, sidewall depletion layer thickness, concentration and position of traps and donor and the distribution of donor charge, which together with the Hartree and the exchange interactions determine the charge distribution of the ground state and the excitation spectrum.

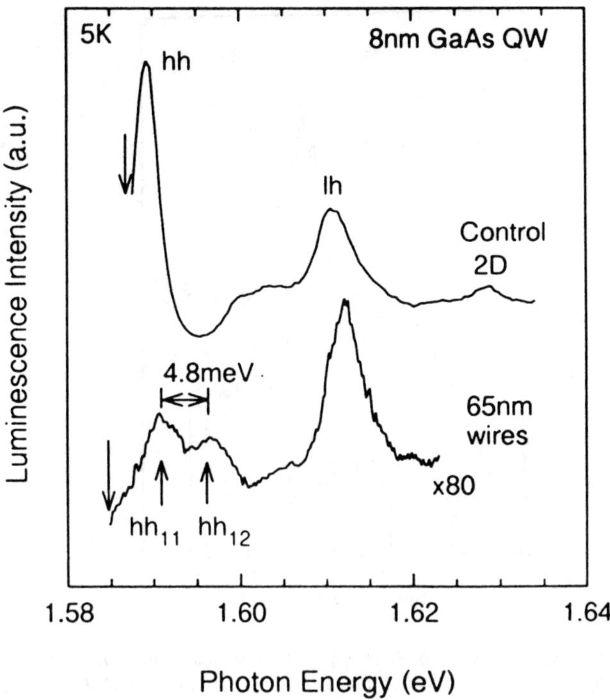

Figure 14. Low temperature PLE of an array of 60 nm quantum wires (bottom spectrum) and its control sample (top spectrum). The heavy-hole exciton line exhibits a splitting due to 1D quantum confinement. The detection energy is marked by downward pointing arrows.

Since these topics are reviewed in this volume in [33, 34, 35, 36], I will only briefly review our recent results in quantum wires and dots.

4.2.1. *Quantum wires*

Figure 15 shows the depolarised spectrum of 100 nm modulation doped wires of GaAs-GaAlAs [37]. The electronic scattering from these wires shows distinct polarisation properties in agreement with general theoretical predictions. The observed electronic excitations are SDE with transition energies of 4.5, 8 and 12.5 meV, corresponding to transitions between confined states due to the 1D potential. These values are in reasonable agreement with calculated SDE energies using the Hartree approximation.

4.2.2. *Quantum dots*

The theoretical formalism for electronic excitations for quantum dots has been developed by, among others, Hawrylak *et al.* [38] using the Hartree approximation. The resulting potential is found to be repulsive in the centre of the dot (as in the 2D case) and the

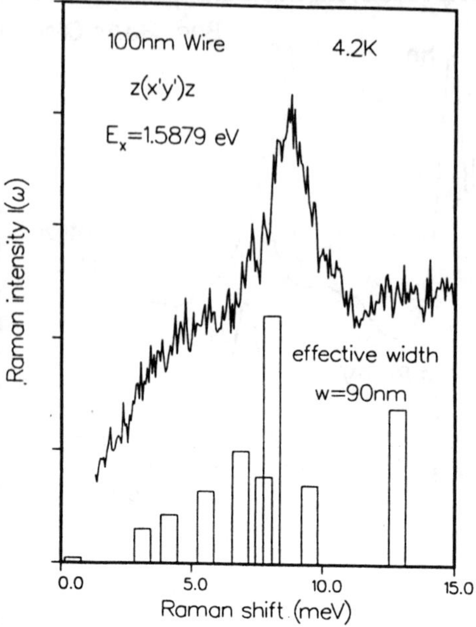

Figure 15. Depolarised resonant Raman spectrum of 100 nm wires. The vertical bars indicate the calculated SPE peaks and the Hartree subbabd spacings [37].

overall arrangement of the energy levels looks like the shell structure of atoms. These calculated values of SPE energies compared favourably to light scattering data from doped quantum dots in a magnetic field obtained by Lockwood *et al.* [39]. Figure 16 shows the measured and the calculated Raman spectra for 150 nm dots at various magnetic fields.

Many body effects are also manifested in the photoreflectance from doped quantum dots as reported in references [40] and modelled by Gumbs *et al.* [41].

5. Conclusions

I have shown that Raman scattering is a valuable non-destructive technique to evaluate nanometer scale pattern transfer processes needed to fabricate quantum nanostructures. The finger prints of typical process-related damages were given in the case of flat etched surfaces and nanostructured surfaces. Surface phonons and multiphonon processes were discussed in the case of GaAs-based nanostructures and the effect of fabrication-induced strain relaxation was mentioned for several semiconductors. Finally a summary is given of our electronic excitation experiments in quantum dots and wires fabricated using the optimised processes discussed in this chapter. Raman scattering is complemented very well by photoreflectance, PLE and X-ray reciprocal space mapping in the study of side-effects arising from nanofabrication techniques.

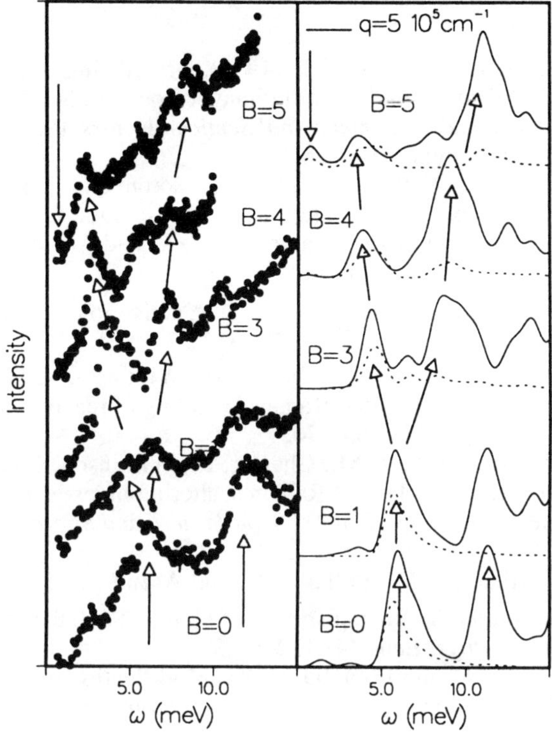

Figure 16. (Left) Measured Raman spectra of modulation doped 150 nm GaAs-GaAlAs quantum dots with nominal carrier density of 8.5 x 10^{11} cm^{-2} up to 5T. (Right) Calculated SPE spectra of a dot with a density of 8.0 x 10^{11} cm^{-2} (124 electrons per dot), 140 nm diameter for the wavevector q= 1 x 10^5 (dashed line) and 5 x 10^5 (solid line). The arrows indicate related peaks in experiment and theory for each magnetic field.

6. Acknowledgements

This work was performed in the author's group while she was at the Nanoelectronics Research Centre of the University of Glasgow, by M. Watt, P.D. Wang, Y.-S. Tang, M.A. Foad, A.P. Smart, R. Cheung, Y.P.Song, C. Guasch, A. Ross and several others. Helpful discussions with C.D.W. Wilkinson are gratefully acknowledged. Fruitful collaborations with the groups of F.H. Pollak (Brooklyn College CUNY) and G.Gumbs (Hunter College CUNY), G. Bauer (University of Linz), A. Cullis (DRA, Malvern), B. Dietrich and W. Kissinger (Institute of Semiconductors, Frankfurt-Oder), D.J. Lockwood and P. Hawrylak (National Research Council, Canada) and A. Pinzuck (Lucent Technologies) have been invaluable and is gratefully acknowledged.

7. References

1. Wilkinson, C.D.W. and Beaumont, S.P. (1990), Dry-etching damage in nanostructures, in S.P. Beaumont and C.M.Sotomayor Torres (eds.), *Science and Engineering of One and Zero-Dimensional Semiconductors*, Plenum Press, New York, pp. 11- 16 and references therein.
2. Cheung, R., Thoms, S., Watt, M., Foad, M.A., Sotomayor Torres, C.M., Wilkinson, C.D.W., Cox, U.J., Cowley, R.A., Dunscombe C. and Williams, R.H. (1992) Reactive ion etching induced damage in GaAs and $Al_{0.3}Ga_{0.7}As$ using $SiCl_4$, *Semicond. Sci. Technol.*, **7**, 1189-1198.
3. Semura S., Saitoh H. and Asakawa, K. (1984) Reactive ion etching of GaAs in CCl_4/H_2 and CCl_4/O_2, *J. Appl. Phys.* **55**, 3131-3135.
4. Watt, M. Sotomayor Torres, C.M., Cheung, R., Wilkinson, C.D.W., Arnot, H.E.G. and Beaumont, S.P. (1988a) Raman scattering of reactive-ion etched GaAs, *Journal of Modern Optics*, **35**, 365-370.
5. Watt, M. Sotomayor Torres, C.M., Cheung, R., Wilkinson, C.D.W., Arnot, H.E.G. and Beaumont, S.P. (1988b) Raman scattering investigations of the damage caused by reactive-ion-etching of GaAs, *Superlattices and Microstructures*, **4**, 243-244.
6. Tiong, K.K., Amirtharaj, P.M., Pollak, F.H. and Aspnes, D. E. (1984) Effect of As^+ ion implantation on the Raman spectra of GaAs: "Spatial correlation" interpretation, *Appl. Phys. Lett.* **44**, 122-124.
7. Chung, Y., Langer, D.W. and Look, D. (1985) Modification of surface characteristics in GaAs with dry processing, *IEEE Transactions on Electron Devices*, **ED-32**, 40-44.
8. Kirillov,D., Cooper, C.B. and Powell, R.A. (1986) Raman scattering study of plasma etching damage in GaAs, *J. Vac. Sci. Technol. B* **4**, 1316-1318.
9. Roughani, B. Jackson, H.E., Jbara, J.J., Mantei, T.D., Hickman, G., Stutz, C.E., Evans, K.R. and Jones, R.L. (1989) Raman scattering studies of Reactive ion-etched MBE <100> n-type GaAs, *IEEE Journal of Quantum Electronics* **25**, 1003-1007.
10. Abstreiter, G., Bauser, E., Fischer, A. and Ploog, K. (1978) Raman spectroscopy - a versatile tool for characterisation of thin films and heterostructures of GaAs and $Al_xGa_{1-x}As$, *Appl. Phys.* **16**, 345-352.
11. M. Cardona and G. Güntherodt (eds.), *Light Scattering in Solids*, Volumes I-VI, Springer Series Topics in Applied Physics, Springer, Berlin (1983-1992).
12. Wang, P.D., Foad, M.A., Sotomayor Torres, C.M., Thoms, S. Watt, M., Cheung, R., Wilkinson, C.D.W. and Beaumont, S.P. (1992) Raman scattering of coupled longitudinal optical phonon-plasmon modes in dry etched n^+-GaAs, *J. Appl. Phys.* **71**, 3754-3759.
13. Murad, S.K., Wilkinson, C.D.W., Wang, P.D., Parkes, W., Sotomayor Torres, C.M. and Camerun, N. (1993) Very low damage etching of GaAs, *J. Vac. Sci. Technol. B* **11**, 2237-2243.
14. Watt, M., Sotomayor Torres, C.M., Arnot, H.E.G. and Beaumont, S.P. (1990) Surface phonons in GaAs cylinders, *Semicond. Sci. Technol.* **5**, 285-290; Watt, M., Arnot, H.E.G, Sotomayor Torres, C.M. and Beaumont, S.P. (1989) Surface phonon studies of nanostructures, in *Nanostructure Physics and Fabrication*, M. Reed and W.P. Kirk (eds.), Academic Press, Boston, pp. 89-96.

15. Hayashi, S. and Kanamori, H. (1982) Raman scattering from the surface phonon mode in GaP microcrystals, *Phys. Rev. B* **26**,7079-7082.
16. Ruppin, R. and Englman, R. (1970) Optical phonons of small crystals, *Rep. Prog. Phys.* **33**, 149-196.
17. Wang, P.D., Cheng, C., Sotomayor Torres, C.M. and Batchelder, D.N. (1993) GaAs micrometer-sized dot imaging by Raman microscopy, *J. Appl. Phys.* **74**, 5907-5909.
18. Tang, Y.S. and Sotomayor Torres, C.M. (1996) Damage, strain and quantum confinement issues in dry etched semiconductor nanostructures, *Mat. Res. Soc. Symp. Proc.* **405**, 99-108.
19. Dietrich, B., Frankenfeld, H., Tang, Y.S., Sotomayor Torres, C.M., Zeindl, H.P. and Wolff, A. (1996) Elastic relaxation of pseudomorphic strain in quantum dots, *Solid State Phenomena*, **47-48**, 535-539.
20. Ni, W.-X., Birch, J., Tang, Y.S., Joelsson, K.B., Sotomayor Torres, C.M., Kvick, A.and Hansson, G.V. (to be published) Lattice distorsion in dry etched Si/SiGe quantum dot arrays studied by 2D-reciprocal space mapping using synchrotron X-ray diffraction, *Thin Solid Films*.
21. Tang, Y.S., Sotomayor Torres, C.M., Ni, W.N. and Hansson, G.V. (1996) Room temperature electroluminescence of nanofabricated Si-$Si_{1-x}Ge_x$ quantum dot diodes, *Superlattices and Microstructures*, **20**, 505-511.
22. Tang, Y.S., Wang, P.D., Sotomayor Torres, C.M., Lunn, B. and Ashenford, D.E. (1995) Process-induced strains in dry etched semiconductor nanostructures studied by photoreflectance, *J. Appl. Phys.* **77**, 6481-6484
23. Qiang, H., Pollak, F.H., Tang, Y.S., Wang, P.D. and Sotomayor Torres, C.M. (1994) Characterisation of process-induced strains in GaAs/GaAlAs quantum dots using room temperature photoreflectance, *Appl. Phys. Lett.* **64**, 2830-2832.
24. Tang, Y.S., Wilkinson, C.D.W., Sotomayor Torres, C.M., Snmith, D.W., Whall, T.E. and Parker, E.H.C. (1993) Optical properties of Si/$Si_{0.87}Ge_{0.13}$ multiple quantum well wires, *Appl. Phys. Lett.* **63**, 497-499.
25. Holy, V., Darhuber, A.A., Bauer, G., Wang, P.D., Song, Y.P., Sotomayor Torres, C.M. and Holland, M.C. (1995) Elastic strains in GaAs/AlAs quantum dots studied by high-resolution x-ray diffraction, *Phys. Rev. B* **52**, 8348-8357, and references therein.
26. Johnson, N.P., Foad, M.A., Murad, S., Holland, M.C. and Wilkinson, C.D.W. (1994) Deep levels induced by $SiCl_4$ reactive ion etching in GaAs, *Mat. Res. Soc. Symp. Proc.* **325**, 443-448.
27. Wang, P.D. and Sotomayor Torres, C.M. (1993) Multi-phonon relaxation in GaAs-AlGaAs quantum well dots, *J. Appl. Phys.* **74**, 5047-5052.
28. Benisty, H., Sotomayor Torres, C.M. and Weisbuch, C. (1991) Intrinsic mechanism for the poor luminescence properties of quantum box systems, *Phys. rev. B* **44**, 10945-10948.
29. Sotomayor Torres, C.M. (in press) Carrier relaxation in 1- and 0-D, in *Hot electrons in semiconductors: Physics and devices*, N. Balkan (ed.), Oxford University Press.
30. Song, Y.P, Wang, P.D., Sotomayor Torres, C.M. and Wilkinson, C.D.W. (1994) Magnetically confined plasma reactive ion etching of GaAs/AlGaAs quantum nanostructures, *J. Vac. Sci. Technol. B* **12**, 3388-3392.

31. Song, Y.P, Wang, P.D., Sotomayor Torres, C.M. and Wilkinson, C.D.W. (1995) Magnetically confined plasma reactive ion etching and photoluminescence of GaAs quantum wires, *Semicond. Sci. Technol.* **10**, 1404-1407.
32. Wang, P.D., Song, Y.P., Sotomayor Torres, C.M., Holland, M.C., Lockwood, D.J., Hawrylak, P., Palacios, J.J., Christianen, P.C.M., Maan, J.C. and Perenboom, J.A. (1994), Optical emission and Raman scattering in modulation-doped GaAs-AlGaAs quantum wires and dots, *Superlattices and Microstructures*, **15**, 23- 27.
33. Cingolani, R., this volume
34. Pinzuck, A., this volume
35. Jusserand, B., this volume
36. Abstreiter, G., this volume
37. Lockwood, D.J., Hawrylak, P., Wang, P.D., Song, Y.P, Sotomayor Torres, C.M., Holland, M.C., Pinczuk, A. and Dennis, B.S. (1996) Inelastic light scattering from electronic excitations in deep-etched quantum dots and wires, *Solid State Electronics*, **40**, 339-342.
38. Hawrylak, P., Wojs, A., Lockwood, D.J., Wang, P.D., Sotomayor Torres, C.M., Pinczuk, A. and Dennis, B.S. (1996) Optical spectroscopies of electronic excitations in quantum dots, *Surface Science* **361/362**, 774-777.
39. Lockwood, D.J., Hawrylak, P., Wang, P.D., Sotomayor Torres, C.M., Pinczuk, A. and Dennis, B.S. (1996) Shell structure and electronic excitations of quantum dots in a magnetic field probed by inelastic light scattering, *Phys. Rev. Lett.* **77**, 354-357.
40. Wang, P.D., Sotomayor Torres, C.M., Holland, M.C., Qiang, H., Pollak, F.H and Gumbs, G. (1994) Photoreflectance study of modulation doped GaAs/GaAlAs quantum dots fabricated by reactive ion etching, *Mat. Res. Soc. Proc.* **324**, 187-193.
41. Gumbs, G., Huang, D, Qiang, H. Pollak, F.H., Wang, P.D., Sotomayor Torres, C.M. and Holland, M.C. (1994) Electromodulation spectroscopy of an array of modulation-doped GaAs/GaAlAs quantum dots: Experiment and theory, *Phys. Rev. B* **50**, 10962-10969.

OPTICAL PROPERTIES OF A LOW DIMENSIONAL SILICON SYSTEM: POROUS SILICON

A. AYDINLI and A. BEK
Department of Physics
Bilkent University
Bilkent, Ankara, 06533
Turkey

1. Introduction

Crystalline silicon is an ineffective light emitter due to its indirect bandgap. Various efforts to obtain efficient room temperature luminescence from Si have so far failed. However, recent discovery of bright photoluminescence from electro-etched silicon leading to the formation of a porous structure has led to intense research activity in this field with the hopes of eventual integration of light emitting silicon devices into all silicon microelectronics circuitry. A main feature of porous Si is its lower dimensionality, which may introduce new mechanisms for luminescence due to revised electronic states. However, none of the several models proposed for the origin of the efficient luminescence has so far been widely accepted.. In this article, we review the current status of the formation and optical characterization of porous silicon.

2. Formation and Microstructure of Porous Silicon

Porous Si can be formed using both n- and p-type Si by either chemical [1-4] or electrochemical [5-7] etching. Luminescence properties similar to porous Si has also been obtained in other materials such as SiGe [8], silicon-on-sapphire [9], and silicon carbide [10]. In addition to chemical

G. Abstreiter et al. (eds.), Optical Spectroscopy of Low Dimensional Semiconductors, 355–373.

methods, techniques as diverse as spark erosion [11] and classical photolithography [12] have been successfully used. Conventional electrochemical etching is monitored either through anodic current or the potential drop across a Pt cathode and a Si anode. Contact to the Si wafer is made through the back side and Pt cathode faces the front face of Si in either aqueous or nonaqueous solutions of HF. Back side is metallized by depositing aluminum followed by thermal annealing at 450 °C to obtain a good ohmic contact, prior to anodization. During the formation of the porous Si, p-type Si wafer forms forward biased Schottky diode with the electrolyte and n-type Si wafer reverse biased. Thus, n-type wafers require heavily doped Si or illumination for the necessary current to flow.

While several models have been proposed for dissolution of Si in fluoride media, the exact mechanism is still not well understood [13]. The proposal by Lehman and Gosele [14] gives the main features of the reaction as;

1. $$=SiH_2 + F^- + h^- \rightarrow =SiFH_2^\circ$$

2. $$=SiFH_2^\circ + F^- \rightarrow =SiF_2 + H_2 + e^-$$

3. $$\begin{matrix} \equiv Si & & \equiv SiH \\ & SiF_2 + HF \rightarrow & + \\ \equiv Si & & \equiv SiSiF_3 \end{matrix}$$

4. $$\equiv SiSiF_3 + HF \rightarrow \equiv SiH + SiF_4$$

5. $$SiF_4 + 2HF \rightarrow 2H+ + SiF_6^{2-}$$

Many such proposals are possible and have been made both for p- and n-type samples [15]. Several models have been proposed for the mechanism of formation as well. Beale et.al. [16] have proposed that under typical conditions, a depletion layer forms at the semiconductor/electrolyte interface limiting current flow. The electrochemical reaction then proceeds by thermionic emission across the barrier (or tunneling in the case of degenerately doped samples). Barrier lowering is expected with applied field suggesting thinner depletion layers at the pore tips due to enhanced electric field there. Hence, dissolution proceeds at the pore tips. Smith and Collins [17] have proposed a diffusion limited model where random walk of a hole (electron) into (away from) the Si surface enhances dissolution at the pore tips. Finally, Lehman and Gosele [14] proposed that quantum effects manifesting themselves in small crystallites making up the porous structure, may govern morphology.

The morphology of porous Si depends on many parameters including the type, doping, crystallographic orientation of the wafer and HF concentration, pH, current density, temperature, duration of dissolution, and illumination during etching. Bulk Si can be made microporous (pore width $\leq$ 2 nm), mesoporous (pore width 2-50 nm), or macroporous (pore width > 50 nm) [18]. Microporous Si is the most interesting material due to its luminescent properties. It may be found in almost all kinds of samples but pure microporous structure is observed only on moderately doped p-type Si. Generally speaking, porosity increases with decreasing HF content. Many techniques have been used to characterize the microstructure of porous Si. Transmission electron microscopy (TEM) shows that [16], layers consist of long pores running perpendicular to the surface with average sizes of ~10 nm for p+ samples while they are a random collection of holes of ~3 nm for p- and n-type samples under illumination. In addition to crystallites [19], an amorphous phase [20] which seems to increase with etching time is also reported. Scanning tunneling microscopy (STM) [21] identified crystallites of 10-2 nm sizes with decreasing sizes leading to a blue shift of observed luminescence, while TEM in combination with atomic force microscopy (AFM) [22] failed to detect any crystalline phase from similar samples. Si crystallites dispersed in an amorphous matrix has been identified using Raman spectroscopy [23]. It is important to note that Si oxidizes rapidly in ambient air. Aging of samples should be taken into account when measurements are made. The use of laser light during experiments affects the samples as well, leading to degradation of sample PL due either to photochemical effects or heating [24].

3. Optical Characterization of Porous Silicon

3.1. FTIR MEASUREMENTS

Fourier Transform Infrared (FTIR) spectroscopy is a powerful tool to identify chemical species that make up the sample. It has been widely used to identify and correlate various optical and electrical properties of porous silicon [25] with chemical species on the surface of porous silicon layers. However, unlike the typical transmission experiments on most IR transparent samples, porous silicon samples require a reflectance measurement due to the Al layer on the back side of the wafers, unless special effort is made to remove this layer prior to transmission measurements. A typical diffuse reflectance FTIR spectrum of a fresh porous silicon sample is shown in Fig. 1a, in the range of 400-4000 cm^{-1}. Several peaks dominate the spectrum at 512 cm^{-1}, 605 cm^{-1}, 655 cm^{-1}, 685 cm-1, 814 cm^{-1}, 920 cm^{-1}, and a relatively broad band centered at 2100 cm^{-1} with three peaks identified at 2090 cm^{-1}, 2115 cm^{-1}, and 2142 cm^{-1}. Exact location of these peaks depend on sample preparation conditions and may vary up to a few 10 cm^{-1} representing the local bonding environment. We start identifying with the peak at 512 cm^{-1} as the F activated Si optic mode. The peak at 605 cm^{-1} is a Si-Si stretching mode and 655 and 685 cm^{-1} as the SiH_2 and SiH wagging modes. The peak at 814 cm^{-1} is due to SiH_2 twisting mode while the peak at 920 cm^{-1} is SiH_2 scissors mode. We also find very weak peaks at 1045, 1106, due to symmetric and asymmetric stretching modes of Si-O-Si. The peaks at 2090, 2215 and 2142 are Si-H, Si-H_2, and Si-H_3 stretching modes broadened into a wide peak indicating surface disorder. We conclude that the surface and the near surface [26] of porous silicon is covered with hydrogen in various forms of hyrides with very little oxygen. Since it is well known from electrochemical studies of Si that even a few ppm of water molecules under illumination can result in oxidation of Si surfaces [13], small amounts of oxygen related features in Figure 1(a) could have formed during the short period between sample preparation and data accumulation. Porous Si with its large surface to volume ratio rapidly oxidizes in air. This is illustrated for the same sample in Figure 1(b). The sample was stored in air for more than a month. Here, we clearly observe Si-O-Si bending and symmetric and asymmetric stretching modes at 428 cm^{-1}, 1070 cm^{-1} and 1153 cm^{-1}, respectively. Also evident are peaks at 2266 cm^{-1}, 2867cm^{-1}, 2926 cm^{-1} and 2959 cm^{-1} which are identified as O-Si-H stretching, CH_2 symmetric and asymmetric stretching and CH_3 asymmetric stretching modes, respectively. CH modes are attributed to hydrocarbon contamination during storage in ambient air [27]. We now observe a broad peak centered at 3400 cm^{-1} due to absorbed water moisture. Maruyama et.al. [28] studied PL spectra of porous Si oxidized in humid and dry ambient air and found that PL efficiency is much higher for samples oxidized in humid air which is attributed to the role -OH groups play in oxidation of porous Si.

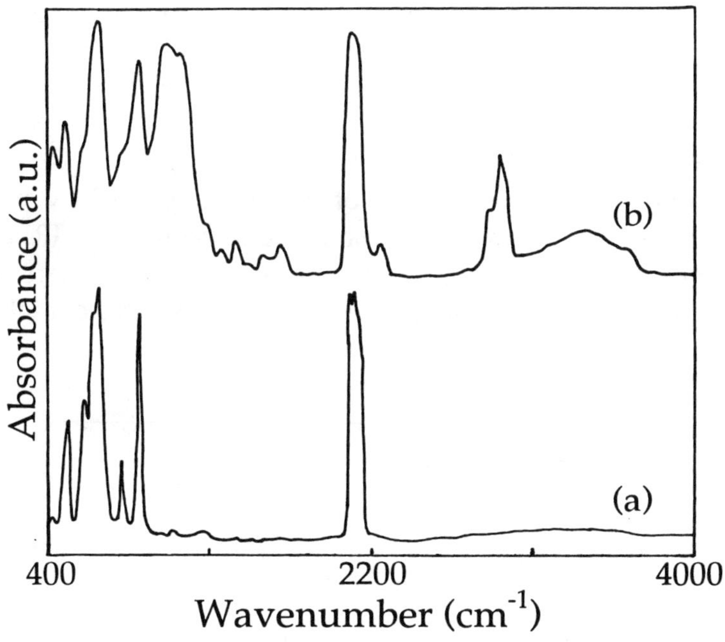

Figure 1. FTIR Diffuse reflectance spectra of (a) fresh and (b) oxidized porous Si. Note the appearance of Si-O-Si bonds after storage in air for one month.

FTIR spectroscopy has been used to characterize a variety of surface and bulk treatments of porous Si. Rehm et.al. [29] followed the effects of methanol post-treatment on porous Si using IR spectroscopy while Xiao et.al. [30] characterized oxygen incorporation with remote plasma treatment. Dubin et. al. [31] performed *in situ* photomodulated IR spectrosocpy during formation of porous silicon. They found that porous Si luminesced even before exposing to air. Extracting the free carrier signal from the IR spectrum, they were able to deduce that photocarriers in wet porous Si in solution and in dry porous Si are delocalized, and localized, respectively.

3.2. PHOTOLUMINESCENCE

Crystalline Si exhibits a weak photoluminescence (PL) peak corresponding to a transition at its indirect band gap energy in the IR whereas PL peaks from below to well above the indirect band gap has been obtained in porous Si [32]. Above the bandgap PL was first observed [33] in 1984 at 4.2K. In 1990, Canham [34] first reported observing room temperature PL in porous Si. The visible PL from porous Si is orders of magnitude stronger than the PL from the band gap PL of c-Si. Considering the exceptionally large parameter space available for fabricating porous Si, numerous studies have been carried out to characterize the PL properties of porous Si. Typically PL spectra consists of a single Gaussian-like peak of approximately 0.3 eV width. It has been possible to tune the PL peak position from near IR to blue using etch times and etchant compositions as control parameters. Figure 2 shows PL peaks of two samples etched in two different chemical solutions, one HF rich and other HF poor.

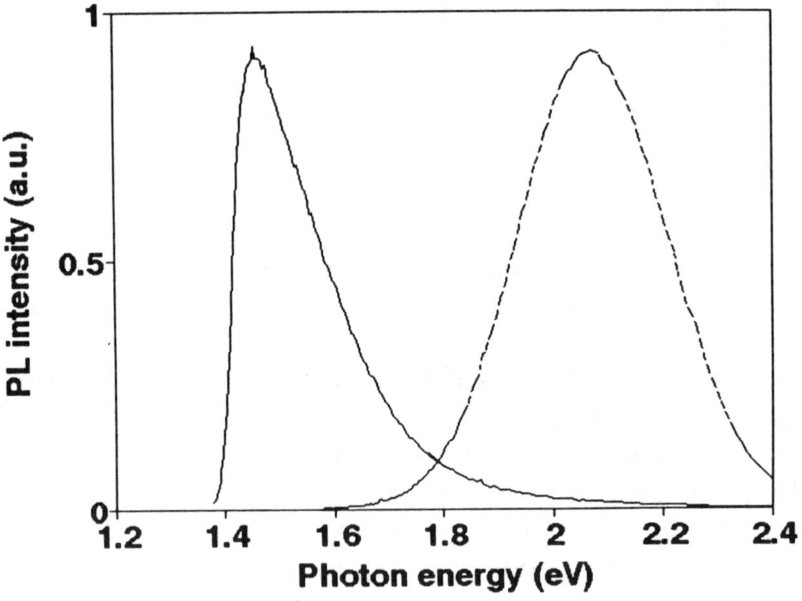

Figure 2. PL spectra of two samples prepared in different ethanoic solutions of HF. Higher energy PL spectra corresponds to lower HF concentration.

In parallel with the increase in porosity with lower HF concentration, we observe the PL peak initially centered in the near IR to shift to blue-green region of the spectrum. The sharp decrease below 1.45 eV is due to rapidly falling detector response. The original paper by Canham [34] shows PL peak positions shifting from ~1.4 eV to 1.55 eV with increasing etch times in the same solution. While Canham used this data to support his model of quantum confinement as the origin of above the band gap luminescence, demonstration of wavelength tunability via PL has important technological implications. Thus, not unexpectedly, wavelength tunability has also been achieved in electroluminescing samples as well [35].

The spectral shape and temperature dependence of PL has also prompted much debate with scattered observations of structure on the PL spectrum of porous Si. Asnin and coworkers have observed [36] fine structure in the low temperature spectra of their porous Si layers. This was explained as recombination due to 1D-excitons in quantum wires. Multiple peak photoluminescence has been inferred also at room temperature [37-38]. These have been taken as evidence of recombination due to quantum confined carriers in Si nanocrytallites as well as being due to compounds of Si. However, in general the presence of Si crystallites with various sizes is expected to smear any sharp features as is observed in majority of the samples.

Temperature dependence of PL spectra may offer important clues as to the origin of luminescence in porous Si. However, there still lacks consistent data in this area. This is in part due to several components of the PL spectra showing different temperature dependence. In fact, PL data obtained by Xu et.al. [39] shows both red and blue shift of PL at different wavelengths upon cooling. On the other hand PL intensity has been observed [40] to increase upon cooling as well as decrease after first rising at around 150 K. Figure 3 shows the temperature dependence of PL spectrum of a porous Si sample prepared under ambient light in an ethaonic solution of HF (1 vol. 48% HF:3 vol. ETH) with a current density of 75 mA/cm^2 for 60 min. As the temperature decreases, PL efficiency increases with decreasing temperature reaching to a maximum at 150 K and thereafter decreases again. Meanwhile, PL peak position continuously shifts towards the blue most of which occurs by 150 K. Total blue shift in this experiment was approximately 139 meV. A summary of temperature dependence of PL spectra of several samples prepared with different current densities is shown in Fig.4.

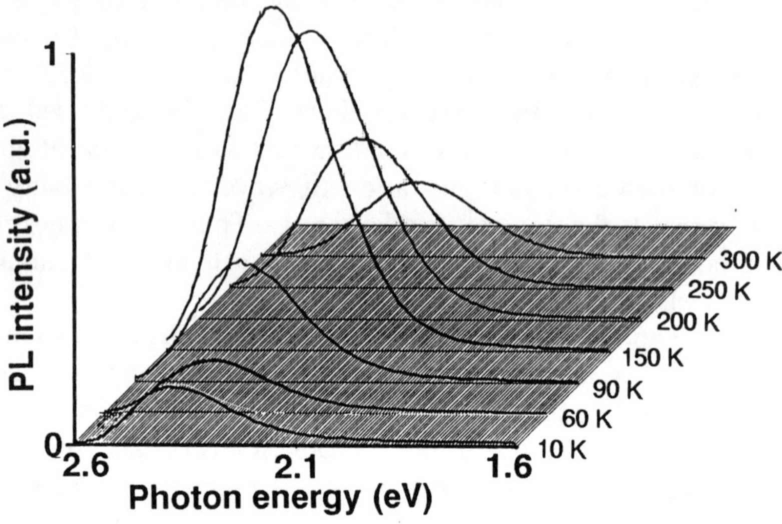

Figure 3. Temperature dependence of PL spectra in porous Si prepared in a solution of 1 vol. %48 HF + 3 vol. Ethanol with 75 mA/cm^2 for 60 min.

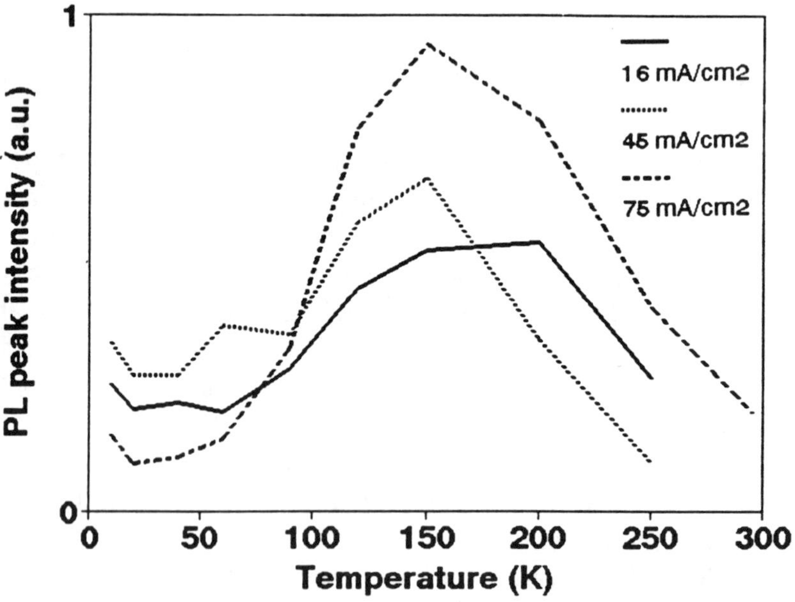

Figure 4. Temperature dependence of three porous Si samples prepared in a solution of 1 vol. %48 HF + 3 vol. Ethanol at three different current densities for 60 min.

Typically, 10 K PL intensities are of the same order of magnitude at room temperature whereas PL intensity at ~150 K is 5-10 times more efficient. Furthermore, with increasing anodic etching current densities, we find that PL efficiency generally increases. As increasing current density results in increased porosity, smaller crystallite sizes and larger surface area, we expect an increase in PL intensity. Assuming radiative recombination taking place at the surface of the crystallites, low temperature dependence of PL peak energy can be explained by invoking thermally activated diffusion of carriers photogenerated at the core of the crystallites to the surface. At higher temperatures, radiative recombination rate at the near surface region decreases with temperature leading to reduction of PL intensity at room temperature [23]. In Figure 5 we also display temperature dependence of the PL peak position for the same set of samples.

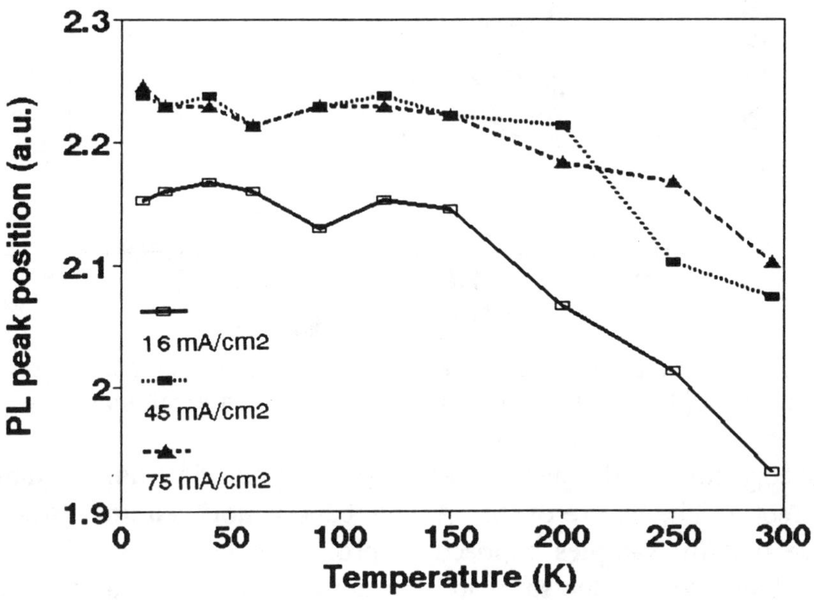

Figure 5. PL peak position of the same samples as in Figure 4.

Several points are to be noted. First, all samples etched at different current densities blue shift as the temperate is lowered. Second, increasing current density also leads to higher PL peak energies. Finally, samples prepared at 45 and 75 mA/cm^2, both blue shift to the same PL peak energy even though their room temperature values are different. The blue shift of the PL peak parallels that of a semiconductor with the band gap widening as temperature falls and hence the resulting blue shift. However, not all samples show this rather simple behaviour. Samples with PL peak position in the red at room temperature (prepared with high HF concentrations) often show double peak behaviour upon cooling with a near infrared peak rising to compete and over take the red PL, Figure 6.

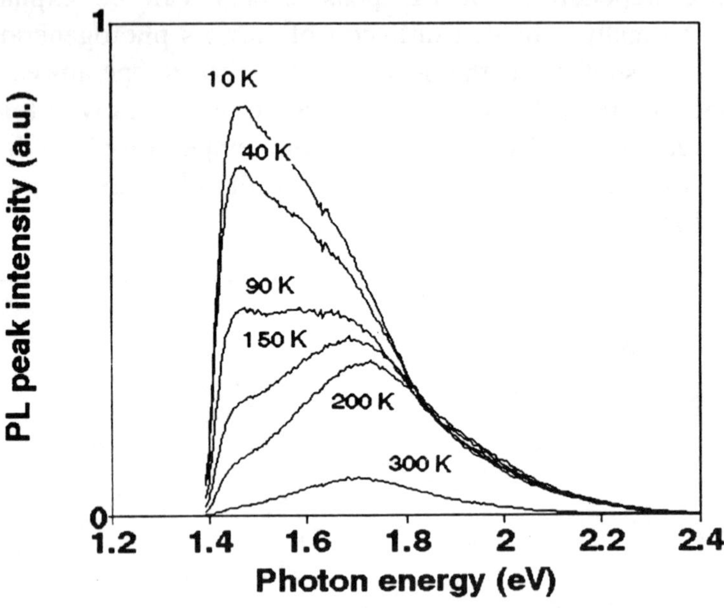

Figure 6. Temperature dependence of PL spectra of a porous Si sample prepared in a solution of 1 vol. %48 HF + 1 vol. Ethanol with 65 mA/cm^2 for 60 min.

The observation of the near infrared peak beyond 850 nm is limited by our detector. The nature of this near IR band is not well understood. We observe that for samples exposed to prolonged laser light illumination, near IR band slowly blue shifts indicating that it may be related to surface species which oxidize or desorp during laser illumination. Koch [32] has pointed out to the universal nature of a near IR peak accompanying the visible peak. However, in contrast with our data they do not observe laser degradation. This may mean that our near IR peak does not necessarily correspond to the one they observe but rather a new peak. Further work is necessary to understand the nature of this peak.

3.3. RAMAN STUDIES

Raman scattering has been used extensively to characterize structural and electrical properties of solids and both crystalline and amorphous Si are not exceptions. The observed Raman shift for crystalline and amorphous Si are at 520 cm^{-1} and ~ 460-480 cm^{-1}, respectively. Room temperature width of triply degenerate crystalline Si phonon band is ~3.6 cm^{-1}. A direct consequence of size reduction is the modification of the wavefunction of the infinite crystal by a weighing function incorporating a correlation length representing the limited size of the crystallites. With decreasing crystallite size a larger portion of the dispersion curve is then sampled. With these assumptions, detailed behaviour of Raman shift in small crystallites have been calculated under strong phonon confinement [41]. In this model, first order Raman spectrum is given by :

$$I(\omega) = \int_0^1 \frac{\exp(-q^2 L^2 / 4a^2) d^3 q}{[\omega - \omega(q)]^2 + (\Gamma_0 / 2)^2} \qquad (1)$$

where q is expressed in units of $2\pi/a$, a is the lattice constant and Γ_0 is the linewidth of the Si phonon in crystalline bulk Si. Dispersion of the optical phonon in crystalline Si is taken as :

$$\omega^2 (q) = A + B \cos(\pi q / 2) \qquad (2)$$

where $A=1.714x10^5$ cm^{-2} and $B=1.000x10^5$ cm^{-2}. It is possible to fit the experimental data with above line shape function and deduce average particle sizes from the data. In Figure 7, we present a Raman spectrum of a free standing porous Si film. We find that peak is downshifted by 5 cm^{-1}. The linewidth of the peak is ~14 cm^{-1}. The calculated lineshape is superimposed on the experimental data. The average size is found to be 4.0 nm. No evidence for an amorphous phase is observed . We find that as the HF concentration decreases, Raman line shape broadens and down shifts further. This is expected since decreasing HF concentration leads to higher porosity, lower crystallite size samples. Several groups have measured Raman spectra of porous silicon [42-43]. however, the most comprehensive Raman work on porous Si is performed by Tanino et. al. [44]. They have measured low temperature Raman spectra on free standing porous films in the range of one- through four-phonon processes.

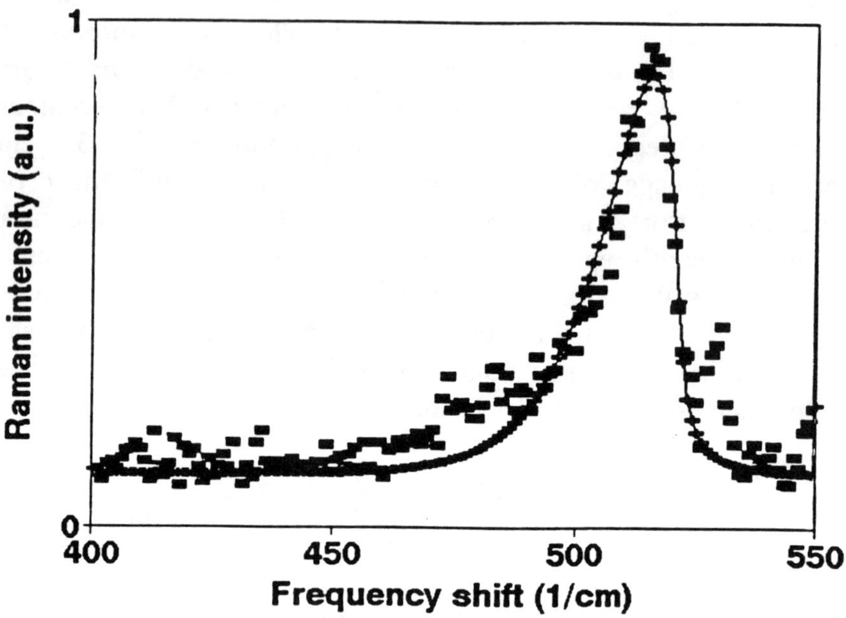

Figure 7. Stokes shift for a free standing porous Si sample prepared in a solution of 3 vol. %48 HF + 1 vol. H_2O with 50 mA/cm^2 for 60 min.

They concluded that aqueous solutions produces smaller crystallite sizes with spherical shapes. Samples prepared with ethaonic solutions resulted in slab like configurations with better ordering of porous Si. Finally, higher order scattering was found to be a very sensitive indicator of surface disorder.

3.4. OPTICAL ABSORPTION

Optical transmission experiments on free standing films provide direct evidence for the absorption processes in porous Si. The earliest such experiment was done by Halimaoui [45]. Their data is shown in Figure 8 with transmission spectra for bulk spectra included for comparison. Clear shift of the absorption edge to the visible is observed which increases as the porosity is increased. They observed that the absorption edge for heavily doped p-Si is red shifted in comparison with a lightly doped sample.

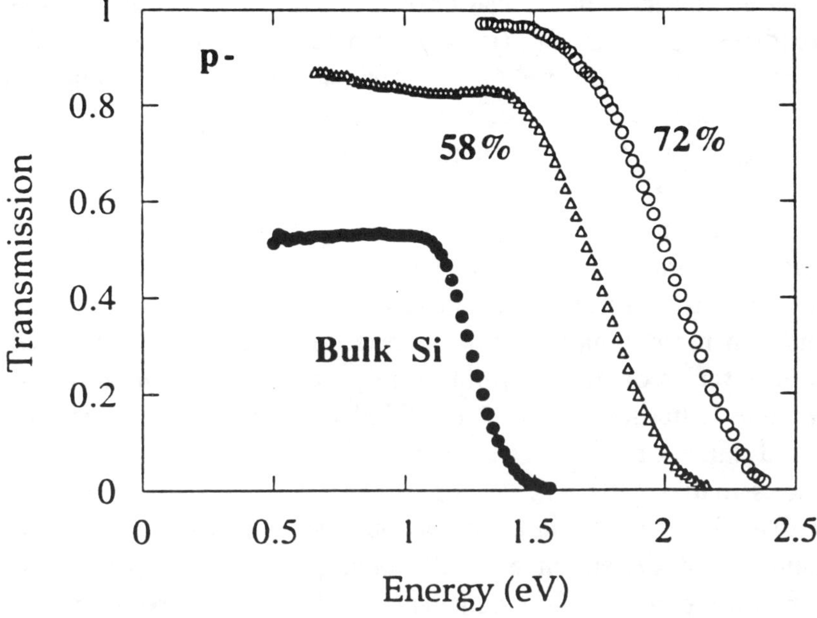

Figure 8. Transmission versus photon energy for two porous Si samples measured at 300 K. From Ref. 4

This behaviour is confirmed by our and other studies [23] and provides clear evidence that absorption takes place through a larger effective band gap in porous Si. However, it is also found that the exact position of the absorption edge is a sensitive function of the surface and red shifts upon hydrogen desorption. We also did transmission experiments as a function of temperature. We find gradual blue shifting of absorption edge without any sharp features as the temperature is lowered, reflecting the crystalline nature of the free standing film.

4. Models of Light Emission in Porous Silicon

Many models have been proposed for the observation of above-the-band gap light emission from porous silicon. Only a few have found strong experimental and theoretical support worthy of further discussion. Among them is the molecular agents theory which attributes luminescence to specific molecules. One such molecule is siloxene ($Si_6O_3H_6$) [46]. While many of the optical features of siloxene mimic that of porous Si, the fact that fresh porous Si does not contain oxygen and that oxidized porous Si (totally replacing hydrogen passivation) luminesces strongly, discredits this proposal. Similarly, luminescence of polysilane chains has also been discounted due to lack of correlation between SiH infrared absorption and PL intensity [31].

The initial proposal of Canham [18] was recombination of carriers across electronic states modified through lowering of dimensionality. Quantum confinement of carriers take place due to very small dimensions of Si columns leading to an larger effective band gap energy and hence above the band-gap recombination light. The large widths of the PL peaks are explained in terms of size distribution of particles. The initial proposal of a columnar structure was first modified to consist of undulating wires [47] and later of crystalline Si particles embedded in an oxide or amorphous matrix [48]. A proof of the quantum confinement model responsible for visible light in porous Si is expected to be the establishment of dependence of PL peak energy on particle size. If quantum confinement is the mechanism of PL, then we expect that the PL energy is inversely proportional to the particle size. Such a proof is not easy and requires the use of several analytical techniques under well controlled conditions. Most of the experiments establishing this relationship have not been found convincing due mainly to incomplete characterization of samples. However, Schuppler et.al. [49] have done extensive work using X-ray absorption measurements which led to an inverse relationship of PL peak energy and particle size. In addition, ours and many other experiments show blue shift of PL peak energy with increasing porosity without direct determination of particle size involved in the luminescence. Unfortunately, experiments to the contrary also exist where blue shifting of the absorption edge with particle size does not lead to any shift of the PL peak energy [23]. We are lead to the conclusion that while photogenerated carriers are created at the crystalline core of a band gap modified particle, strong PL originates from near surface region of small crystallites. Low temperature PL data [50] provided further evidence for inhomogenous distribution of indirect band gap energy of Si particles. Luminescence occurs within an entity with a phonon spectrum very similar to crystalline Si. Using an effective mass approach, Hybertsen [51] calculated both zero phonon and phonon assisted radiative transition rates in Si crystallites stressing the importance of surface states. He finds that phonon assisted transitions dominate above 15-20 A sizes. He also suggests that PL from smaller size particles should decay faster. Thus, time resolved studies of PL decay at different wavelengths (assumed to be originating from different size crystallites) is expected to show multiexponential decay. Indeed, this is what is observed experimentally [52]. Both the phonon sidebands and the multiexponential decay time can also be understood within the framework of a surface modified quantum picture advocated by Koch [32]. As opposed with the pure quantum confinement picture where a perfect particle in a box model is assumed, surface modified model invokes states with a large amplitude in the near surface region leading to a tail of energy states which are absent in the pure confinement picture. Surface states model shares several features of the pure quantum confinement model, such as absorption and high energy luminescence thorough bulk-like states. Considering the vast surface area and the large number of broken bonds present in porous Si, it claims to present a more comprehensive picture explaining the position and decay time of red luminescence through surface states.

5. Conclusions

Porous Si provides a serious model for efficient room temperature light emission in Si. While the peak quantum efficiency is lower than GaAs integrated efficiency is comparable. However, both the physical understanding of the light emission mechanism as well as technological problems of stability needs to be addressed further before it becomes a serious candidate for a practical light emitting Si device.

6. Acknowledgments

It is a pleasure to acknowledge the contributions of M. Gure during sample preparations. This work is supported by Scientific and Technical Research Council of Turkey (TUBITAK) under grant no: TBAG-1244.

7. References

1. Sarathy, J. et.al. (1992) Demonstration of photoluminescence in nonanodized silicon, *Appl. Phys. Lett.* **60**, 1532-1533.
2. Noguchi, N. and Suemune, I. (1993) Luminescent porous silicon synthesized by visible light irradiation, *Appl. Phys. Lett.* **62**, 1429-1431.
3. Kidder, J.N. et.al. (1992) Comparison of light emission form stain-etch and anodic-etch silicon films, *Appl. Phys. Lett.* **61**, 2896-2898.
4. Fathauer, R.W. et.al. (1992) Visible luminescence from silicon wafers subjected to stain etches, *Appl. Phys. Lett.* **60**, 995-997.
5. Xie, Y.H. et.al. (1992) Luminescence and structural study of porous silicon films, *J. Appl. Phys.* **71**, 2403-2407.
6. Xu, Z.Y., Gal, M., and Gross, M. (1992) Photoluminescence studies on porous silicon, *Appl. Phys. Lett.* **60**, 1375-1377.
7. Shih, S., Jung, K.H., and Kwong, D.L. (1993) Photoluminescence study of anodized porous Si after HF vapor phase etching, *Appl. Phys. Lett.* **62**, 1904-1906.
8. Ksendzov, A. et.al. (1993) Visible photoluminescence of porous $Si_{1-x}Ge_x$ obtained by stain etching, *Appl. Phys. Lett.* **63**, 200-202.
9. Dubbelday, W.B. et.al. (1993) Photoluminescent thin film porous silicon on sapphire, *Appl. Phys. Lett.* **62**, 1694-1696.
10. Konstantinov, A.O. et.al. (1995) Photoluminescence studies of porous silicon carbide, *Appl. Phys. Lett.* **66**, 2250-2252.
11. Hummel,R.E., and Chang, S.S., (1992) Novel technique for preparing porous silicon, *Appl. Phys. Lett.* **61**,1965-1967.
12. Nassiopoulos, A.G. et.al. (1995) Visible luminescence from one and two dimensional silicon structures produced by conventional lithographic and reactive ion etching techniques, *Appl. Phys. Lett.* **66**, 1114-1116.
13. Chazalviel, J.N. (1994) The silicon/electrolyte interface, in Vial J.C. and Derrien J.(eds.), *Porous silicon Science and Technology*, Springer -Verlag, p.17.
14. Lehmann, V. and Gosele U. (1991) Porous silicon formation: A quantum wire effect, *Appl. Phys. Lett.* **58**, 856-858.
15. Searson, P.C., Macaulay, J.M., and Ross, F.M. (1992) Pore morphology and the mechanism of pore formation in n-type silicon,

J. Appl. Phys. **72**, 253-258.
16. Beale, M.I.J. et. al. (1985) An experimental and theoretical study of the formation and microstructure of porous silicon, *J. of Crys. Growth* **73**, 622-636.
17. Smith, R.L., and Collins, S.D., (1992) Porous silicon formation mechanisms, *J. Appl. Phys. 71, R1-R22.*
18. Canham, L.T. (1990) Silicon quantum wire array fabrication by electrochemical and chemical dissolution of wafers, *Appl. Phys. Lett.* **57**, 1046-1048.
19. Takasuka, E. and Kamei, K. (1994) Microstructure of porous silicon and its correlation with photoluminescence, *Appl. Phys. Lett.* **65**, 484-486.
20. Shih, S., Jung, K.H., Qian, R.Z., Kwong, D.L. (1993) Transmission electron microscopy study of chemically etched porous Si, *Appl. Phys. Lett.* **62**, 467-469.
21. Enachescu, M., Hartmann, E., and Koch, F. (1994) Correlation of surface morphology with luminescence of porous Si films by scanning tunneling microscopy, *Appl. Phys. Lett.* **64**, 1365-1367.
22. George, T. et.al.(1992) Microstructural investigations of light emitting porous Si layers, *Appl. Phys. Lett.* **60**, 2359-2361.
23. Kanemitsu,Y. et.al. (1993) Microstructure and optical properties of free standing porous silicon films: Size dependence of absorption spectra in Si nanometer sized crystallites, *Phys. Rev. B* **48,** 2827-2830.
24. Mariotto, G., Ziglio, F., and Freiere Jr., F.L. (1995) Raman scattering and nuclear surface characterization of aged porous silicon, *J. Appl. Phys.* **78**, 3335-3341.
25. Porous Silicon, Eds. Z.C. Feng and R. Tsu, World Scientific, 1994.
26. Allongue, P. et.al. (1995) Evidence for hydrogen incorporation during porous silicon formation, *Appl. Phys.Lett.* **67**, 941-943.
27. Teschke, O. et.al.(1994) Photoluminescence spectrum red shifting of porous silicon by a polymeric carbon layer, *Appl. Phys. Lett.* **64**, 3590-3592.
28. Maruyama T. and S. Ohtani S. (1994) Photoluminescence of porous silicon exposed to ambient air, *Appl. Phys. Lett.* **65**, 1346-1348.
29. Rehm, J.M. et.al.(1995) How methanol affects the surface of blue and red emitting porous silicon, *Appl. Phys. Lett.* **66**,3669-3671.
30. Xiao,Y., et.al.,(1993) Enhancement and stabilization of porous silicon

photoluminescence by oxygen incorporation with a remote plasma treatment, *Appl. Phys. Lett.* **62**, 1152-1154.
31. Dubin, V.M., Ozanam, F., and Chazalviel, J.N., (1994) Electronic states of photocarriers in porous silicon studied by photomodulated infrared spectroscopy, *Phys. Rev. B* **50**, 14867-14880.
32. Koch, F. and Petrova-Koch V. (1994) The surface state mechanism or light emission from porous silicon, in Z.C. Feng and R. Tsu (eds.), *Porous Silicon,* World Scientific, Singapore, pp. 133-147.
33. Pickering, C. et.al. (1984), Optical studies of porous silicon films formed in p-type degenerate and non-degenerate silicon, *J. Phys. C: Solid State Physics* **17**,6535-6553.
34. Canham, L.T. (1990) Silicon quantum wire array fabrication by electrochemical and chemical dissolution of wafers, *Appl. Phys. Lett.* **57**, 1046-1048.
35. Lang, W., Steiner, P., and Kozlowski, F. (1994) Optoelectronic properties of porous silicon-The electroluminescent devices, in V.C. Vial, J. Derrien (eds.), *Porous silicon Science and Technology,* Springer-Verlag., Berlin, pp. 293-305.
36. Asnin, V.M. et.al. (1993) Quantum size effect in the photoluminescence of porous silicon layers, *Solid State Commun.* **84**, 817-820.
37. Cheah, K.W. et.al. (1993) Multiple peak photoluminescence of porous silicon, *Appl. Phys. Lett.* **63**, 3464-3466.
38. Zhang, Z.L. et.al. (1995) Two peak photoluminescence and light emitting mechanism of porous silicon, *Phys. Rev. B* **51**, 11194-11197.
39. Xu, Z.Y., Gal, M., and Gross, M. (1992) Photoluminescence studies on porous silicon, *Appl. Phys. Lett.* **60**, 1375-1377.
40. Zheng, X.L., Wang, W., and Chen, H.C. (1992) Anomalous temperature dependencies of photoluminescence for visible light emitting porous Si, *Appl. Phys. Lett.* **60**, 986-988.
41. Campbell, I.H. and Fauchet, P.M. (1984) The effects of microcrystal size and shape on the one phonon Raman spectra of crystalline semiconductors, *Solid State Commun.* **58**, 739-742.
42. Sui, Z. et.al. (1992) Raman analysis of light emitting porous silicon, *Appl. Phys. Lett.* **60**, 2086-2088.
43. Tsu, R., Shen, H., and Dutta, M. (1992) Correlation of Raman and photoluminescence spectra, *Appl. Phys. Lett.* **60,** 112-114.
44. Tanino, H. et.al. (1996) Raman study of free-standing porous silicon,

Phys. Rev. B **53**, 1937-1947.

45. Halimoui, A., Porous silicon; material processing properties and applications, in J.C. Vial and J. Derrien (eds.), *Porous silicon Science and Technology,* Springer -Verlag, pp. 33.
46. Brandt, M.S. et .al. (1992) The origin of visible luminescence from porous silicon: A new interpretation, *Solid State Commun.* **81**, 307-312.
47. Cullis, A.G. and Canham, L.T. (1991) Visible light emission due to quantum size effects in highly porous crystalline silicon, *Nature* **353**, 335-337.
48. Teschke, O. et.al. (1993) Nanosize structures connectivity in porous silicon and its relation to photoluminescence, *Appl. Phys. Lett.* **63**, 1927-1929.
49. Schuppler, S. et.al. (1994) Dimensions of Luminescent oxidized and porous silicon structures, *Phys. Rev. Lett.* **72**, 2648-2651.
50. Suemoto, T. et.al. (1993) Observation of phonon structures in porous Si luminescence, *Phys. Rev. Lett.* **70**, 3659-3662.
51. Hybertsen, M.S. (1994) Absorption and emission of light in nanoscale silicon structures, *Phys. Rev. Lett.* **72**, 1514-1517.
52. Kovalev, D.I. et.al. (1994) Fast and slow visible luminescence bands in oxidized porous Si, *Appl. Phys. Lett.* **64**, 214-216.

Ankara

Antalya

PARTICIPANTS

LECTURERS

Prof. Gerhard Abstreiter
Walter Schottky Institut
Technische Universität München
Am Coulombwall
D-85748 Garching
Germany

Prof. Yasuhiko Arakawa
Institute of Industrial Science (IIS)
University of Tokyo
7-22-1 Roppongi
Minato-ku, Tokyo 106
Japan

Prof. Atilla Aydinli
Bilkent University
Dept. of Physics
Bilkent, 06533 Ankara
Turkey

Prof. Dieter H. Bimberg
Institut für Festkörperphysik
Technische Universität Berlin
Hardenbergstraße 36
D-10623 Berlin
Germany

Prof. Salim Çiraci
Bilkent University
Dept. of Physics
Bilkent, 06533 Ankara
Turkey

Prof. Roberto Cingolani
Universita di Lecce
Dept. Materiali
Via Arnesano
I-73100 Lecce
Italy

Dr. Karl Eberl
MPI für Festkörperforschung
Heisenbergstraße 1
D-70569 Stuttgart
Germany

Dr. Francois H. Julien
Institut d'Electronique Fondamentale
Univ. Paris-Sud
URA 022 du CNRS
F-91405 Orsay Cedex
France

Dr. Bernard Jusserand
France Telecom - CNET-CNRS
196, Avenue Henri-Ravera
F-92225 Bagneux Cedex
France

Prof. Eli Kapon
Swiss Federal Institute of Technology
Inst. of Micro and Optoelectronics
CH-1015 Lausanne
Switzerland

Prof. Jean Pierre Leburton
University of Illinois
Gen. Phys. Adv. Abs.
405 N. Mathews Ave.
Urbana, IL 61801
USA

Prof. Roberto D. Merlin
Department of Physics
University of Michigan
Ann Arbor, MI 48109-1120
USA

Dr. Aron Pinczuk
AT&T Bell Labs
Room 1D-433
600 Mountain Avenue
Murray Hill, NJ 07974-0636
USA

Prof. Michael L. Roukes
Condensed Matter Physics
Caltech 114-36
Pasadena, CA 91125
USA

Prof. Clivia M. Sotomayor Torres
LS für Materialwissenschaften in der
Elektrotechnik
Bergische Univ. GH Wuppertal
Fuhlrottstr. 10
D-42097 Wuppertal
Germany

Prof. Bilal Tanatar
Bilkent University
Dept. of Physics
Bilkent, 06533 Ankara
Turkey

Dr. Werner Wegscheider
Walter Schottky Institut
Technische Universität München
Am Coulombwall
D-85748 Garching
Germany

ASI STUDENTS

Abay Bahattin
Atatürk Üniversitesi
Fen Edebiyat Fakültesi
Fizik Bölümü
25240 Erzurum
Turkey

Akbas Hasan
Trakya Üniversitesi
Fen Fakültesi
Fizik Bölümü
22030 Edirne
Turkey

Aksay Sabiha
Anadolu Üniversitesi
Fen Fakültesi
Fizik Bölümü
26470 Eskisehir
Turkey

Aktas Saban
Trakya Üniversitesi
Fen Fakültesi
Fizik Bölümü
22030 Edirne
Turkey

Arzberger Markus
Walter Schottky Institut
Technische Universität München
Am Coulombwall
D-85748 Garching
Germany

Askenasy N.
Department of Physical Electronics
Faculty of Engineering
Tel-Aviv University
Ramat Aviv 69978
Israel

Bairamov Bahish
A.F. Ioffe Physico-Technical Institute
Russia Academy of Sciences
Polytechnicheskaya 26
St. Petersburg 194021
Russia

Bandoyopadhy Supriyo
Dept. of Electrical Engineering
University of Nebraska
Lincoln, NE 68588-051
USA

Basaran Engin
Gebza Yüksek Teknoloji Ens.
Fizik Bölümü
41400 Gebze-Kocaeli
Turkey

Bennett Colin
Department of Physics
University of Essex
Colchester CO4 3SQ
U.K.

Blanton Sean
The University of Chicago
The James Frank Institute
5640 S Ellis Ave
Chigago, IL
USA

Cerqueira Maria
University of Minho
Physics Department
Campus de Gualtar
4709 Braga Codex
Portugal

Doll Monica
Department de Fisica
U. de Santiago de Chile
Avda Ecuador 3493
Casilla 307 Santiago
Chile

Eitel Sven
Paul Scherrer Institut
Badenerstr. 569
CH-8048 Zürich
Switzerland

Elsass Chriss
Materials Dept.
Eng. 3 Bldg
UCSB
Santa Barbara, CA 93106
USA

Filoramo Arianne
L.P.M.C.
Ecole Normale Superieure
24, Rue Lhomond
F-75005 Paris
France

Frank Wolfgang
LMU München
Sektion Physik
Geschwister-Scholl-Platz 1
D-80539 München
Germany

Gehrsitz Stefan
Paul Scherrer Institut
Badenerstr. 569
CH-8048 Zürich
Switzerland

Givant Amichai
Division of Applied Physics
Hebrew University
Jerusalem 91904
Israel

Glutsch Stephan
Friedrich-Schiller-Universität
Institut für Festkörpertheorie und
Theoretische Optik
Max-Wien-Platz 1
D-07743 Jena
Germany

Govorov Andrei
LMU München
Sektion Physik
Geschwister-Scholl-Platz 1
D-80539 München
Germany

Guasch Cathy
Dept. of Electronics and Electrical Engineering
University of Glasgow
Rankine Bldg.
Glasgow, G12 8QQ
Scotland, U.K.

Güven Kaan
Bilkent University
Faculty of Science
Physics Dept.
06533 Bilkent, Ankara
Turkey

Hagn Marcus
Walter Schottky Institut
Technische Universität München
Am Coulombwall
D-85748 Garching
Germany

Kalem Seref
Marmara Arasurma Merkezi
Elektronik Bölümü
P.O. Box 21
41470 Gebze-Kocaeli
Turkey

Katz Matty
Department of Physics
Solid State Institute
Technion, Haifa 32000
Israel

Kaya Ismet
Bilkent University
Faculty of Science
Physics Dept.
06533 Bilkent, Ankara
Turkey

Kronik Leor
Department of Physical Electronics
Faculty of Engineering
Tel-Aviv University
Ramat Aviv 69978
Israel

Kulik Igor
Verchenko Institute of Physics and Technology at Low Temperatures
Kharkov
Ukraine

Kumru Mustafa
Selcuk Üniversitesi
Fen Fakültesi
Fizik Bölümü
42079 Konya
Turkey

Lavon Yoav
Division of Applied Physics
Hebrew University
Jerusalem 91904
Israel

Lavrova Olga
Dept. of Electrical Engineering
University College London
Torrington Place
London WC1Y 7JE
U.K.

Loser Falk
Institut für Angewandte Photophysik
Technische Universität Dresden
Mommsenstraße 13
D-01062 Dresden
Germany

Malajovich Irina
Materials Dept.
Eng. 3 Bldg
UCSB
Santa Barbara, CA 93106
USA

Marks Daniel
Beckman Institute
University of Illinois
405 N. Mathews Ave.
Urbana, IL 61801
USA

Martinet Eric
EPFL
DP Inst. Microoptoelectronique
Ecublens
CH-1015 Lausanne
Switzerland

Martinez Jorge Garcia
Materials Dept.
Eng. 3 Bldg
UCSB
Santa Barbara, CA 93106
USA

Maximov Mikhail
A.F. Ioffe Physico-Technical Institute
Russian Academy of Sciences
26, Polytehnicheskaya
St.-Petersburg 194021
Russia

Pugh Loretta
Optoelectronics Group
Cavendish Laboratory
Madingley Road
Cambridge, CB3 OHE
U.K.

Ribayrol Ann
Dept. of Electronics and Electrical
Engineering
University of Glasgow
Rankine Bldg.
Glasgow, G12 8LT
U.K.

Rinaldi Ross
Dipart. di Scienza dei Materiali
Universita di Lecce
Via Arnesano
I-73100 Lecce
Italy

Rolo Anabela
University of Minho
Physics Department
Campus de Gualtar
4709 Braga Codex
Portugal

Schrottke Lutz
Paul-Drude-Institut für
Festkörperelektronik
Hausvogteiplatz 5-7
D-10117 Berlin
Germany

Schumovsky Alexander
Bogoliubov Lab. of Theoretical
Physics
Joined Inst. for Nuclear Research
Dubna
Russia

Senger Tugrul
Bilkent University
Faculty of Science
Physics Dept.
06533 Bilkent, Ankara
Turkey

Serpengüzel Ali
Bilkent University
Faculty of Science
Physics Dept.
06533 Bilkent, Ankara
Turkey

Siarkos Anastassios
Friedrich-Schiller-Universität
Institut für Festkörpertheorie und
Theoretische Optik
Max-Wien-Platz 1
D-07743 Jena
Germany

Stevens Troy
University of Michigan
Randall Lab. of Physics
500 East University Ave.
Ann Arbor, MI 48109-1120
USA

Theodorou George
Dept. of Physics
Solid State Section
54006 Thessaloniki
Greece

Tsatsul'nikov Andrei
A.F. IOFFE Physical-Technical Institute
Russian Academy of Sciences
26, Polytehnicheskaya
St.-Petersburg 194021
Russia

Tüzemen Sabahattin
Atatürk Universitesi
Fen Edebiyat Fakültesi
Fizik Bölümü
25240 Erzurum
Turkey

Vittorio Massimo
Dipart. Di Scienza die Materiali
Universita di Lecce
Via Arnesano
I-73100 Lecce
Italy

Vouilloz Fabrice
EPFL
DP Inst. Microoptoelectronique
Ecublens
CH-1015 Lausanne
Switzerland

Yalcin Nuh
Gebza Yüksek Teknoloji Ens.
Fizik Bölümü
41400 Gebze-Kocaeli
Turkey

Zaluzny Miroslaw
Institute of Physics
M. Curie-Sklodowska University
pl. M. Curie-Sklodowskiej 1
20-031 Lublin
Poland

Zimmermann Stefan
LMU München
Sektion Physik
Geschwister-Scholl-Platz 1
D-80539 München
Germany

Zor Muhsin
Anadolu Üniversitesi
Fen Fakültesi
Fizik Bölümü
26470 Eskisehir
Turkey

AUTHOR INDEX

SUBJECT INDEX